NDT in Canada Conference 2019

AA004996

Inform-Engage-Advance

Edmonton, Canada
18-20 June 2019

ISBN: 978-1-5108-9277-4

Printed from e-media with permission by:

Curran Associates, Inc.
57 Morehouse Lane
Red Hook, NY 12571

Some format issues inherent in the e-media version may also appear in this print version.

NDT.net
Rheintalstr 3
53498 Bad Breisig
Germany

Phone: + 49 2633 4756670
Fax: + 49 3212-1030513

info@ndt.net

Additional copies of this publication are available from:

Curran Associates, Inc.
57 Morehouse Lane
Red Hook, NY 12571 USA
Phone: 845-758-0400
Fax: 845-758-2633
Email: curran@proceedings.com
Web: www.proceedings.com

TABLE OF CONTENTS

Artificial Intelligence
Advanced NDT Data Analysis

Neil Harrap
Principal Project Leader

CINDE
CANADIAN INSTITUTE FOR NON-DESTRUCTIVE EVALUATION

NDT in Canada
NDTiC 2019
Inform - Engage - Advance
June 18 - 20
River Cree Resort & Casino
Edmonton, Alberta

Complex Ultrasonic Inspection and Analysis

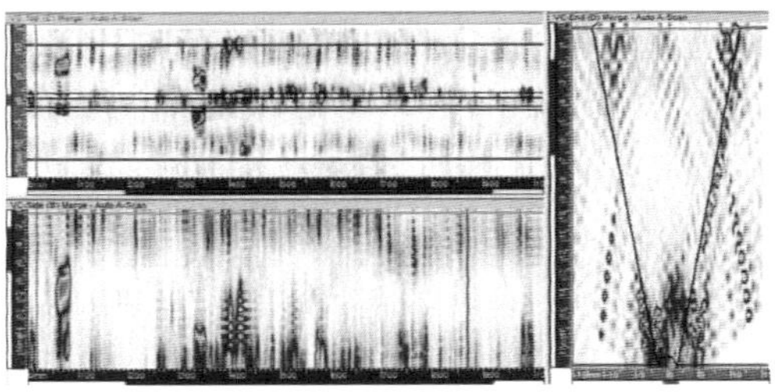

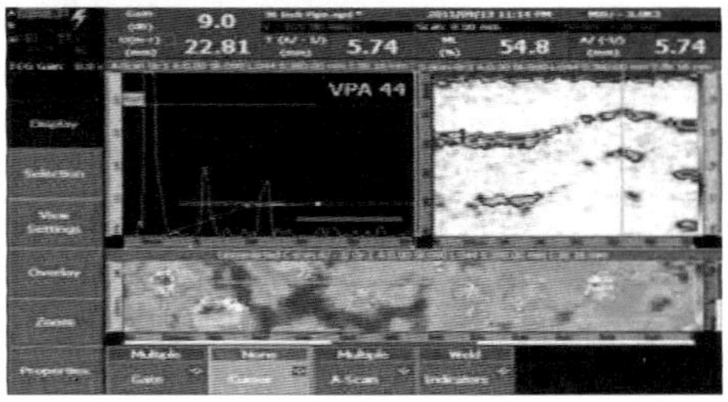

NDT in Canada 2019 | June 18-20 | Edmonton, AB

Data Analysis Using Artificial Intelligence

Why?

1) The volume of complex data is ever increasing
2) The number of competent data analysts is not !!
3) The need for consistent high quality ‘Data Analysis,’ to provide accurate quantification for Engineering Critical Assessment and Asset Integrity Management

The Solution

1) A Powerful, scalable, Cloud-based analysis software
2) Utilising 3D Artificial Intelligence algorithms
3) The ‘Artificial Intelligence (AI)’ becomes an invaluable assistant to the ‘Data Analyst’

Software Description

1) Cloud-based data analysis software
2) Browser-based interface
3) Automated data analysis using Artificial Intelligence (AI)
4) Manual data analysis using 3D visualisation and measuring tools
5) Full reports for automated and manual analysis
6) Comparison of ‘Manual Analysis’ versus ‘AI Analysis’, and External Report, in order to provide final ‘Audit Report’

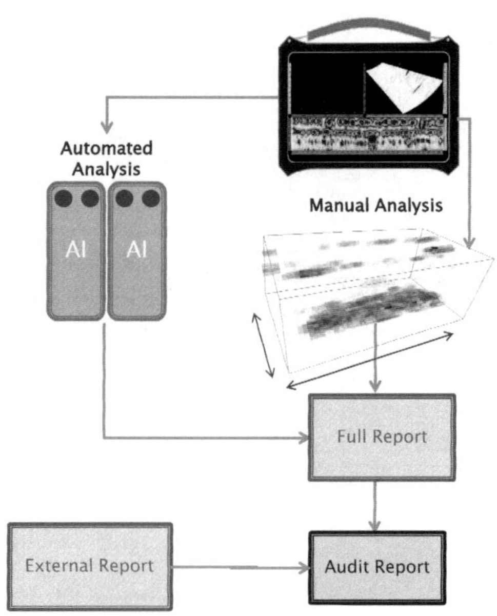

Software Operation

1) Data is acquired by the operator
2) Data is sent to cloud-based servers
3) Data is analysed for quality (lack of data, lack of couplant)
4) A SMS or email is sent to the operator and the supervisor
5) Data is fully analysed by Artificial Intelligence
6) A full report is sent to the Supervisor

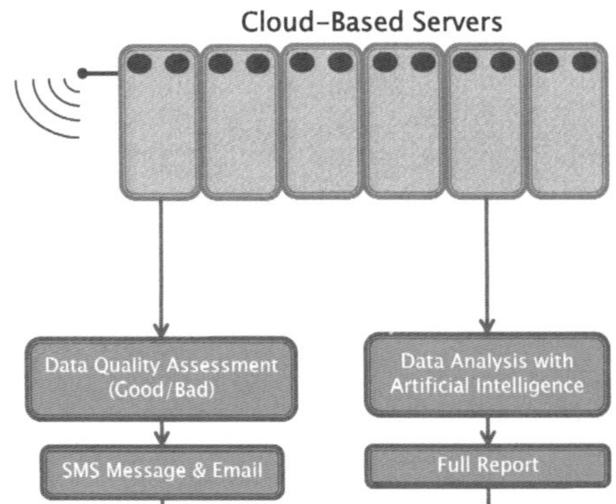

NDT in Canada 2019 | June 18-20 | Edmonton, AB

Reporting

- Data Quality Assessment
- Defect Detection and Measurement
- Location and Characterisation of Geometric Indications

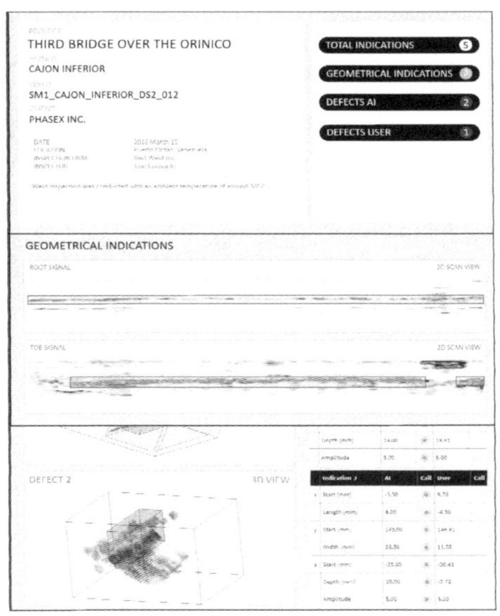

NDT in Canada 2019 | June 18-20 | Edmonton, AB

Summary

- Audits
  - External audits: Checking the quality of external inspection company's work

  - Internal audits: Checking of inspectors competency

- Efficiency tool
  - More analysis per hour
  - Higher quality analysis

- Follow-up tool
  - Monitor the evolution of indications
  - Checking previous inspections

- Future Aspiration: Auto Defect Recognition(ADR)

Case Study

- Third Party Cross Check

  - Fabrication on High Pressure, High Temp System
  - Suspect quality issues with subject contractor
  - Initial investigation indicated that PAUT Data Files were being duplicated

- Using Conventional OEM 2D Software

  - Many Hours spent attempting to identify copied files - this proved inconclusive

- Using AI 3D Software

  - Random selection of 650 data files uploaded to cloud AI processed in minutes and identified that 40% of files proved to be duplicated

NDT in Canada 2019 | June 18-20 | Edmonton, AB

9

Factors Affecting Spatial Resolution in Pulsed Eddy Current Inspection of Pipe

Yuewen Fu[1], *P. Ross Underhill*[2] *and Thomas W. Krause*[2]

[1]Nanchang Hangkong University, China
[2]*Department of Physics and Space Science*
Royal Military College of Canada

CINDE
CANADIAN INSTITUTE FOR NON-DESTRUCTIVE EVALUATION

NDT in Canada
NDTiC 2019
Inform - Engage - Advance
June 18 - 20
River Cree Resort & Casino
Edmonton, Alberta

Outline

- Motivation
- Basics
- Probe Geometry
- Resolution
- Summary
- Acknowledgements

Motivation

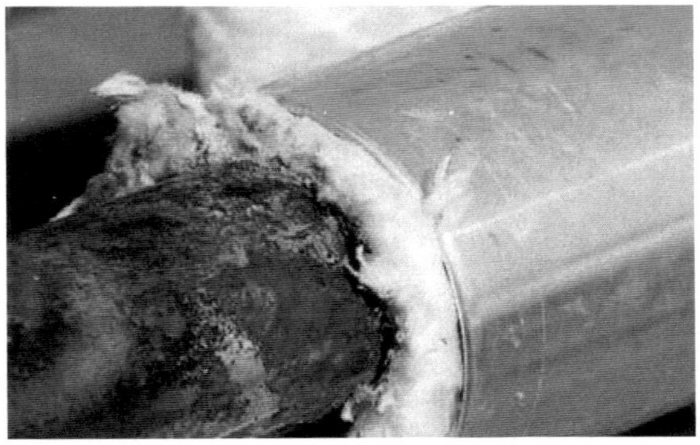

NDT in Canada 2019 | June 18-20 | Edmonton, AB

3

Basics

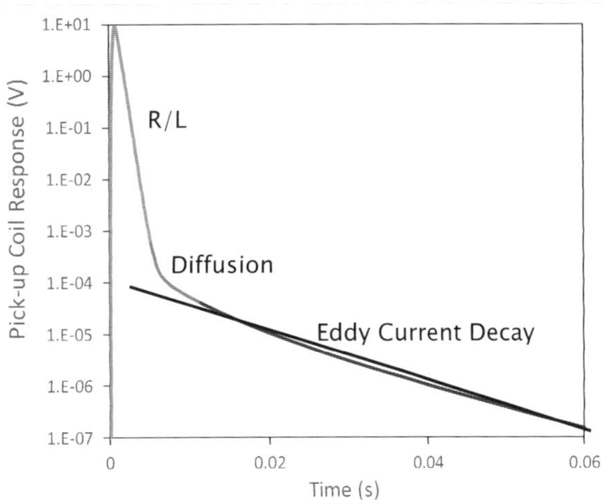

NDT in Canada 2019 | June 18-20 | Edmonton, AB

4

Basics II

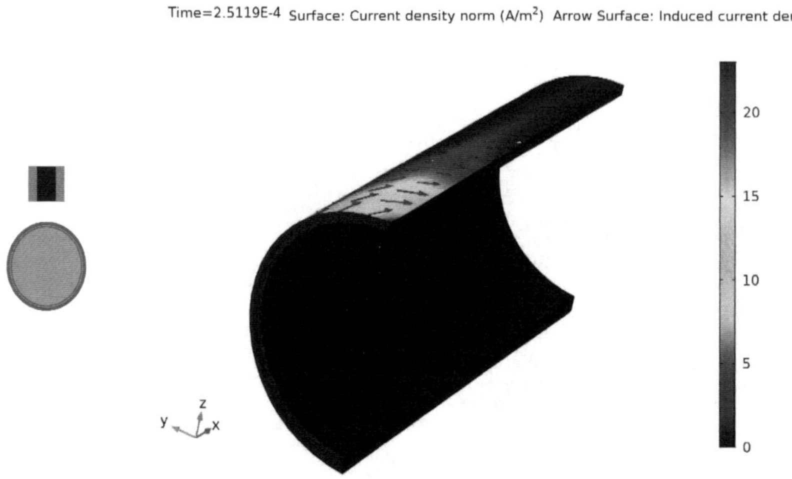

Basics

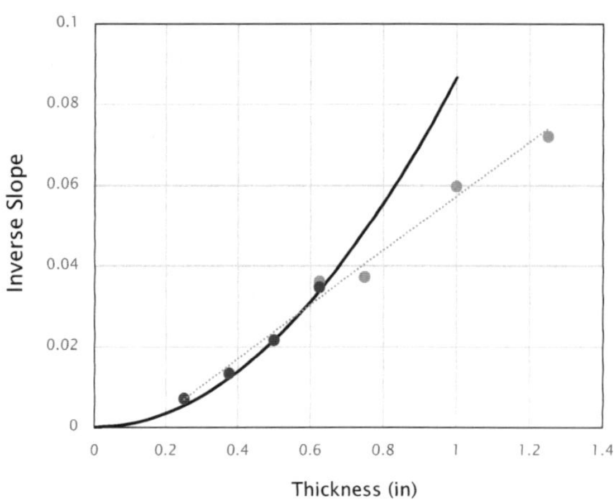

Core

Flat Plates

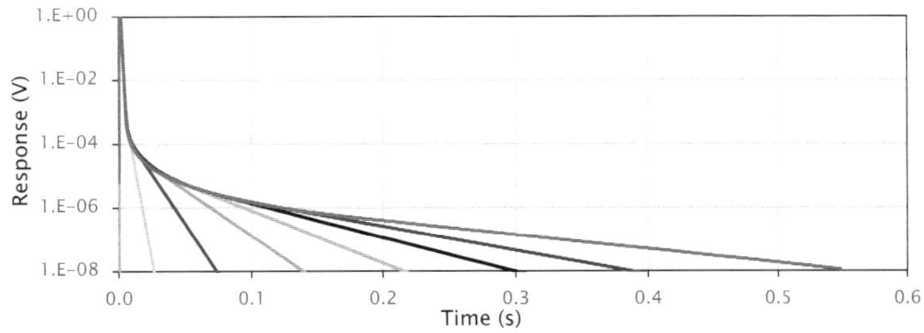

Air

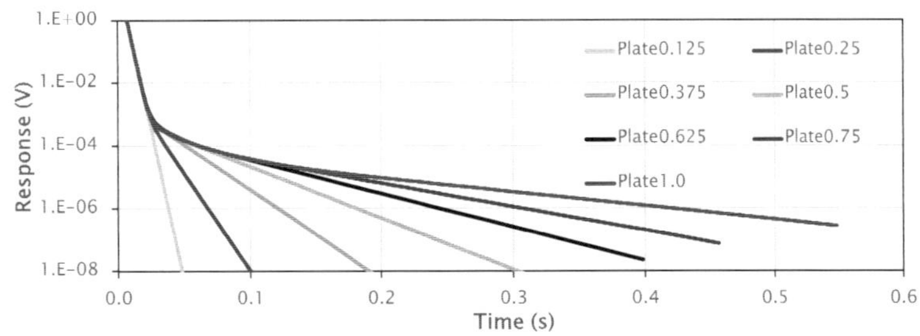

Ferrite
$\mu_r = 50$

NDT in Canada 2019 | June 18-20 | Edmonton, AB

7

Probe Geometry – ¼ model

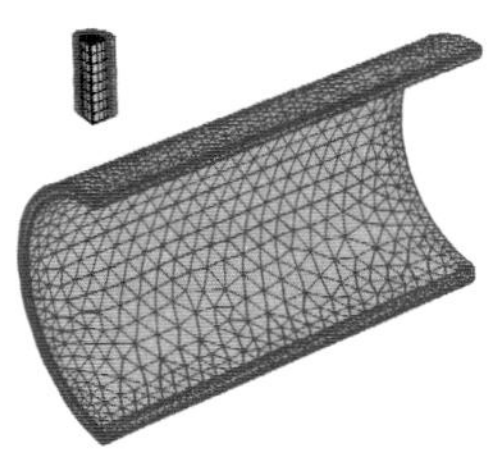

Radial

Transverse

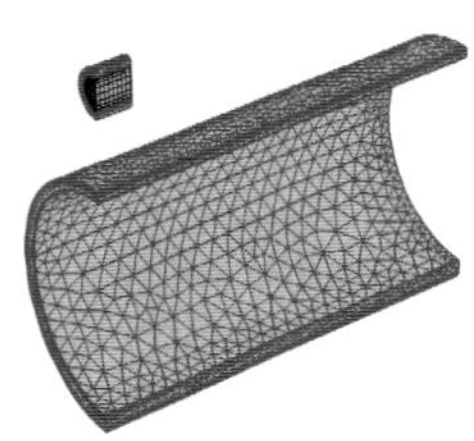

Longitudinal

Symmetry provides the rest of the model

NDT in Canada 2019 | June 18-20 | Edmonton, AB

8

Model Validation

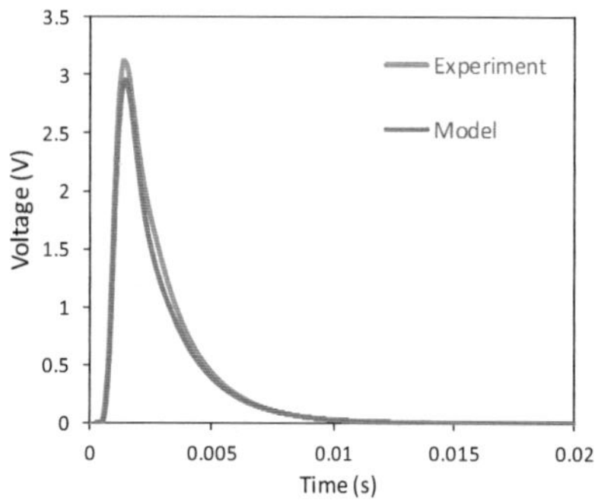

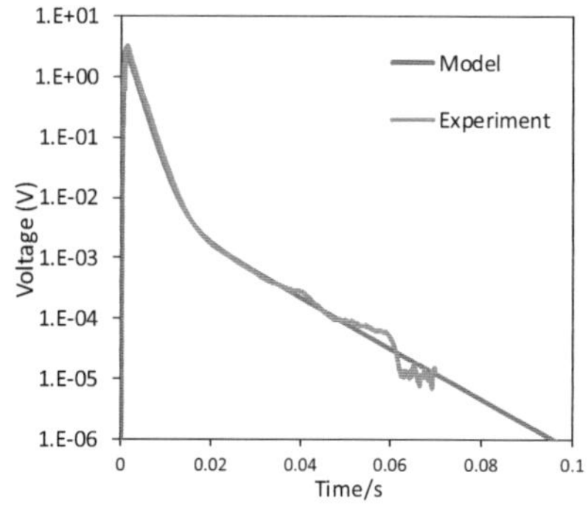

NDT in Canada 2019 | June 18-20 | Edmonton, AB 9

Probe Geometry II - Radial

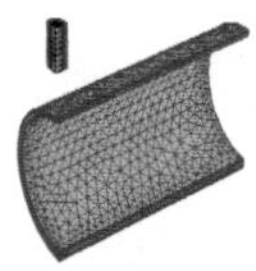

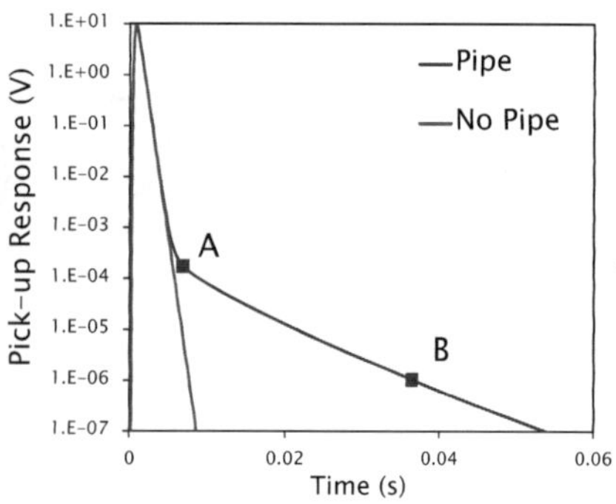

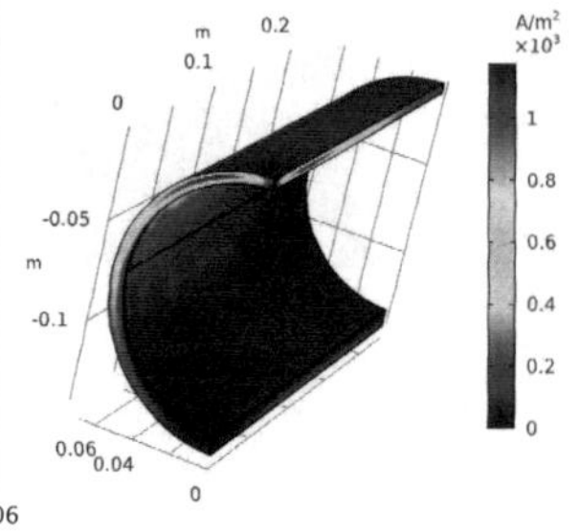

A

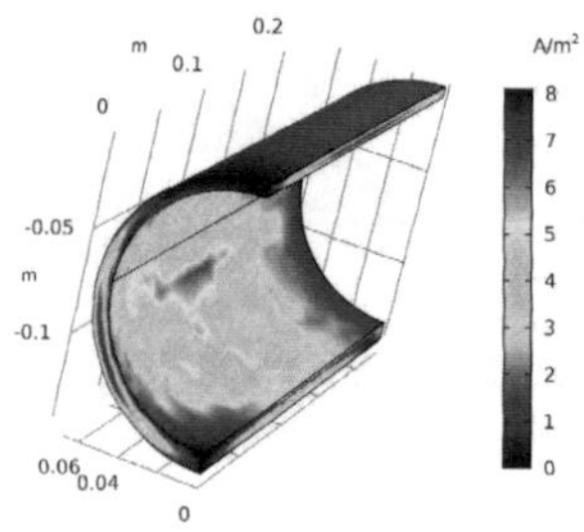

B

NDT in Canada 2019 | June 18-20 | Edmonton, AB 10

Probe Geometry III – Longitudinal

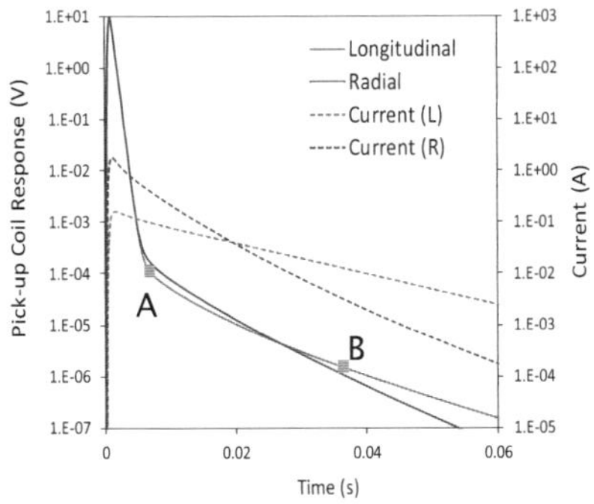

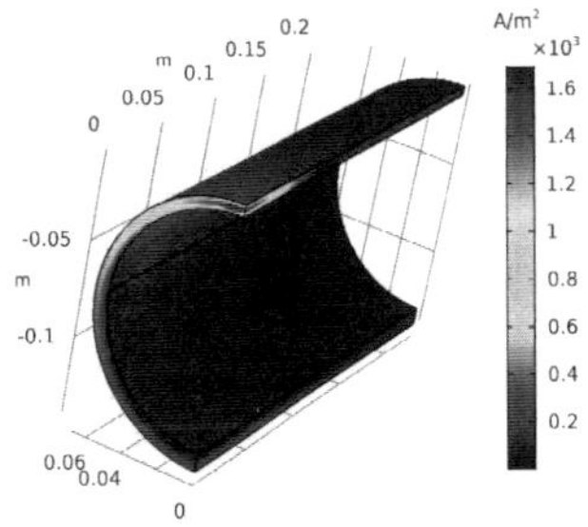

A

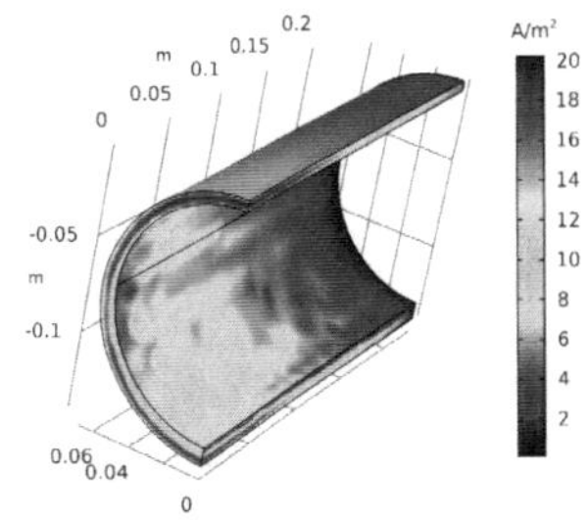

B

Probe Geometry IV - Longitudinal

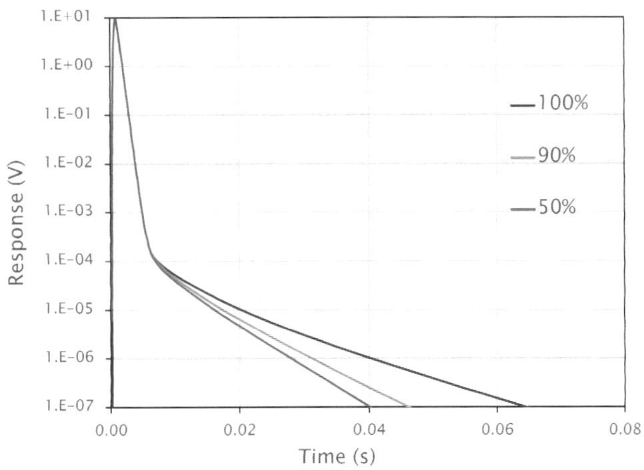

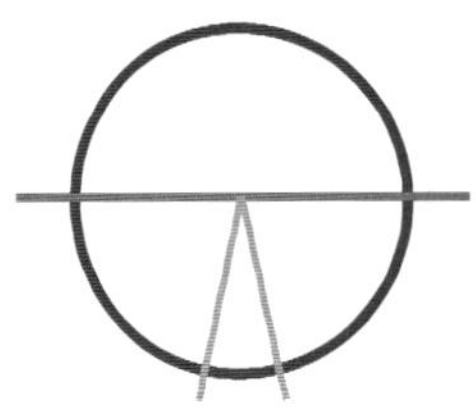

Probe Geometry V - Transverse

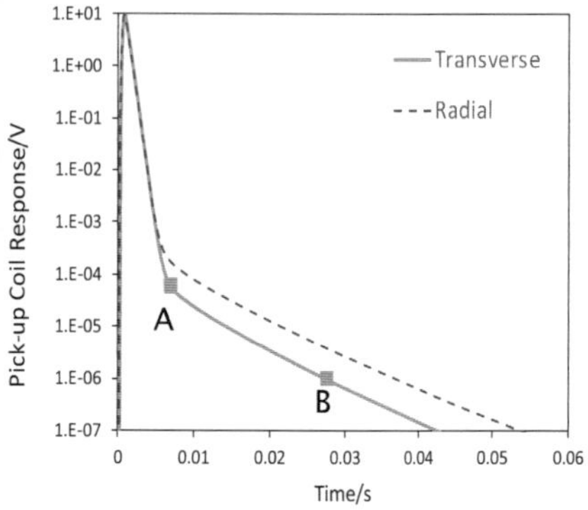

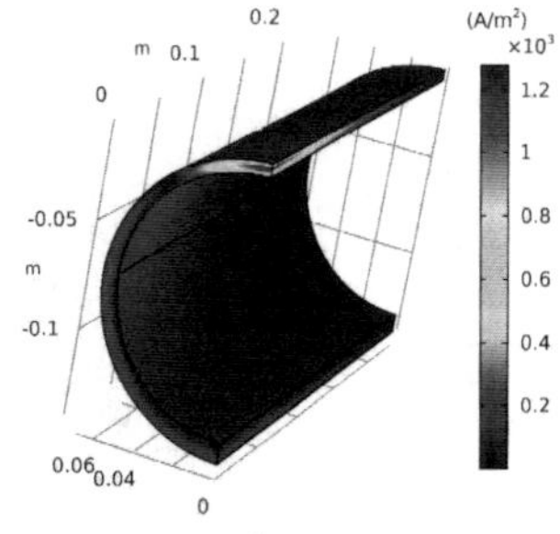

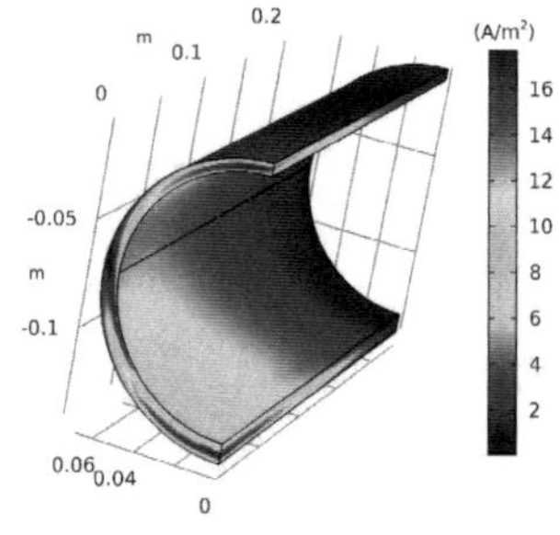

Probe Geometry VI - Summary

All probes
- Get information from delocalised eddy currents

Longitudinal Probe provides
- Information from directly below the probe
- Most confined eddy currents
- Best intensity

Resolution I

ProbeLength	0.04[m]
DriverID	0.020 [m]
DriverTurns	750
PickUpID	18[mm]
PipeOD	140.5[mm]
LiftOff	50[mm]
PipeLength	18.25*2[in]

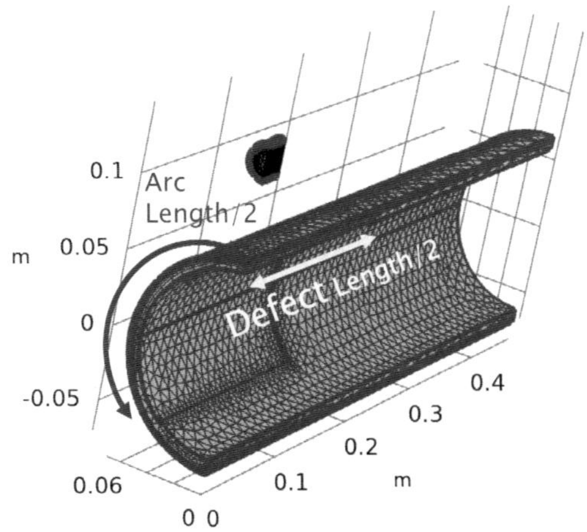

Resolution II - Angular Length

Thickness	10 mm
Defect Depth	2 mm
Defect Length	100 mm

Longitudinal probe

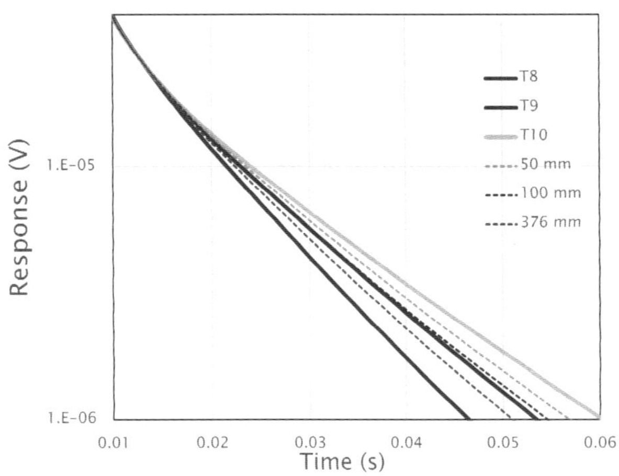

Resolution III - Defect Length

Thickness	10 mm
Defect Depth	2 mm
Angular Length	376 mm

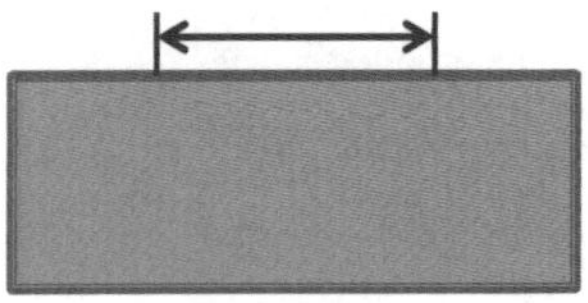

Longitudinal probe

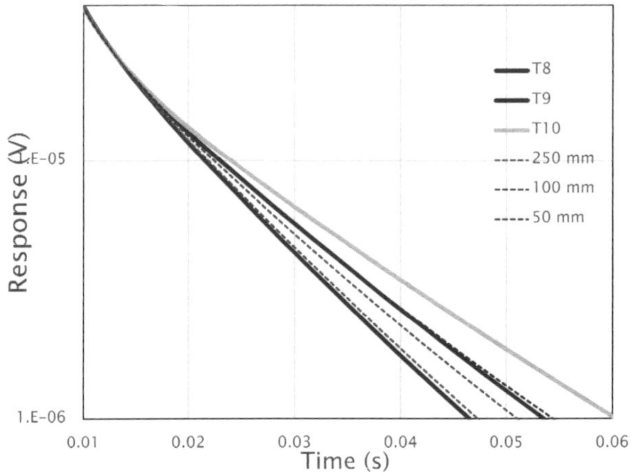

Resolution IV - Summary

Longitudinal probe

- Footprint covers nearly the whole circumference
  More sensitive to arc length than longitudinal length
  Wall thickness has little effect on footprint

Resolution V – Defect Length

Thickness	10 mm
Defect Depth	2 mm
Angular Length	376 mm

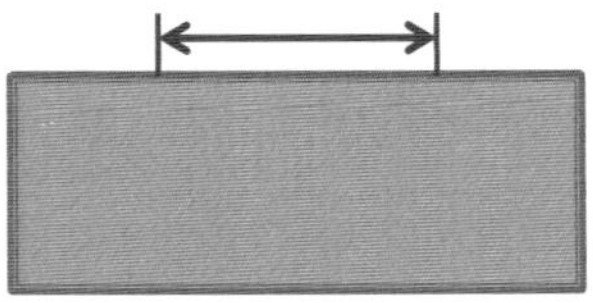

Radial probe

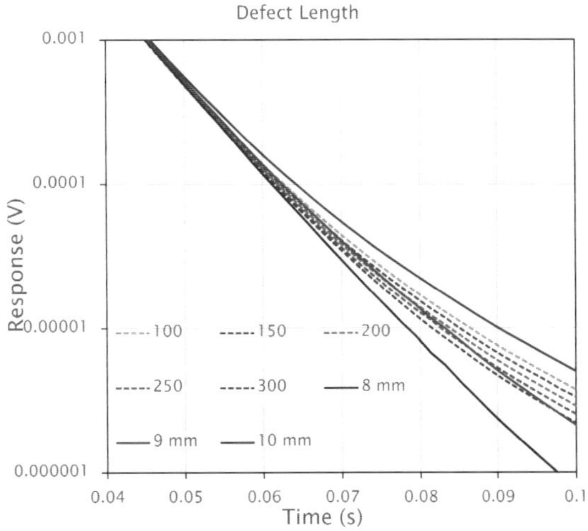

Summary

- Best resolution obtained with longitudinal probe
- Eddy currents go all the way around the pipe (at least for diameters 6” or less)
- Wall thickness has little effect on footprint
- More sensitive to defect arc length than longitudinal length

Summary

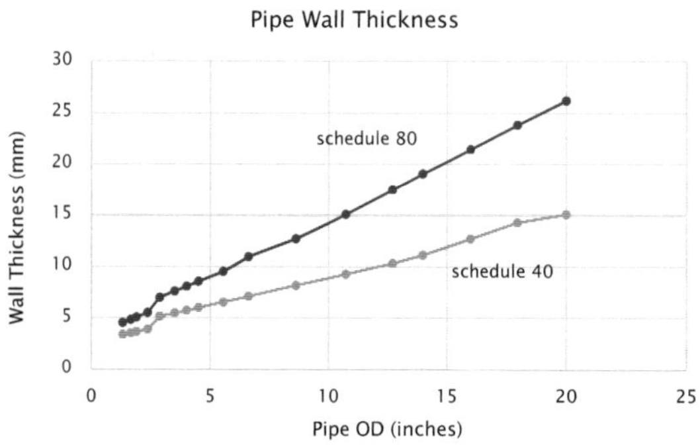

Low inductance
air core
low # of turns on a ferrite core)

NDT in Canada 2019 | June 18-20 | Edmonton, AB 21

Acknowledgements

- NSERC-CRD
- Mequaltech

NDT in Canada 2019 | June 18-20 | Edmonton, AB 22

Questions

Pipe Diameter

10 mm wall

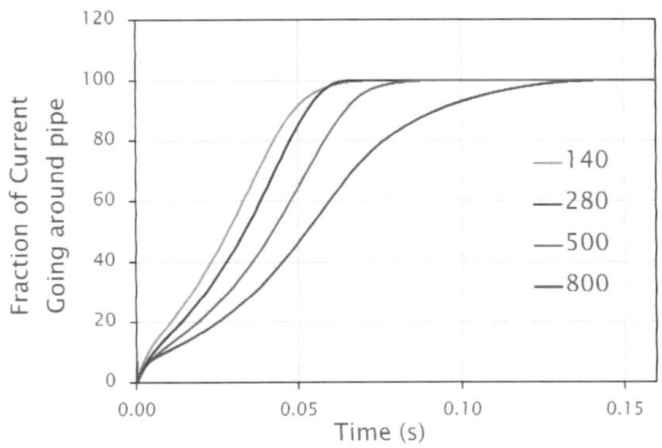

Increasing Uptime and Plant Reliability by Deploying Asset Monitoring Solutions

Eloy Gasperin, Technical Leader
MISTRAS Group

CINDE
CANADIAN INSTITUTE FOR NON-DESTRUCTIVE EVALUATION

NDT in Canada
NDTiC 2019
Inform - Engage - Advance
June 18 - 20
River Cree Resort & Casino
Edmonton, Alberta

Agenda

- ✓ Benefits of Monitoring
- ✓ Acoustic Emission
  - ✓ Transformers
  - ✓ Valves
  - ✓ Boiler tube leaks
  - ✓ Pressure Vessels
  - ✓ Stator Vane Cracking
  - ✓ High Energy Piping
- ✓ Ultrasonic Thickness Monitoring
- ✓ Metal Loss Detector

NDT in Canada 2019 | June 18-20 | Edmonton, AB

Benefits Of Monitoring

- ✓ More data available, better idea of condition of the asset
- ✓ Allows the user to see correlation with operating parameters
- ✓ Reduces the costs of multiple trips to evaluate the asset
- ✓ Information provided in real time, early detection of potential problems allows better planning

Acoustic Emission

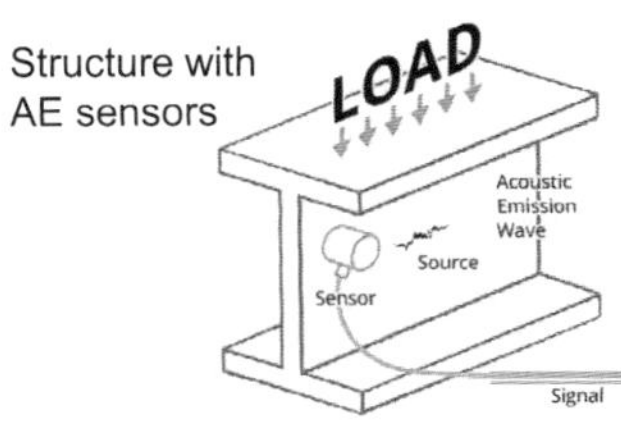

Signal collected by AE sensor

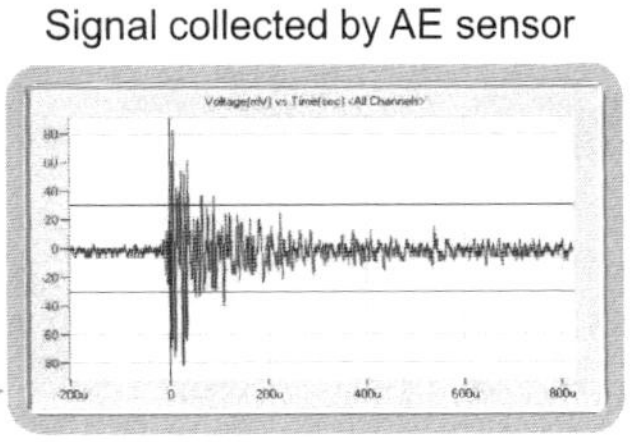

Shape and characteristics (ranges for specific AE features) vary depending of the mechanism producing the emission.

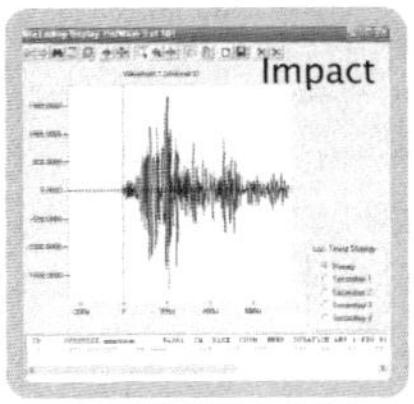

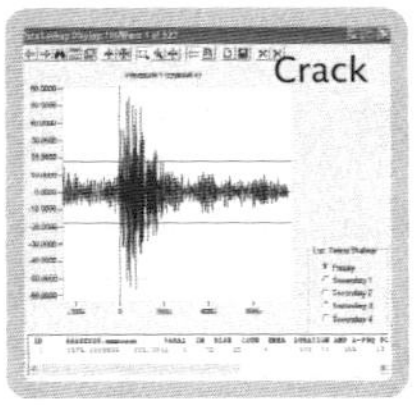

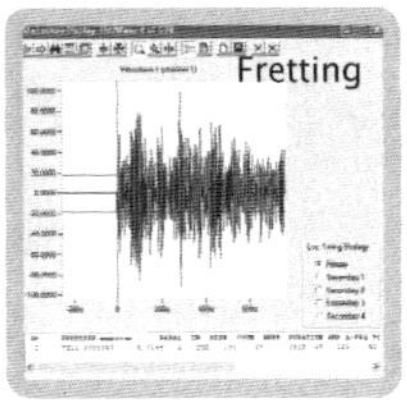

Asset Monitoring Installations ~1700

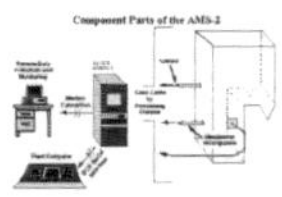

1,040
Valves

230
Transformers
- ✓ GSU
- ✓ Transmission
- ✓ Distribution

55
ACTMS
- ✓ Combustion Turbines

323
AMS
- ✓ Power Boilers
- ✓ HRSG
- ✓ Feed water heaters
- ✓ Recovery Boilers

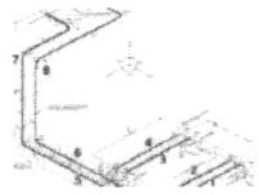

25
Bridges

8
Pressure Vessels
- ✓ Reactors
- ✓ Columns
- ✓ Separators

4
High Energy Piping

Other applications:
- ✓ PIPES – Thickness, corrosion
- ✓ Cranes
- ✓ Blast Furnace
- ✓ Flexible risers
- ✓ Above Ground Storage tank
- ✓ Generators, motors, HV Cables - **NEW**

NDT in Canada 2019 | June 18-20 | Edmonton, AB

What Are We Detecting?

TRANSFORMERS
- Electrical faults (Partial Discharge and/or Arcing)
- Thermal faults
- Mechanical Faults

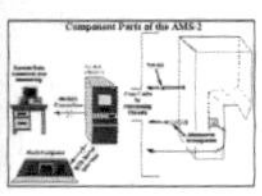

BOILER/HRSG
- Tube Leaks
- Sootblower effectiveness, stuck sootblowers
- Leaking valves
- Slag/pluggage
- Damaging backpass vibration
- Aspirating and cooling leaks
- SCR horn operation

COMBUSTION TURBINES
- Crack growth on stationary blade
- Rubbing
- Clashing
- Foreign object damage

BRIDGES
- Wire break in post tensioned concrete and cable structures.
- Strain, rotation, acceleration, displacement. Steel fatigue, crack detection.
- Concrete micro fracture detection.
- Structural health.

PRESSURE VESSELS
- Local overstressed areas
- Crack growth and/or corrosion products de-cohesion

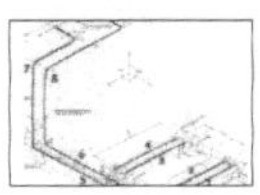

HIGH ENERGY PIPING
- Micro and macro cracks

VALVES
- Through-valve loss

NDT in Canada 2019 | June 18-20 | Edmonton, AB

Power Transformers

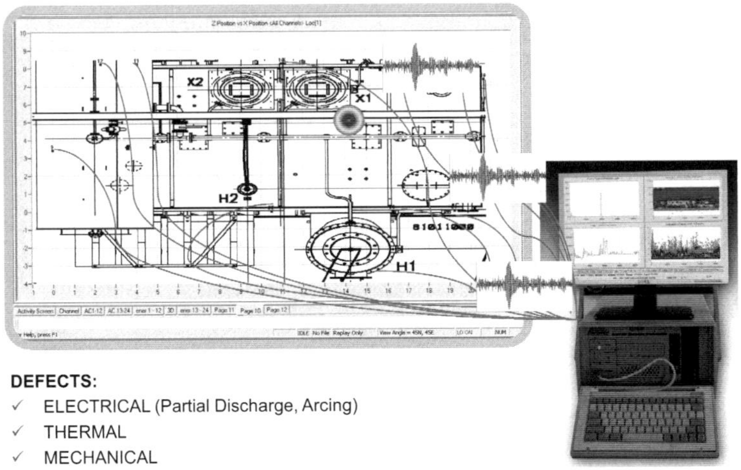

DEFECTS:

- ✓ ELECTRICAL (Partial Discharge, Arcing)
- ✓ THERMAL
- ✓ MECHANICAL

Source Location

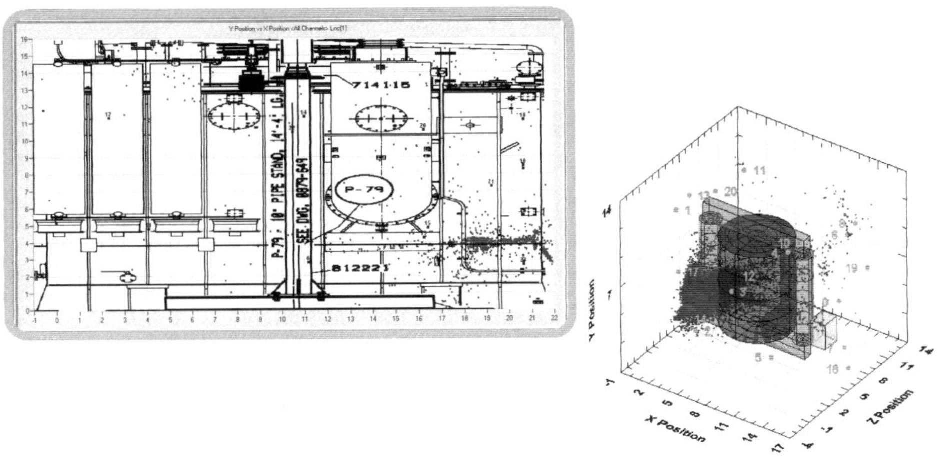

NDT in Canada 2019 | June 18-20 | Edmonton, AB

Acoustic Emission On-Line Monitoring System

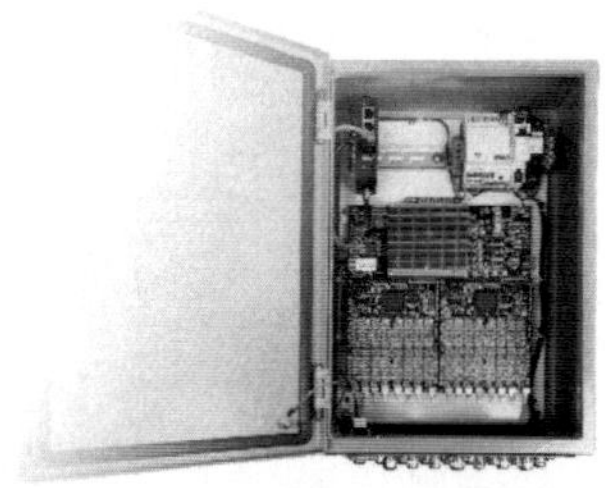

Power Transformers

- ✓ 24/7 Monitoring
- ✓ Short term or permanent

OLM Webpage

Utility's Data Center: OPC or MODBUS

Remote Access

http://monitor.transformer.clinic/

Case Study

Case 6 On IEEE STD. C57.127, 2018

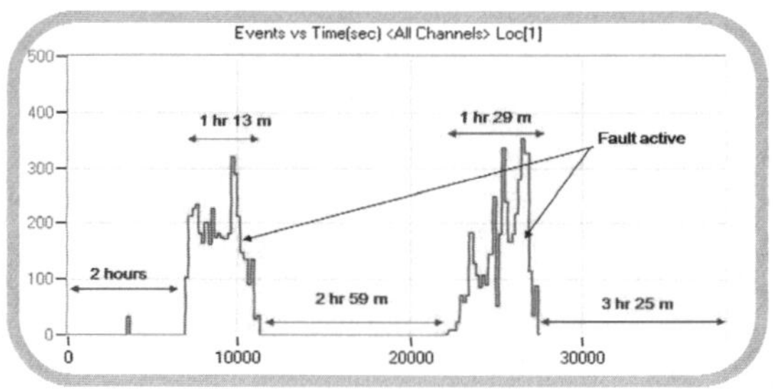

Date	H2	CO	CO2	CH4	C2H6	C2H4	C2H2	TDCG
5/31/05	192	187	728	17	5	3	0	404

Case Study

Case 6 On IEEE STD. C57.127, 2007

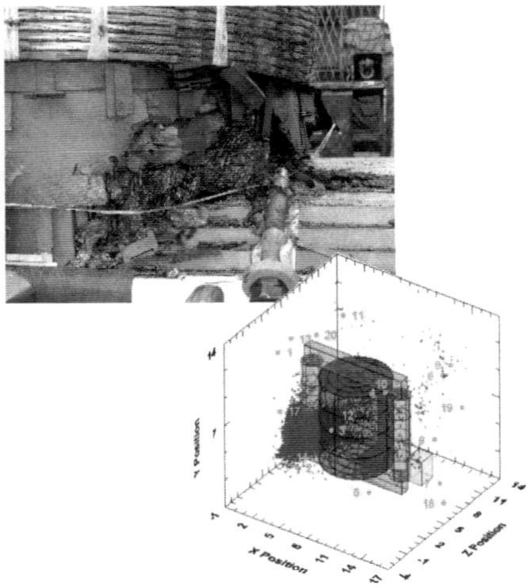

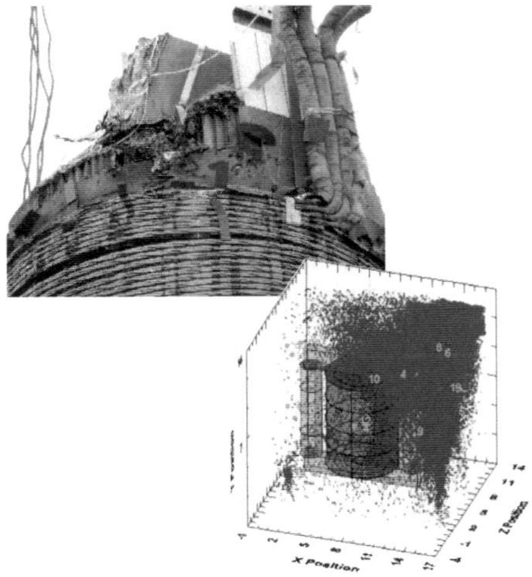

Valves

- ✓ 5-10% of valves suffer through-valve leakage.
- ✓ 1-2% of these passing valves are responsible for 75% of total losses.
- ✓ Real-time on-line valve leak detection and quantification.
- ✓ Instant response when a valve leaks or sticks.
- ✓ Enables monitoring of difficult to reach valves.
- ✓ Can be connected to the plant DCS with a suitable galvanic or zener barrier.

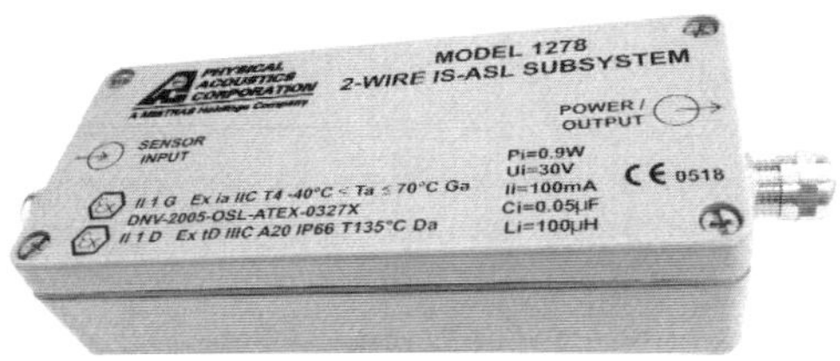

NDT in Canada 2019 | June 18-20 | Edmonton, AB

Boiler Tube Leak Monitoring

Acoustic signals are obtained from processes such as formation\growth of cracks, leaks, and variables such as load.

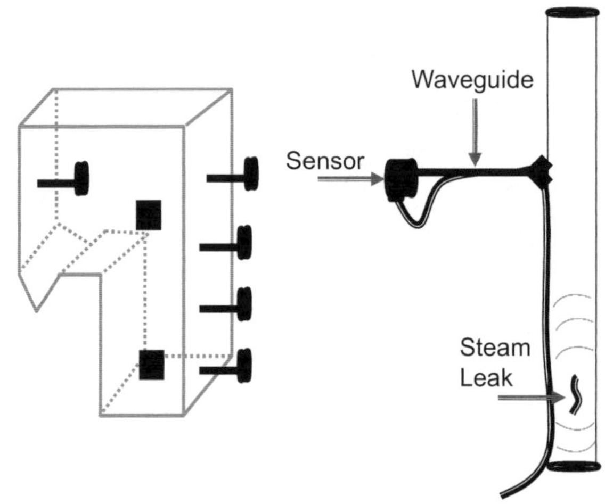

Sensor Location Layout

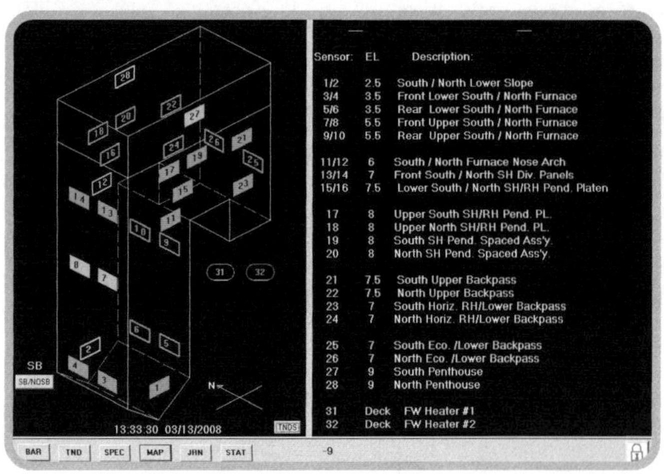

NDT in Canada 2019 | June 18-20 | Edmonton, AB

Finishing Superheat Leak

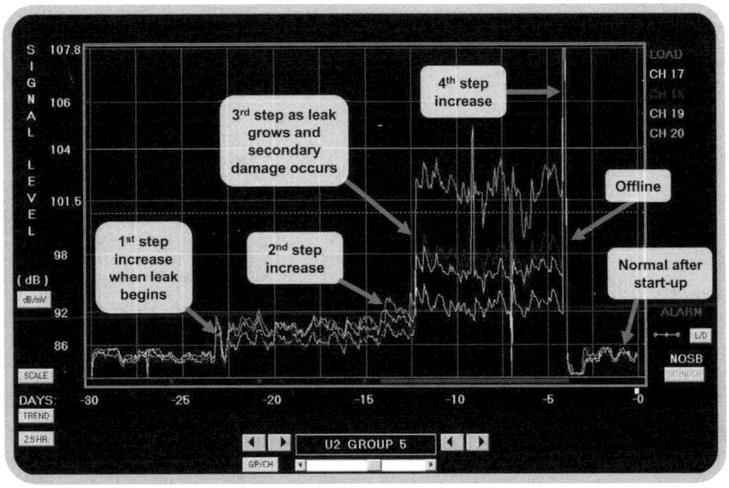

NDT in Canada 2019 | June 18-20 | Edmonton, AB

Finishing Superheat Leak

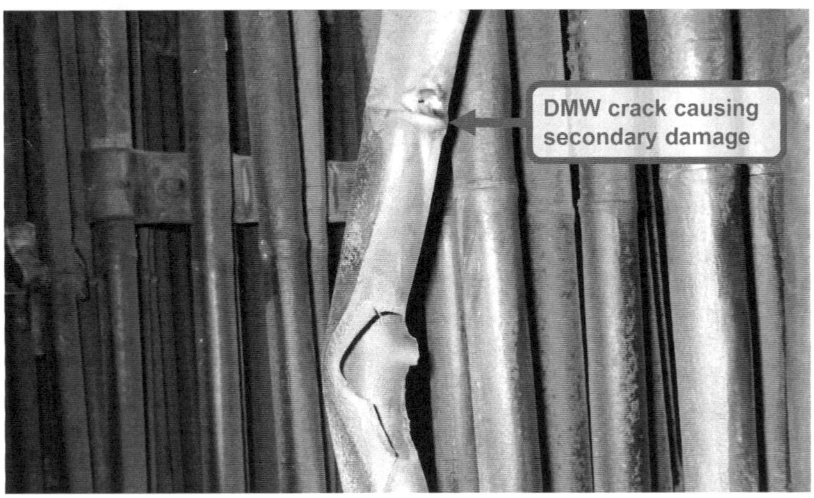

NDT in Canada 2019 | June 18-20 | Edmonton, AB

Pressure Vessels

- ✓ Safety
- ✓ Reliability
- ✓ Longevity
- ✓ Economics
- ✓ Compliance

The Procedure:

- ✓ Techniques for Sensing
- ✓ Techniques for Analyzing
- ✓ Guidelines to Determine Nature, Severity and Location of Structural Defects
- ✓ Criteria to use as a Basis for Assessing Structural Integrity
- ✓ Guidelines for Loading (depending on vessel type and history)

Pressure Vessel Data Interpretation

Intensity Analysis

- ✓ Expanded Data Base
- ✓ Signature Analysis
- ✓ Source Correlation
- ✓ Individual Channel Evaluation

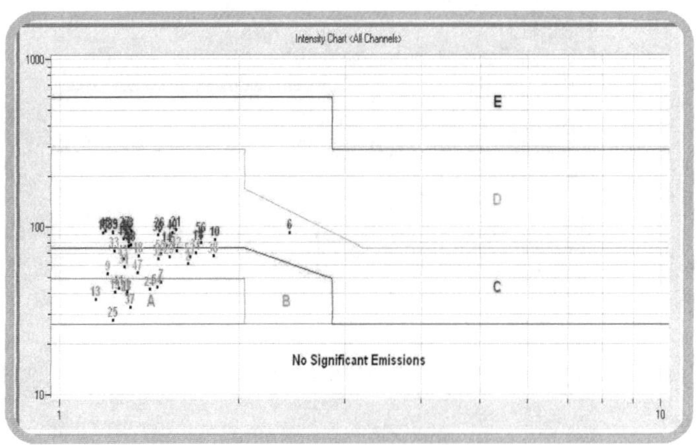

Pressure Vessels

Crack Locations

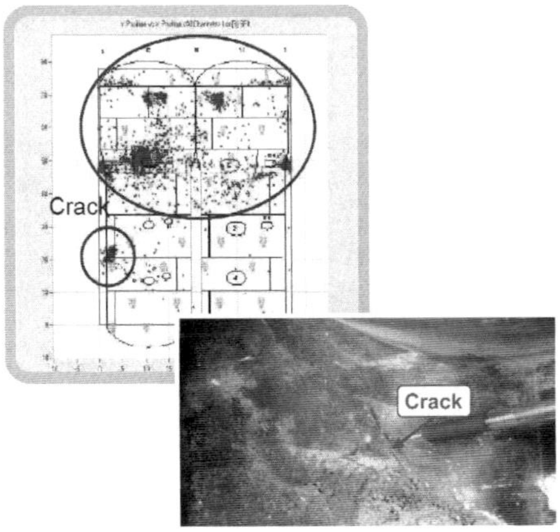

Sensor Positions

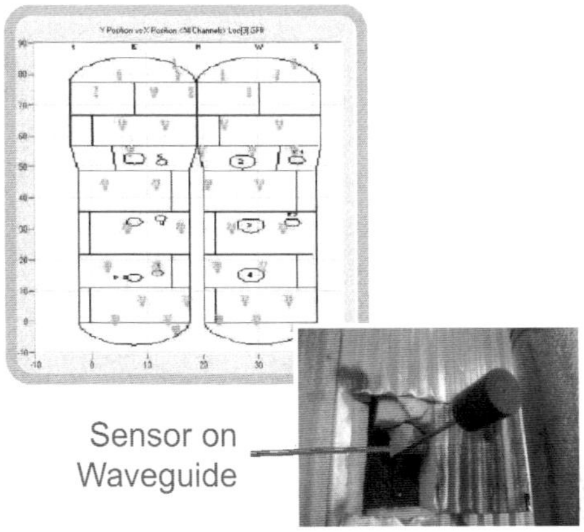

Online Detection of Stator Vane Cracking Using AE

GE F-Class Gas Turbine

- ✓ Compressor Stator vanes
  - ✓ Alternate with Rotor blades
  - ✓ Largest in the front
  - ✓ Numbered from 0 through 17
    - ✓ i.e. S0, S3

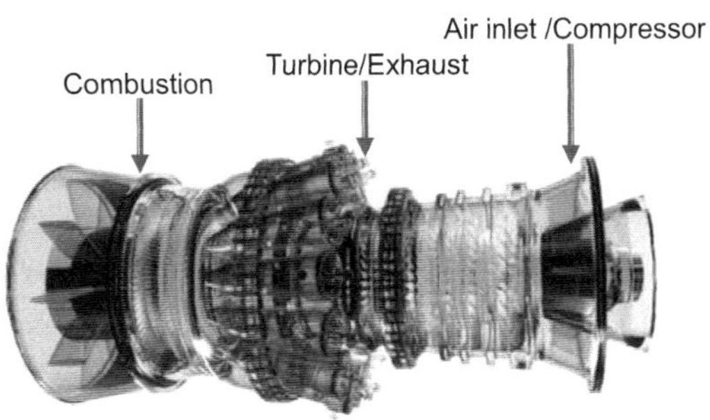

Stator Vane Cracking Using AE

Cluster of Activity over S1 row

Sensor Locations

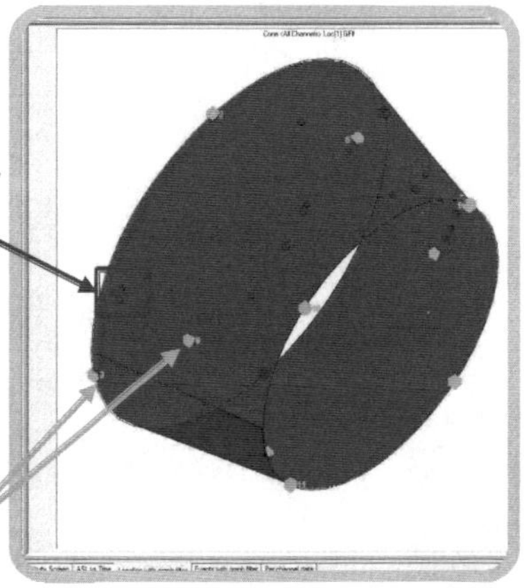

Crack Detection, Fretting and Clashing

S1 Vane crack found in location of ACTMS indication

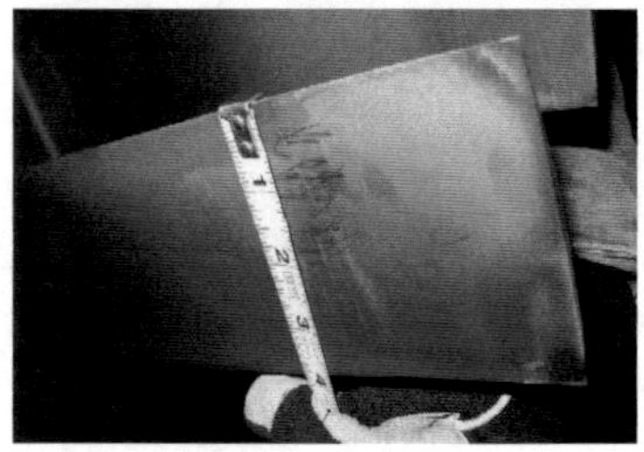

Turbine B crack cluster location

NDT in Canada 2019 | June 18-20 | Edmonton, AB

High-Energy Piping (HEP)

- ✓ Creep
- ✓ Fatigue
- ✓ Thermal fatigue
- ✓ Creep-fatigue
- ✓ Microstructural instability
- ✓ Flow-accelerated corrosion

AE is the best possible technology to:

- ✓ Detect signals as micro and macro cracking occur
- ✓ Easily deployed for continuous monitoring

NDT in Canada 2019 | June 18-20 | Edmonton, AB

High-Energy Piping (HEP)

- ✓ All channels receive data all the time
- ✓ Less likely to loose sudden crack jumps
- ✓ It requires higher count of channels in the system

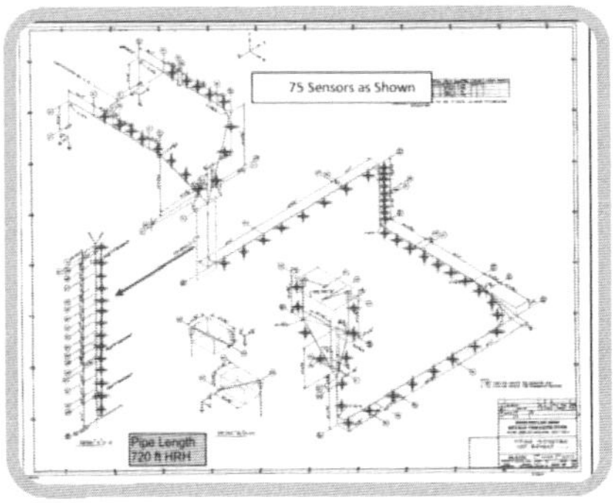

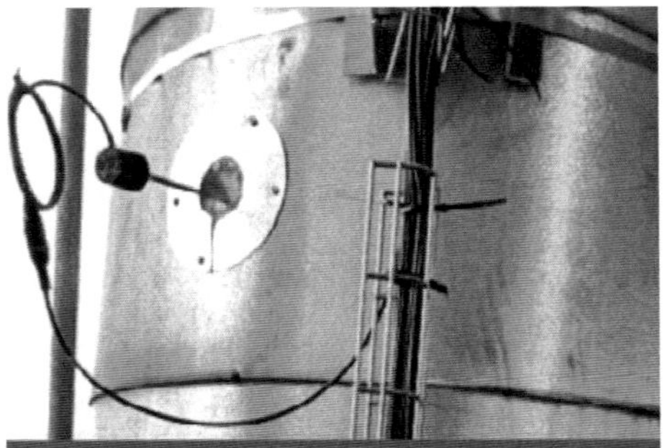

Permanently-mounted sensor

Ultrasonic Thickness Monitoring

- ✓ Very good accuracy of readings
- ✓ Not necessary to have personnel access
- ✓ Can be used at High Temperature (up to 550 °C)
- ✓ Identify corrosion issues at an early stage
- ✓ Improved corrosion management

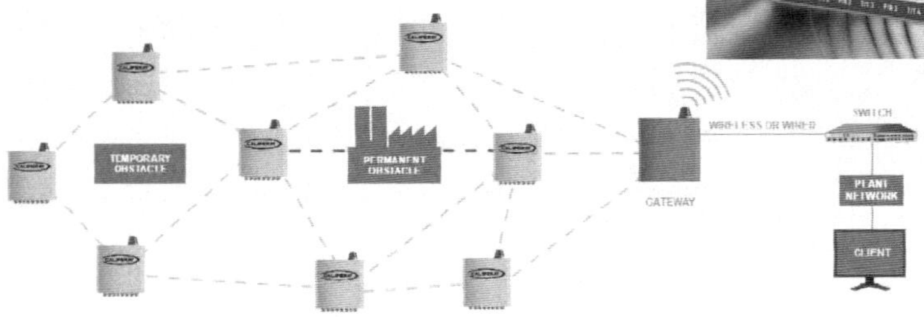

Ultrasonic Thickness Monitoring

ADVANTAGES
Non-intrusive design and installation
Highly precise readings
High temperature functionality
Self-organizing and self-healing network
Increased reading intervals
Improved corrosion management & detection equiptment
Reduced installation costs (no welding or wires required)

KEY FEATURES
Single Channel Pulser and Receiver
Four Channel Multiplexer for 4 Single or Dual Crystal Transducers
Receiver: 20 - 70 dB Gain, 1 MHz - 8 MHz Bandwidth
Pulser: Square Wave
Smart Sensor Network Wireless Communication
Ultra Low Power - 5 year Battery Life
Intrinsically Safe Certified

Ultrasonic Thickness Monitoring

Transducers

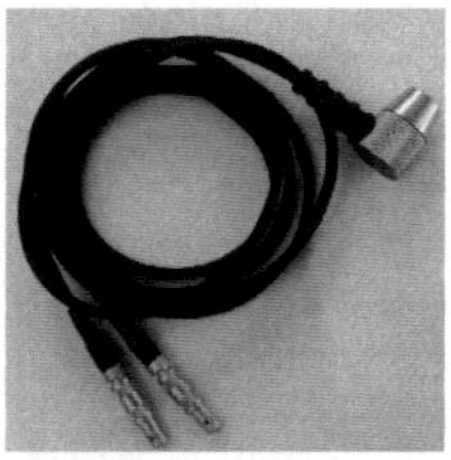

- Dual crystal transducer
- Operating temperature up to 150 °C
- Thickness measurement range between 0.1” to 2.5”
- Gel or glue coupling

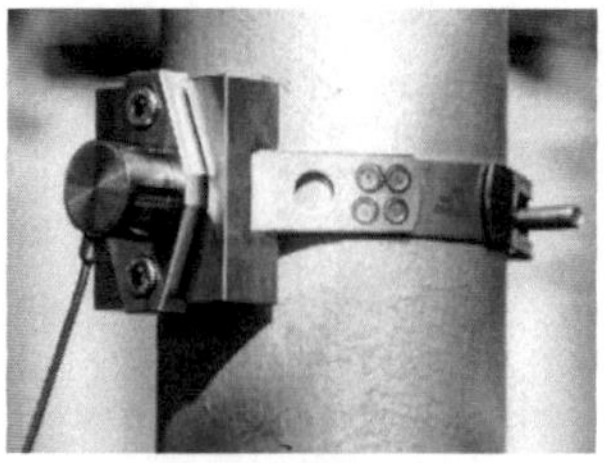

- Single element transducer
- Built-in thermocouple for temperature compensation
- Operating temperature up to 350 °C
- Thickness measurement range between 0.1” to 0.9”
- Dry coupling (metal foil)

NDT in Canada 2019 | June 18-20 | Edmonton, AB

Ultrasonic Thickness Monitoring

Web Application Management Portal (WAMP)

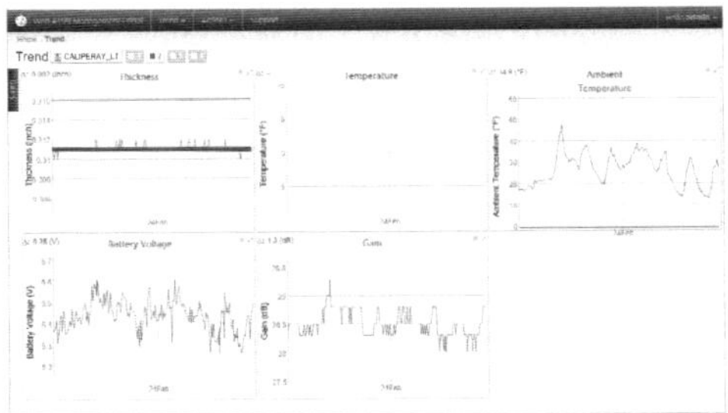

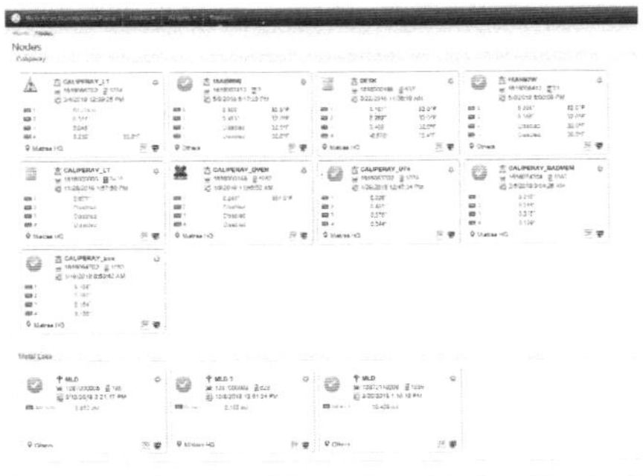

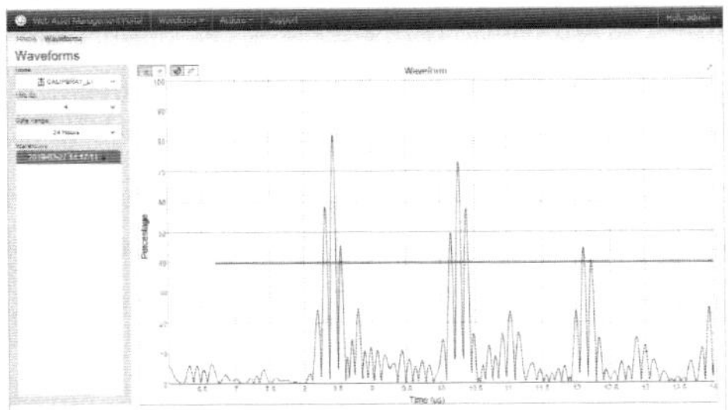

NDT in Canada 2019 | June 18-20 | Edmonton, AB

Metal Loss Detector (MLD)

- A novel type of sensor for monitoring and measuring on-line corrosion in process piping and vessels with high sensitivity
- System connects to existing or new WirelessHART networks.
- Measurements are made available for analysis through the monitoring host software suite
- Provides continuous automated monitoring of piping in corrosive environments and early detection of unexpected corrosion rates
- Allows rapid evaluation of risk while piping systems remain online

NDT in Canada 2019 | June 18-20 | Edmonton, AB

Metal Loss Detector

- Based on Tuning Fork Principle
- Two probes are required, a reference probe and a measurement probe.
- Probes compensate for temperature, flow, and pressure variations, so measurements are not affected by non-corrosion environmental conditions
- High Temperature – up to 450° C.
- High Pressure Operation
- Long Life – 6 years at corrosion rate of 10 mpy
- High Resolution – 0.3 mil
- Usable in conductive and non-conductive environments
- Almost 6 years of in-service experience & data

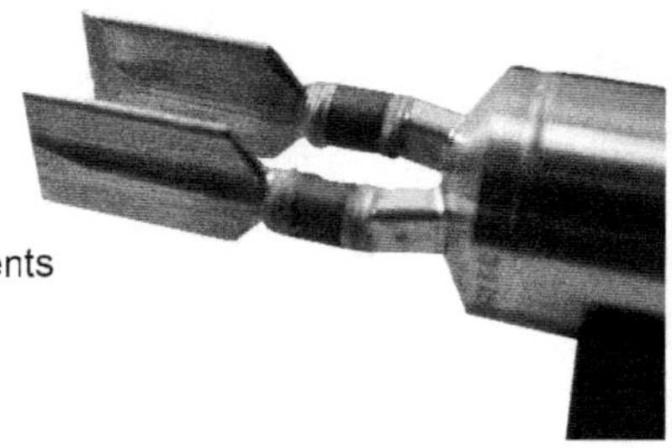

Metal Loss Detector (MLD)

Basic Configuration

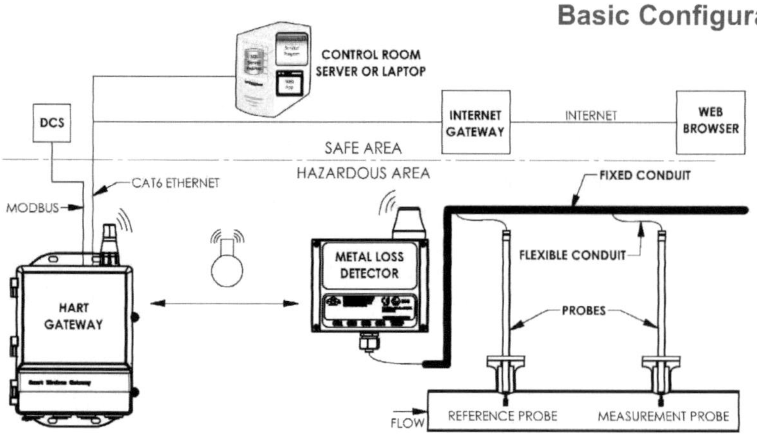

Metal Loss Detector

Trending of Corrosion Rates

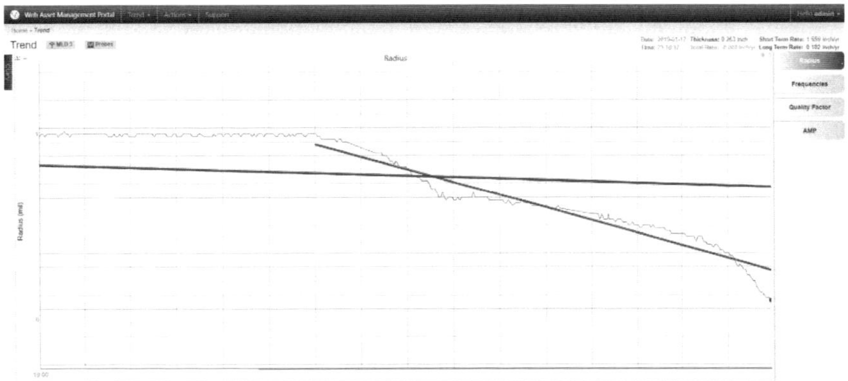

Summary

Multiple applications have been developed in collaboration with asset owners, in order to:

- ✓ Ensure safe operation of assets
- ✓ Determine the condition of the asset at the time of evaluation
- ✓ Extend inspection intervals
- ✓ Identify necessity and level of inspection follow up

THANK YOU

Eloy Gasperin
Technical Leader MISTRAS Group
Eloy.Gasperin@mistrasgroup.com

NDT in Canada 2019 | June 18-20 | Edmonton, AB

Worldwide patented PA UT

Automated UT Industrial applications

FAAST-PA delivers high speed inspection for Tubes, Plates, Bars/Billets, Rail, Turbine discs and more…

CINDE
CANADIAN INSTITUTE FOR NON-DESTRUCTIVE EVALUATION

Speaker: Chris Chollet

NDT in Canada
NDTiC 2019
Inform - Engage - Advance
June 18 - 20
River Cree Resort & Casino
Edmonton, Alberta

SUMMARY

I

Socomate International Overview

The Phased Array FAAST-PA Technology

II

Inspection of Bars & Billets

Inspection of Titanium & Aluminium Plates

III

Industrial references using FAAST-PA

Socomate International overview

- Created in 1977, based in France
- 40 years experience in NDT environment for UT inspection

- Core Business: development of high-end UT Electronics
- Worldwide Sales: over 40 countries

- Design, manufacture and provide OEM UT Electronics
- Conventional UT and PA UT Instruments for custom systems manufacturers

Socomate International product range

SOCO-I-UT | SOCO-8S-UT | SOCO-8P-UT | SOCOSCAN-PA | FAAST-PA

Common DLL for Conventional and PA UT working with any languages on Windows 7 and 10

OEM stand alone products, working with Ethernet connection

Worldwide patent on FAAST-PA

What are the differences between Conventional PA and FAAST-PA ?

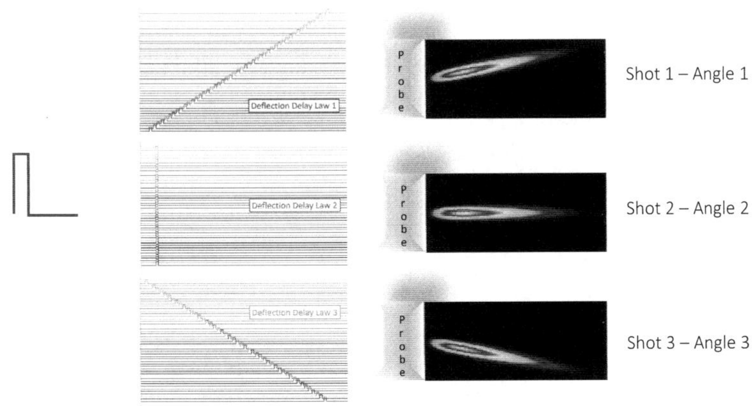

What are the differences between Conventional PA and FAAST-PA ?

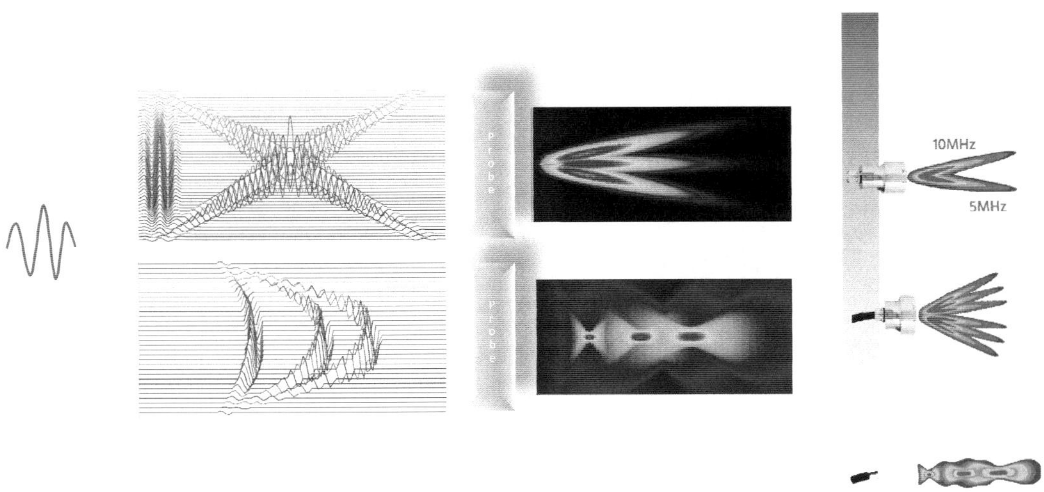

Bars and Billets Inspection using FAAST-PA

7

Bars/Billets Inspection - Context

INTERNAL FLAWS

SURFACE FLAWS (LONGITUDINAL CRACKS)

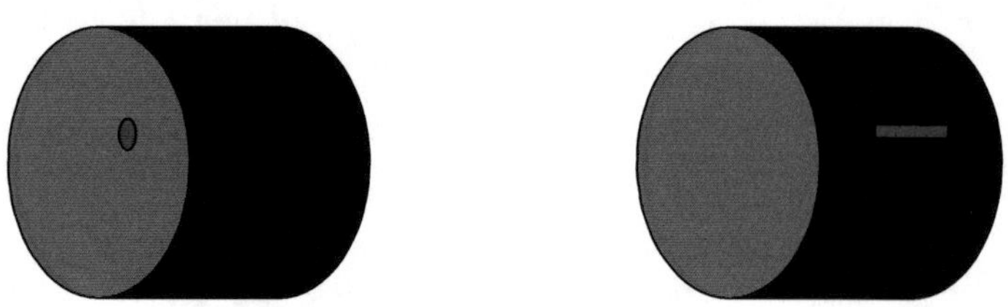

NDT in Canada 2019 | June 18-20 | Edmonton, AB

8

Bars/Billets Inspection – Inspection's Principles

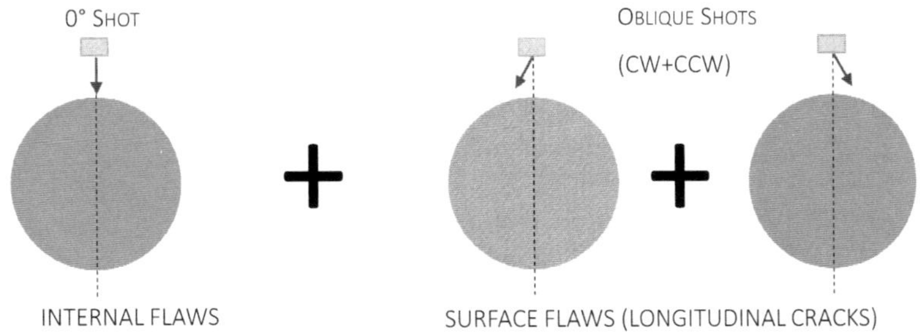

- A CONVENTIONAL PHASED ARRAY INSTRUMENT WILL REQUIRE 3 SEQUENTIAL SHOTS TO COVER THE FULL INSPECTION

Bars/Billets Inspection – FAAST-PA Solution

- FAAST-PA INSPECTION : ONLY **ONE SHOT** IS REQUIRED FOR MULTIPLE ANGLES

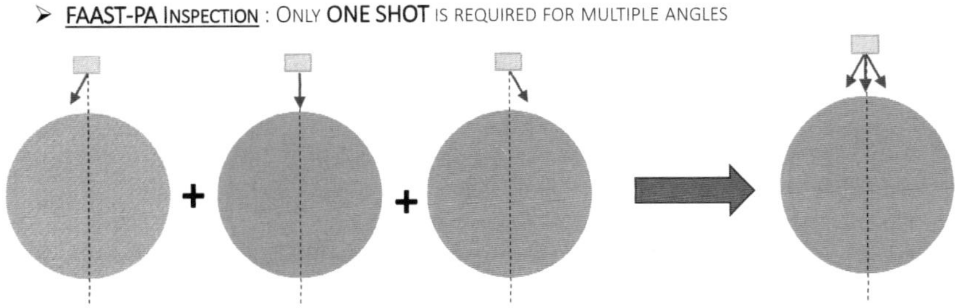

- IN THIS CONFIGURATION, **FAAST** ACQUISITION IS **3 TIMES FASTER** THAN CONVENTIONAL PA

Bars/Billets Inspection – Comparison between solutions

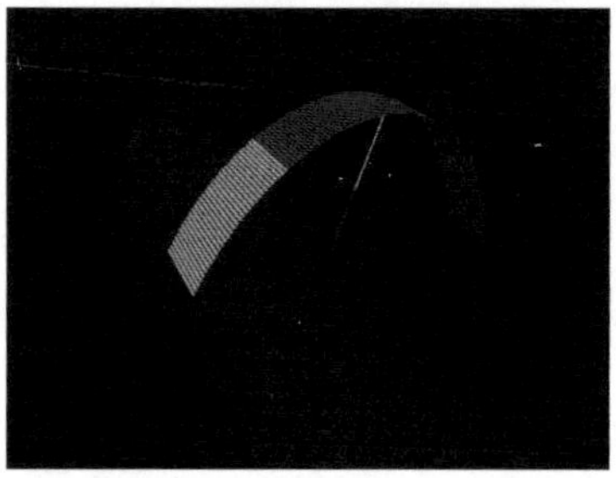

USE OF A 1D LINEAR PROBE R60

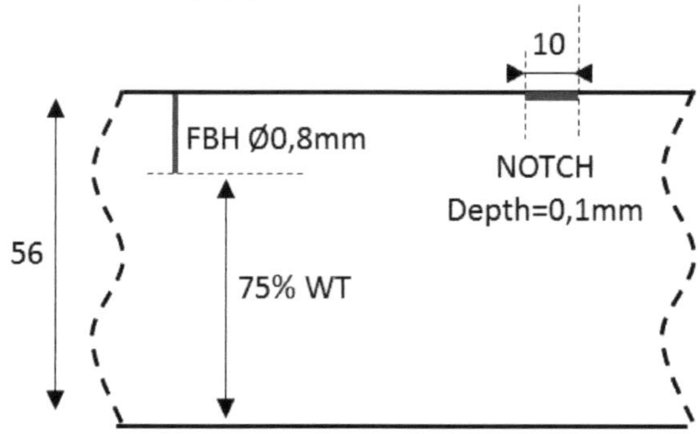

Bars/Billets Inspection – Comparison between solutions

FBH detection

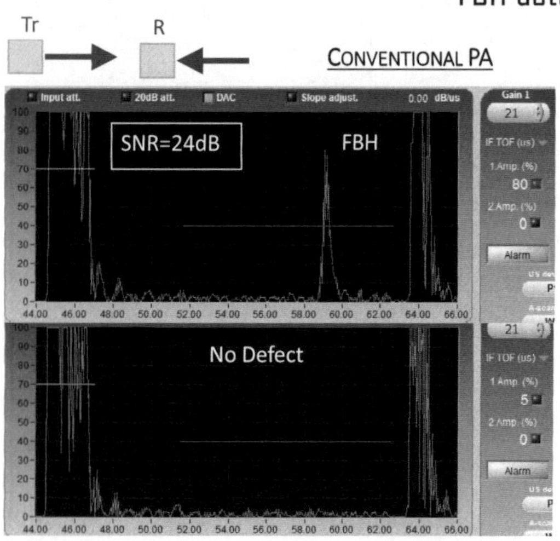

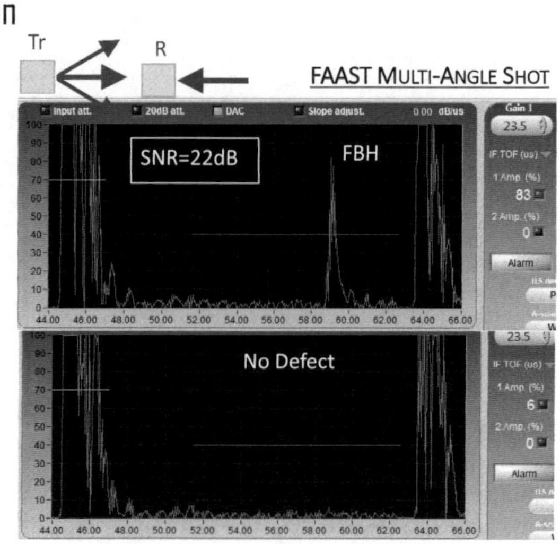

Bars/Billets Inspection – Comparison between solutions

Notch detection

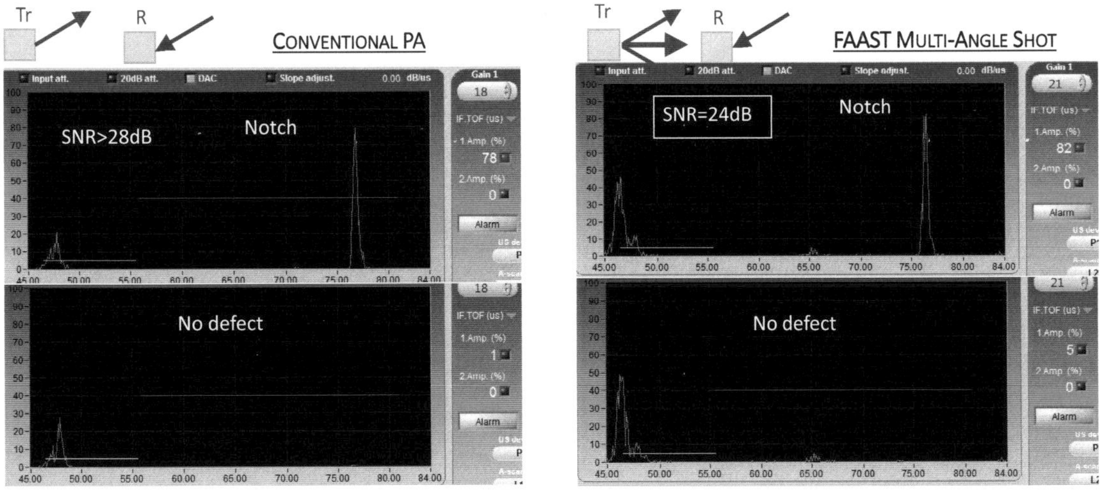

NDT in Canada 2019 | June 18-20 | Edmonton, AB

13

Titanium Plate Inspection using FAAST-PA

14

Titanium Plate– Context

- Detection of flaws at different depths
- Looking for 0,8mm FBH
- From 15mm to 185mm with defects at : 15, 25, 40, 80, 135, 185

- Material: Titanium TA6V
- Probe: 1D Linear PA, 10MHz

Titanium Plate– Inspection using conventional PA

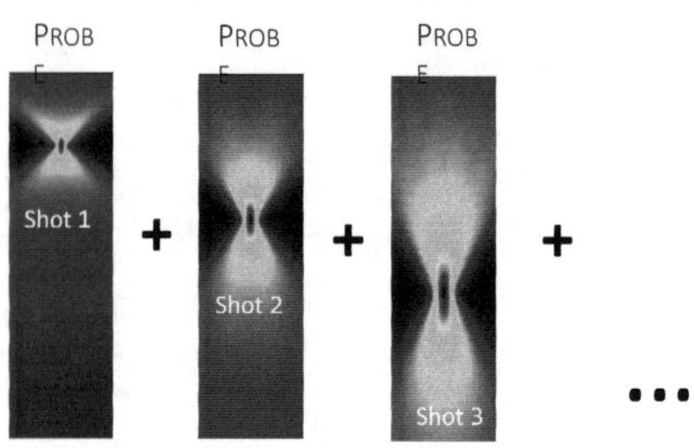

- The principle of conventional PA is the use of different US focused beam for each zone

- This require several sequential shots which lead to time consuming

Titanium Plate– Inspection using conventional PA

SEQUENTIAL UNIFOCAL SHOTS

FI = OPTIMAL FOCALE FOR FBH N° I

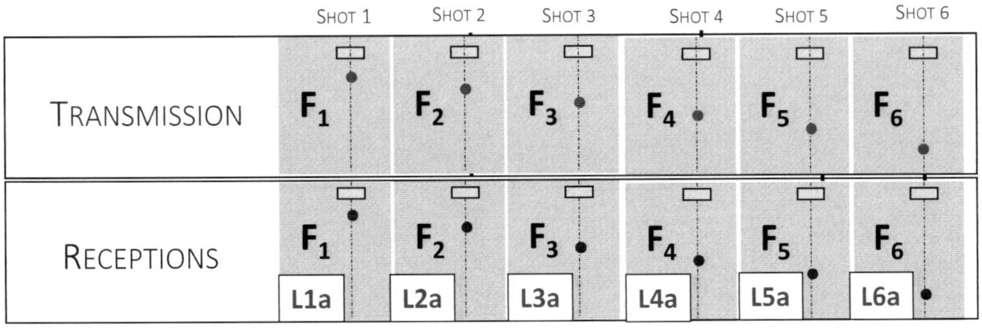

Titanium Plate– Inspection using FAAST-PA

- BASED ON FAAST-PA, ONE SHOT IS REQUIRED FOR MULTIPLE DEPTH FOCUSING

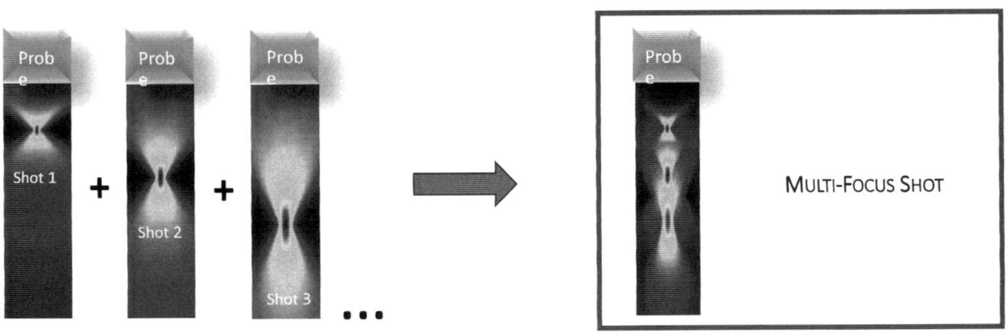

Titanium Plate– Inspection using FAAST-PA

- Example of FAAST-PA using Multiple focal within 1 US Shot
- Multi-Focus Transmission + DDF Reception ▶ **Advanced DDF**

Multi-Focus Shot

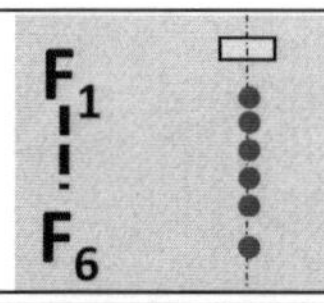

DDF Receptions

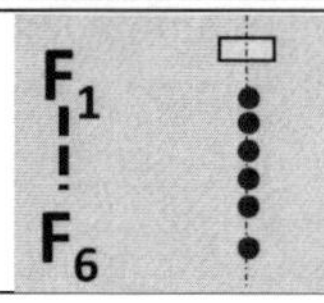

Titanium Plate– Results Comparison

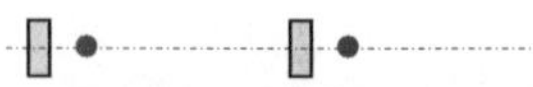

Conventional PA

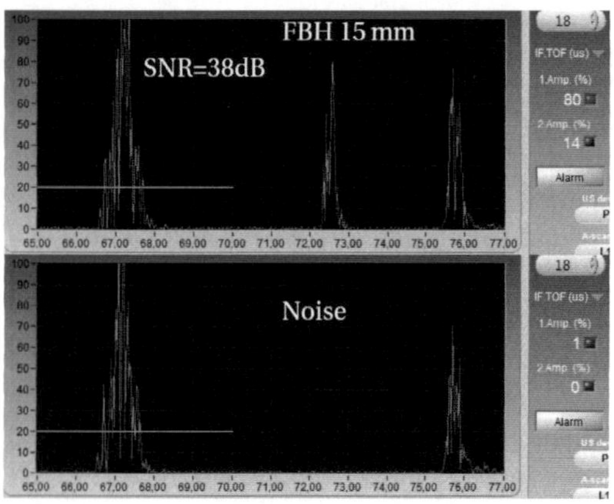

FAAST Advanced DDF

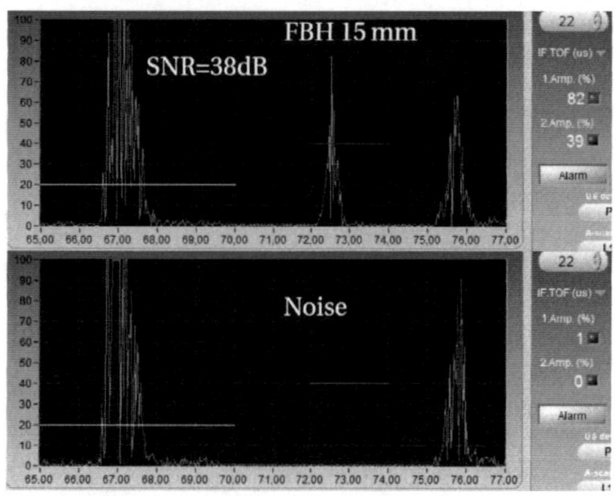

Titanium Plate– Results Comparison

CONVENTIONAL PA

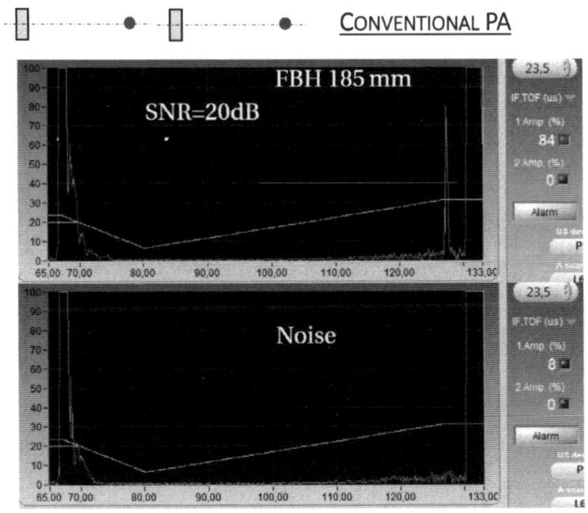

FAAST ADVANCED DDF

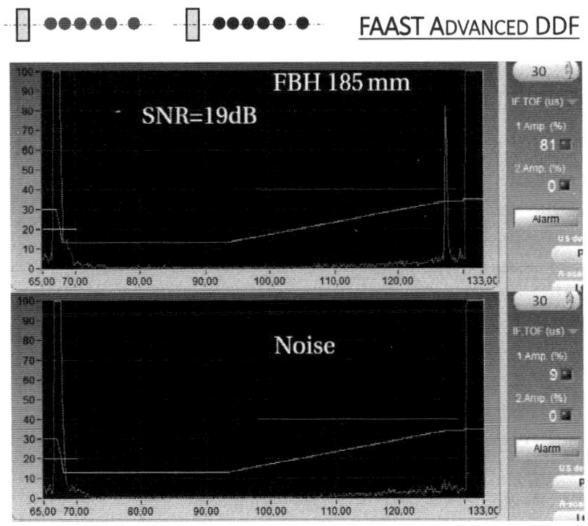

NDT in Canada 2019 | June 18-20 | Edmonton, AB

21

Aluminium Plate Inspection using FAAST-PA

22

Aluminium Plate– Context

MECHANICAL SCANNING SPEED: 700MM/S

MAXIMUM PRF 800HZ (LIMITED BY GHOST ECHOES)

PITCH: 1MM WITH 3 DB REPEATABILITY

SPECIFICATION:

1,2MM FBH

FROM 1,5MM TO 190MM

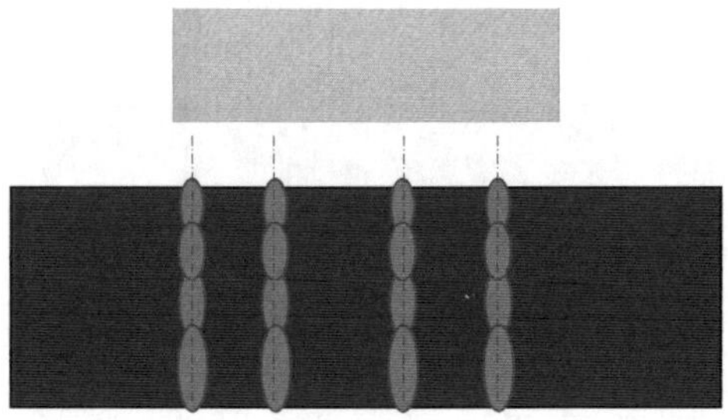

Aluminium Plate– Laboratory results

- ALUMINIUM BLOC TEST WITH FBH Ø1.2MM

NEAR ZONE DETECTION

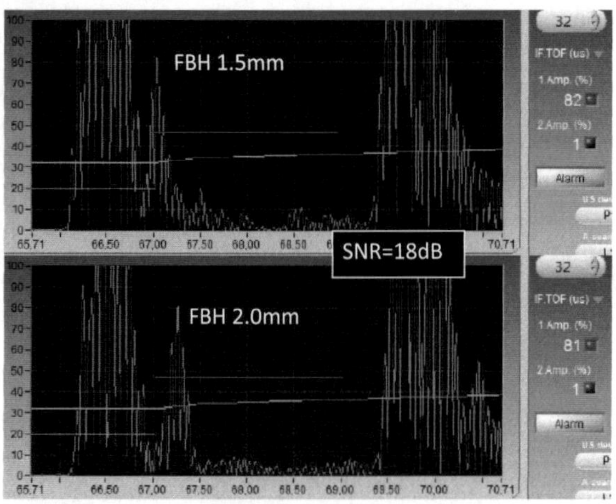

Aluminium Plate– Laboratory results

- Aluminium bloc test with FBH Ø1.2mm

Deep zone detection

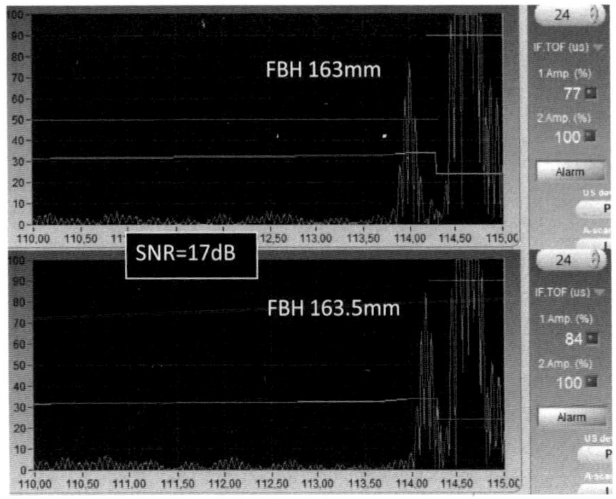

Aluminium Plate – On-site results

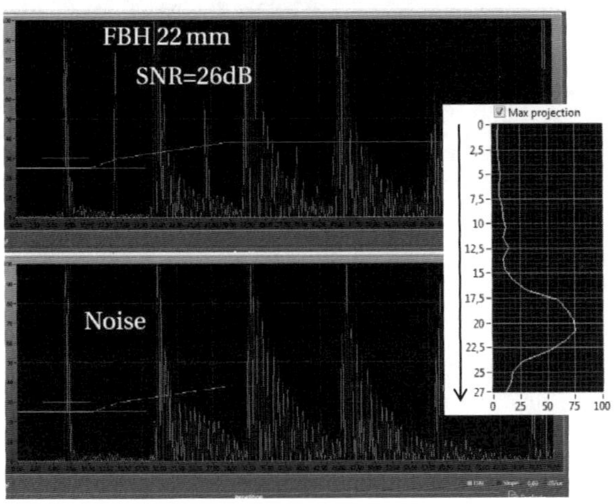

- Aluminium plate with FBH Ø1.2mm

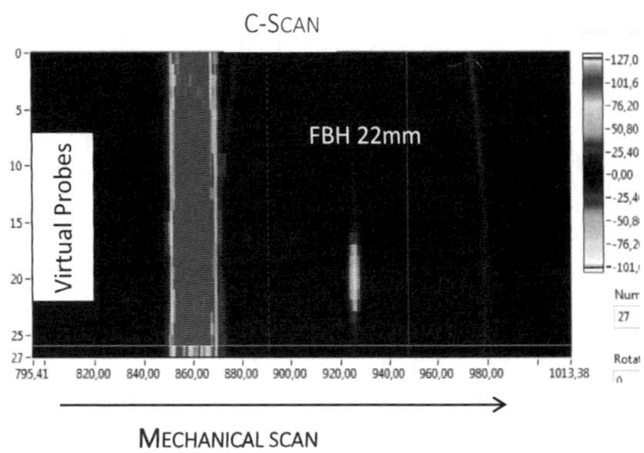

Aluminium Plate – On-site results

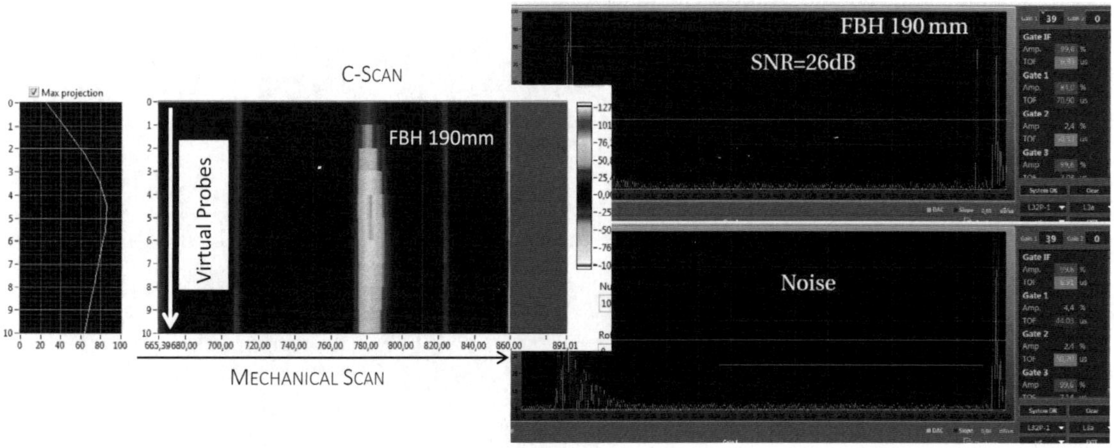

NDT in Canada 2019 | June 18-20 | Edmonton, AB

INDUSTRIAL REFERENCES USING FAAST-PA

Aircraft turbine discs using Multi-zone process

- Use of multi-focus and multi frequency with only one 128 elements 2D matrix PA probe 10MHz
- High speed rotating inspection
- Specification:
  - Titanium
  - 0,4mm FBH
  - From 2,5mm to 140mm
  - Inspection of 7 zones in 2 shots

29

OCTG seamless tubes overview

- Use of Multi-angle, Multi-frequency and Multi-beam acquisition mode using 1D & 2D Matrix PA probe
  - 1D: Lamination and Wall thickness
  - 2D: Flaw detection
- Specification:
  - Transverse, Longitudinal, all oblique flaws inner and outer
  - Real time to inspect 100%
  - 1,5m/s rotating tube
  - 58mm pitch

30

Rail inspection at high speed

- Use of Multi-angle acquisiton mode using 1D Linear probe and conventional
- 1 probe, up to 16 directions in one single shot and processing signals in real time
- Specification:
  - 72 km/h
  - Inspection pitch: 4mm
  - 1 Linear probe for 8 directions,
  - 1 conventional for 0°, 55° and 70° offset

http://cutewallpaper.org

31

High precision tube

- Use of Multi-angle acquisition mode using Bi-linear Phased Array ring probe
- 1 probe for Transverse, Longitudinal, inner and outer flaw detection
- Specification:
  - Minimum flaw depth of 5% of Wall thickness
  - Range from 15 to 50mm
  - Thickness range from 1 to 4mm
  - Linear speed of 15m/min
  - 4 ring probes to fully inspect the tube

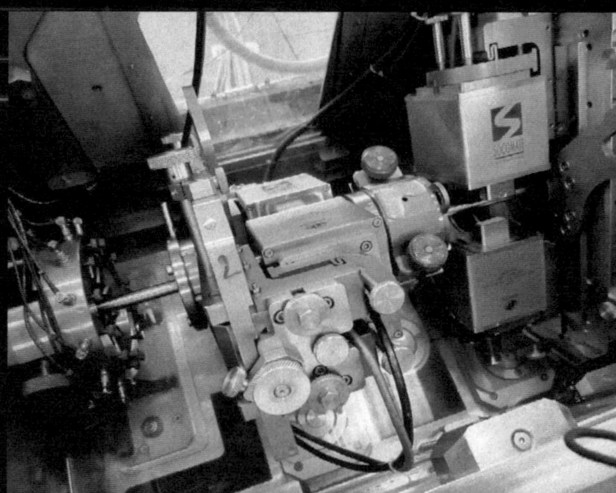

http://cutewallpaper.org

32

Thank you for your attention & participation
Any
Questions ?
33

Machine Learning in Pipeline Inspection: Applications of supervised learning in non-destructive evaluation

Roberto H. HERRERA [1], Paul CHRISTENSEN [1], Adrianus ELVERS [1]

[1] UT Quality Inc. a RAE Energy Company; Edmonton, Canada
Phone: +1 780 930 2802; Fax: +1 780 484 5664; e-mail: h.herrera@rae.com, p.christensen@rae.com, a.elvers@rae.com

Abstract
The large volume of data generated during pipeline inspection is usually reduced to key indicators while valuable information remains hidden in intricate relations and trends between these data. Machine Learning (ML) is a branch of Artificial Intelligence (AI) that helps to discover these hidden patterns. ML is becoming a ubiquitous tool in data science. It provides meaningful information beyond traditional statistical analysis. In this paper, we present an application of supervised learning, a method where the algorithm learns from previous classifications, and then independently performs classifications in new datasets. First, we use a training dataset from a known database with flaws classified by operators and validated by supervisors. Then, we apply the trained algorithm in new datasets from a large inspection database. Our method shows an accuracy above 97% in both flaw type and acceptance criteria classifications.

Keywords: Machine learning, pipeline inspection, supervised learning, flaw type classification, acceptance criteria.

1. Introduction

The classification of indications in non-destructive testing (NDT) is a labor-intensive task, subject to operator bias. The advancements in the electronics and inspection techniques are ultimately bounded by a human interpretation step that can be improved by using machine learning algorithms.

Machine learning (ML) is a subfield of Artificial Intelligence (AI) which uses statistical techniques to give computer systems the ability to learn from data [1]. Two conventional methods in ML are supervised and unsupervised learning. In the former, we need the input data, and some output data to learn the correct answers (e.g., classification and regression) in the latter the algorithm learns from the input data without knowing the output data (e.g., cluster analysis and association).

Supervised learning simulates the NDT process where the operator labels the indications according to parameters like shape, location, and inspection method. Multiple variables are influencing this labeling: operator's experience, inspection technique, material properties, etc. Supervised learning works on labeled data, which is the case of ultrasonic inspection by learning the patterns and associations between those labels. Once the algorithm has learned the classification model from a known dataset, it can be used in unclassified data. The known dataset is used to train the model and learn the classifications. In simple words, the algorithm learns from the indications classified by the operators, and then it could be applied to an unlabeled dataset and do the job with high accuracy.

In this paper with first describe the data preparation and the database system used by the UTScan® system, which allows the direct application of ML techniques to our current and past inspection databases. We then discuss the algorithm selection. The results section describes the accuracy in the classification with different datasets, and finally, we provide conclusions and future work.

2. Getting your data ready for Machine Learning

Data collected during the NDT inspection is crucial to ensure the safety of pipelines. However, the consistency of the results and the reliability of the findings are heavily dependent on the personnel performing the inspections and evaluation of the results [2]. The good news is that ML algorithms can operate over large datasets and learn the classification for multiple indications from different operators. Even, if some indications have been incorrectly classified, thanks to the central limit theorem, the error in large numbers will significantly reduce. Hence the classification accuracy will improve.

Pipeline inspection data is the perfect case to apply a supervised learning algorithm. The database holds information about the indications, such as depth, height, length, etc., together with the identification, e.g., incomplete fusion, porosity, lack of fusion, etc. These are the labels the operator used to classify the indications along with their evaluation as acceptable or rejectable.

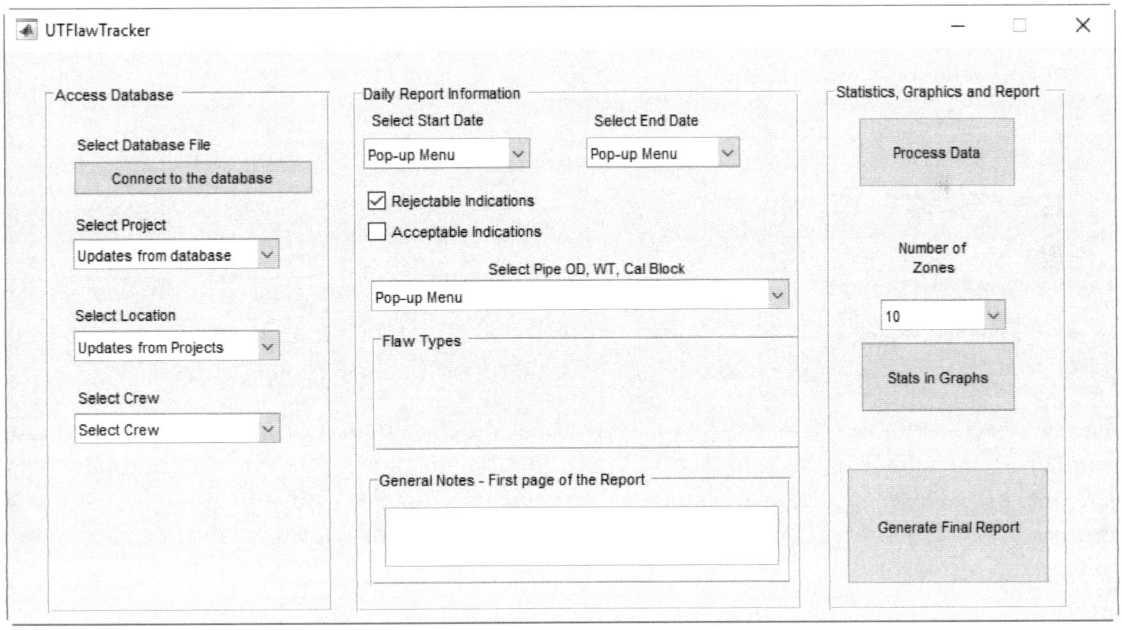

Figure 1: UT Flaw Tracker©, filters the dataset to feed the proper data into the ML algorithm.

In principle, ML learns from data, but you have to feed the right data into the algorithm to get valuable results. We have built an in-house statistical tool (UT Flaw Tracker©) which filters the data before the ML step. The UT Flaw Tracker© connects to the database and sequentially filters out the data by selecting: Project, Location, Crew, Dates, Evaluation (Rejectable, Acceptable), Pipe OD, Wall Thickness and Cal Block (you can have different pipe diameters in the same project, the same with WT and calibration blocks). After selecting these parameters, you are prompted to choose the flaw types or identified indications. A sample filtered looks like Table 1.

Table 1: A subset of the UT Flaw Tracker© output table used to perform statistical calculations and to feed the ML algorithm.

WeldNum	Result	Class	Ident	Label	WeldSide	Start	End	Length	Depth	Height	WallThickness	ScanDateTime
Weld01	Accepted	LB	IF	VOL-3	UpStream	274.0	282.0	8.0	4.3	1.0	19.1	9/28/2017 0:05
Weld02	Rejected	LB	IF	VOL-Cap	UpStream	366.0	475.0	109.0	1	1.0	19.1	9/28/2017 0:35
Weld03	Accepted					0.0	0.0	0.0	0	0.0	19.1	9/28/2017 0:38
Weld04	Accepted					0.0	0.0	0.0	0	0.0	19.1	9/28/2017 0:50
Weld05	Accepted					0.0	0.0	0.0	0	0.0	19.1	9/28/2017 1:34
Weld06	Rejected	LB	IF	TOFD	N/A	331.0	358.0	27.0	16.1	1.0	19.1	9/28/2017 1:54
Weld07	Accepted	LB	IF	TOFD	N/A	281.0	286.0	5.0	15.4	1.0	19.1	9/28/2017 2:07
Weld08	Accepted	LB	IF	TOFD	N/A	433.0	437.0	4.0	13	1.0	19.1	9/28/2017 2:24
Weld09	Rejected	LB	IF	TOFD	N/A	160.0	182.0	22.0	16.3	1.0	19.1	9/28/2017 2:45
Weld10	Rejected	VR	VOL	TOFD	N/A	238.0	250.0	12.0	19.1	3.5	19.1	9/28/2017 3:25
Weld11	Rejected	VI	P	TOFD	N/A	451.0	495.0	44.0	16.5	1.3	19.1	9/28/2017 3:37
Weld12	Accepted					0.0	0.0	0.0	0	0.0	19.1	9/28/2017 3:42
Weld13	Accepted					0.0	0.0	0.0	0	0.0	19.1	9/28/2017 4:18
Weld14	Accepted					0.0	0.0	0.0	0	0.0	19.1	9/28/2017 4:34
Weld15	Accepted					0.0	0.0	0.0	0	0.0	19.1	9/28/2017 4:55
Weld16	Accepted					0.0	0.0	0.0	0	0.0	19.1	9/28/2017 5:13
Weld17	Accepted					0.0	0.0	0.0	0	0.0	19.1	9/28/2017 5:27
Weld18	Accepted	LB	IF	VOL-Cap	UpStream	517.0	8.0	20.0	1.6	1.0	19.1	9/28/2017 5:55
Weld19	Accepted	LB	IF	TOFD	N/A	370.0	375.0	5.0	16	1.0	19.1	9/28/2017 6:12
Weld20	Accepted	LB	IF	VOL-1	UpStream	121.0	127.0	6.0	13.5	1.0	19.1	9/28/2017 6:27
Weld21	Rejected	LB	IF	TOFD	N/A	182.0	243.0	61.0	16.2	2.5	19.1	9/28/2017 6:46
Weld22	Accepted	LB	Geo	CL	UpStream	524.0	50.0	55.0	0	0.0	19.1	9/28/2017 7:02
Weld23	Accepted	LB	Geo	CL	DownStre	493.0	50.0	86.0	0	0.0	19.1	9/28/2017 7:28
Weld24	Accepted	LB	IF	VOL-4	UpStream	252.0	254.0	2.0	4.3	1.0	19.1	9/28/2017 7:42
Weld25	Accepted	LB	Geo	CL	DownStre	472.0	518.0	46.0	0	0.0	19.1	9/28/2017 7:54
Weld26	Rejected	VR	VOL	TOFD	N/A	258.0	270.0	12.0	19.1	3.5	19.1	9/28/2017 8:18
Weld27	Accepted	LB	Geo	CL	UpStream	426.0	451.0	25.0	0	0.0	19.1	9/28/2017 8:49
Weld28	Accepted					0.0	0.0	0.0	0	0.0	19.1	9/28/2017 9:11
Weld29	Accepted					0.0	0.0	0.0	0	0.0	19.1	9/28/2017 9:28
Weld30	Accepted	LB	Geo	CL	DownStre	483.0	62.0	108.0	0	0.0	19.1	9/28/2017 9:47
Weld31	Accepted	LB	Geo	CL	DownStre	509.0	57.0	77.0	0	0.0	19.1	9/28/2017 10:27
Weld32	Accepted	LB	Geo	CL	DownStre	445.0	33.0	117.0	0	0.0	19.1	9/28/2017 10:44
Weld33	Rejected	VR	VOL	TOFD	N/A	202.0	219.0	17.0	19.1	2.2	19.1	9/28/2017 11:13
Weld34	Accepted	LB	IF	Cap2	UpStream	250.0	279.0	29.0	2	1.2	19.1	9/28/2017 11:42
Weld35	Accepted	LB	Geo	CL	DownStre	467.0	520.0	53.0	0	0.0	19.1	9/28/2017 12:05
Weld36	Accepted	LB	Geo	CL	DownStre	476.0	14.0	67.0	0	0.0	19.1	9/28/2017 12:26
Weld37	Accepted	LB	Geo	VOL-4	DownStre	414.0	468.0	54.0	0	0.0	19.1	9/28/2017 12:43

The next steps involve the statistical analysis about the distribution of indications in the pipe, number of indications in the wall thickness and the frequency of occurrence of each flaw type and the percentage of the total amount. The statistical analysis already provides valuable information, but ML could be an additional tool for validation. Thus, we use the generated table to feed the ML algorithms.

3. Algorithm Selection

The algorithm selection is entirely data-driven. As there is no previous study on the dataset, the best option is to run different algorithms and compare their performances, to select the best-suited algorithm for your application — usually, the one consistently producing the best results.

We train the algorithm to learn from the inspection data to infer the classification of the indications with the same accuracy as the operators do or higher. The task can be described as:

I- Given the measured data from the UT system, classify the Flaw Types as Geometry, Incomplete Fusion, Porosity, etc. and evaluate the indication into Accepted it or Rejected.

The parameters we use to train the algorithm are called predictors; in this case:

- Flaw Identification
- Channel
- WeldSide
- Start
- End
- Length
- Depth
- Height
- WallThickness – Physical data

Another condition to impose to the algorithm is to be consistent and with an accuracy close to 100%; independently of your mood, the weather, your vision, or any other external situation affecting the inspection result and its reproducibility.

For the algorithm selection, we use the Classification Learner App in Matlab® [3] over a dataset of 766 indications.

We use the nine predictors listed above to train multiple algorithms, as is shown in Figure 2.

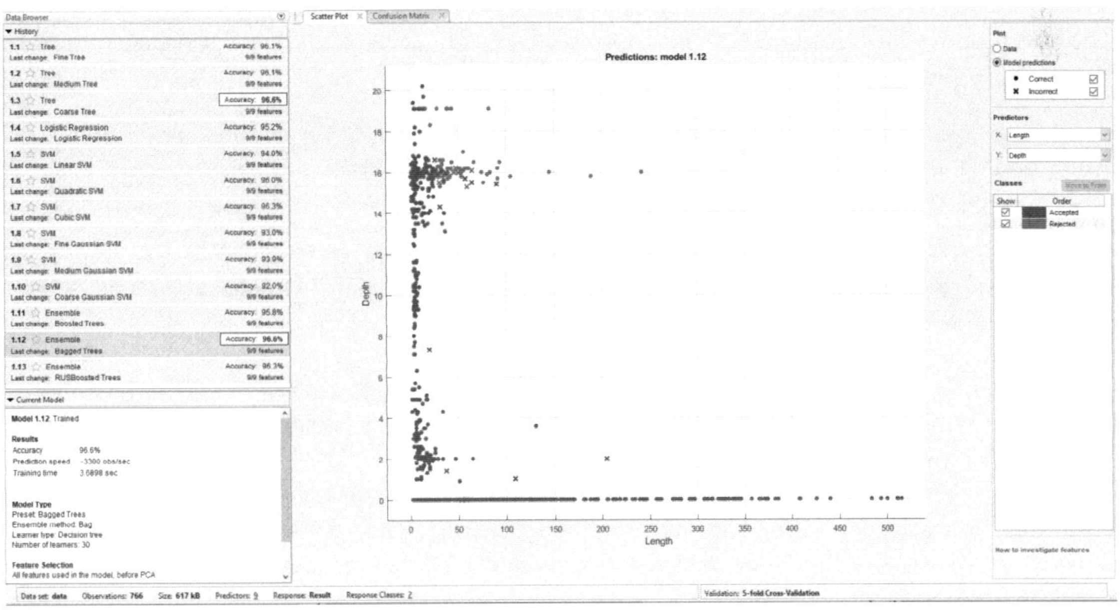

Figure 2: Training the multiple algorithms on the training dataset. An Ensemble method produces the highest accuracy of 96.6%.

The Confusion Matrix shows the true positive rate and false negative rate. We can interpret from the confusion matrix that for the accepted indications, the algorithm predicted class matches 99% of the true class, while only the 78% matches the rejected, meaning that we should expect more false negatives in rejected indications.

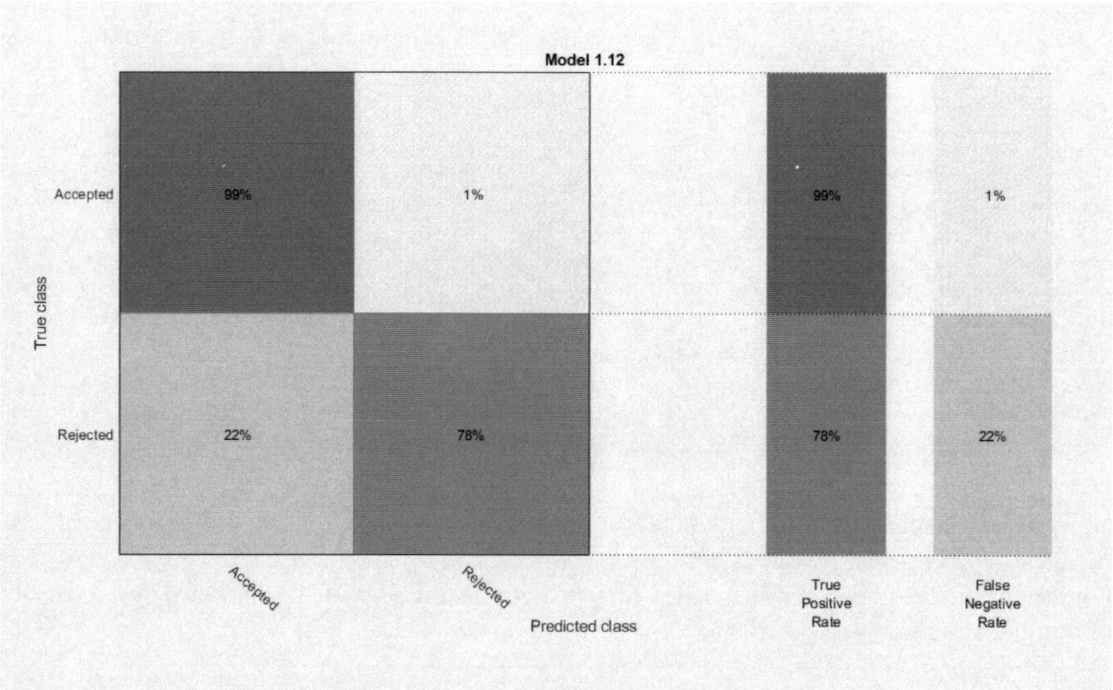

Figure 3. The Confusion Matrix.

The best performing algorithm was the ensemble Bagged Trees. With this algorithm, we create the model to run over the new unlabeled datasets.

4. Results

Once the Machine Learning algorithm produces a reasonable result, we process a new dataset with the Trained Algorithm.

A database from a recent project (2018) was used as a testing dataset. We ran the trained model over the 657 indications in this new dataset, hiding the labels and evaluation performed by the operators (flaw types and evaluation: accepted or rejected).

From 657 Indications, the ML algorithm correctly classified 654 in Accepted or Rejected for an accuracy of 99.5%. Only three indications were accepted that the algorithm thinks we should reject. After consultation with the specialist, these were close-calls. These indications were correctly classified by the operators but were so close to the acceptable limit that another operator could have had a different opinion.

For the flaw type, the accuracy was 97.56%, the algorithm misclassified 16 indications, but after comparing the results in Table 2, many of the indications were labeled as non-classified

(NC) by the ML algorithm. This NC label comes from the training dataset where operators marked but did not label the indication. In practice, only two indications (Geo as IF and IF as Non-Classified), were wrongly classified by the algorithm leading to an accuracy of 99.69%.

Table 2: Results of the ML Algorithm. Misclassified labels, 16 out of 657 indications. See the explanation on the text for the NC label.

Operator	ML Algorithm
{'Geo'}	{'NC' }
{'Geo'}	{'NC' }
{'Geo'}	{'IF' }
{'Geo'}	{'NC' }
{'Geo'}	{'NC' }
{'Geo'}	{'NC'}
{'Geo'}	{'NC' }
{'Geo'}	{'NC'}
{'Geo'}	{'NC' }
{'Geo'}	{'NC' }
{'Geo'}	{'NC' }
{'Geo'}	{'NC' }
{'Geo'}	{'NC' }
{'Geo'}	{'NC' }
{'Geo'}	{'NC' }
{'IF' }	{'Geo' }

3. Conclusions

The machine learning algorithm classified the new dataset with high accuracy, yet, at this stage, we advise against completely unsupervised classifications, because the method is intended as a guide instead of a fully automated blind approach.
We see the ML algorithm as a tool for the operator in the classification of flaw types and evaluation. Also, the algorithm could be used to validate the acceptance criteria.
We are working to integrate the Machine Learning functionality into the current version of the UTScan® system as a validating method.

References

[1] A. Ng, *Machine Learning Yearning*. Online Draft. [book.ml], 2017.

[2] A. Gonzalez Rodriguez, A. Gosselin, R. Rheaume, and N. Harrap, "Computational intelligence approaches for data analysis, the next step of innovation for advanced UT techniques in NDT," in *ECNDT 2018: 12th European Conference on Non-Destructive Testing (ECNDT 2018), Gothenburg 2018, June 11-15*, 2018, pp. 1–2.

[3] MATLAB, *version 9.5.0.1033004 (R2018b)*. Natick, Massachusetts: The MathWorks Inc., 2018.

PULSED EDDY CURRENT RESPONSE TO GENERAL CORROSION IN CONCRETE REBAR

I.C. Eddy[1,2], P.R. Underhill[1], J. Morelli[2], T.W. Krause[1]

[1] Royal Military College of Canada, Dept. of Physics and Space Science, Kingston ON
[2] Queen's University, Dept. of Physics, Engineering Physics and Astronomy, Kingston ON

CINDE CANADIAN INSTITUTE FOR NON-DESTRUCTIVE EVALUATION

NDT in Canada
NDTiC 2019
Inform - Engage - Advance
June 18 - 20
River Cree Resort & Casino
Edmonton, Alberta

Overview of Presentation

- Motivation & Background
- Theory
- Experimental Setup
- Test Methods & Results
  - Varying Rebar Diameter
  - Varying Liftoff Distances
  - Varying Angle
  - Inspection of Junction Nodes
- Conclusions
- Future Work

NDT in Canada 2019 | June 18-20 | Edmonton, AB

2

Rebar Motivation

- Generally a carbon steel
- Used in structures to stabilize concrete and maintain structural integrity
- Usually layered and cross-hatched
- Steel can corrode over time, causing damage to concrete and degrading tensile strength
- Goal is to determine reduced diameter due to corrosion, when compared to original diameter, independent of rebar depth into concrete structure

[1] https://ridgeend.files.wordpress.com/2014/10/ridge-end-2-224a.jpg

Theory

Magnetic field diffusion equation is given by:

$$\nabla^2 \boldsymbol{B} = \mu\sigma\left(\frac{\partial \boldsymbol{B}}{\partial t}\right) \qquad (1)$$

where B is the magnetic field, μ is material permeability and σ is material electrical conductivity.

Equation (1) has general solution of [1],

$$\mathbf{B}(t) = f\left(e^{-t/\tau_D}\right)$$

where τ_D is the diffusion time.

Magnetic flux, Φ, through a coil loop is given by:

$$\Phi = \int B \cdot da$$

where 'a' is the cross-sectional area of the pickup coil.

Theory Continued

- Following Faraday's Law [2], where V is voltage generated, and N is the number of coil turns:

$$V = -N\left(\frac{\partial \Phi}{\partial t}\right)$$

- The diffusion time for cylindrical geometry of radius R is given by [3]

$$\tau_D \sim \mu \sigma R^2,$$

which leads to the following relationship in semi-log space:

$$\log(V) = \left(-\frac{M}{\tau_D}\right) * t + \log(C)$$

where 't' is time, 'M' and 'C' are constants dependent on probe parameters

Transient Response due to PEC

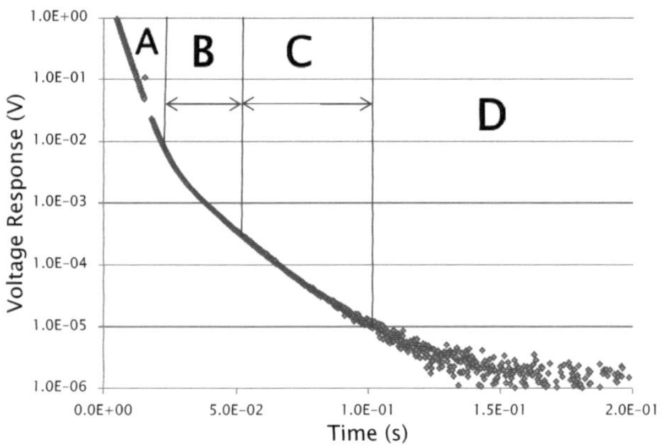

Figure 1 - Regions of the transient response worth note

- A: Decay of signal dominated by probe parameters
- B: The transition area where the response changes from exponential decay dominated by probe parameters to a decay dominated by sample parameters
- C: Useful long-time decay region that is analyzed
- D: Noise floor reached (1E-5V)

Experimental Setup

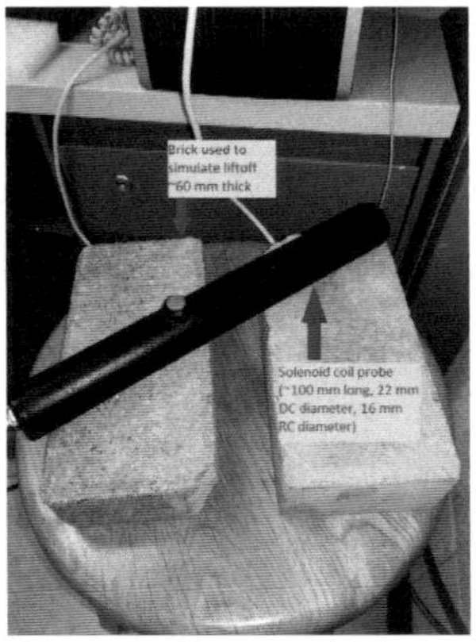

Figure 2 - Data acquisition setup

- As seen in Figure 2 on the left, the solenoid coil is within the upper half of the black rod
- The green button activates data acquisition
- Response from rebar samples, Figure 3 on right, was obtained at varying liftoff

Figure 3 - From left to right diameters (closest fractional inches): 1", 3/4", 5/8", 2/5" (untouched rebar samples), 7/8" (lathed rebar), 1 ½" (smooth rod)

Permeability, μ, and Conductivity, σ

- Expression for diffusion time:
  $\tau_D \sim \mu\sigma R^2$
- Eddy-current indicated all samples had similar permeabilities.
- 4-point measurement used to determine conductivity

Table 1 - Parameters of rebar, including conductivity (green)

Label	Radius(mm)	Conductivity (S/m)	Error (S/m)
1 1/2"	19.0	6.90E+06	3E+04
1"	12.7	6.84E+06	3E+04
7/8"	11.1	6.84E+06	2E+04
3/4"	9.6	6.68E+06	2E+05
5/8"	7.8	5.92E+06	4E+05
2/5"	5.6	6.69E+06	3E+05

Varying Diameter

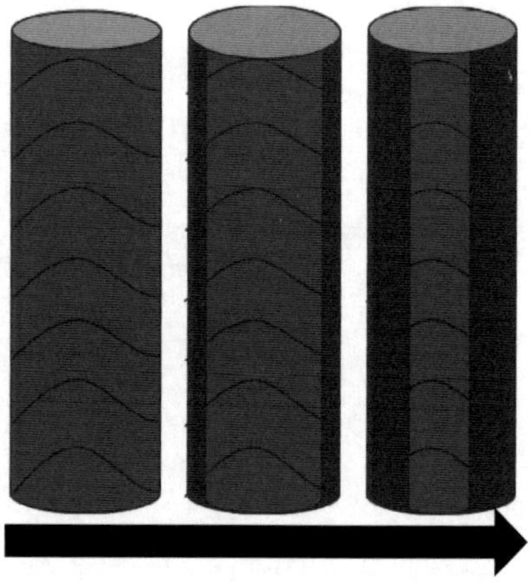

Figure 4 -Corrosion progression over time

- Ingress of water and salt into concrete may deteriorate rebar over time
- Effective strength relies on non-corroded areas
- Effective diameter picked up by PEC is the metallic middle unaffected by corrosion
- Fig 4 shows effective inner ferromagnetic diameter

Varying Rebar Diameter Data

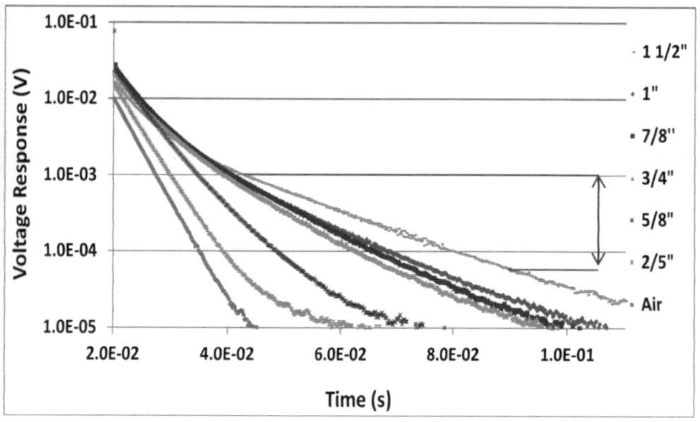

Figure 5 - The transient response to varying rebar diameters from a pulsed Eddy current solenoid coil.

- Noise floor reached at 1.00E-05V
- Solenoid used couples better with diameters close to 1" (~25 mm)
- For sake of initial data analysis, long term decay slopes were taken over the voltage response range of 1E-3 V to 7E-5 V.
- It is noted that the long-term decay transition is present for ½" & 5/8" at 0.048 s & 0.06 s, respectively, but approaches the noise floor.

Relationship Between Slope & Diameter(mm)

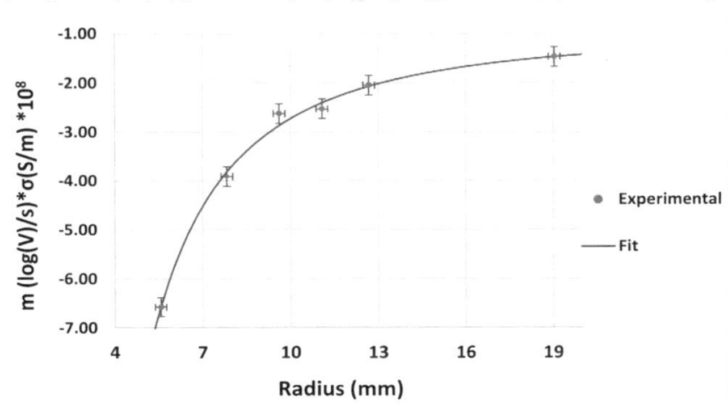

Figure 6 -Power law relationship between diameter of rebar and long time decay slope

- Values found through Matlab curve fitting tool
- y-axis values were graphed/fitted through $\log_{10}$ conversion
- Slope continues to increase in magnitude
- When slope is multiplied by conductivity, then relationship shows R^{-2} relationship
- Permeabilities were similar (except for carburized surface of 1" Sample)

$$m\sigma = (-173.2 * \boldsymbol{R}[mm]^{-2} - 0.9902)x10^8$$

Introducing Liftoff (Concrete Modeling)

- Signal strength is affected by distance of probe from rebar
- Noise floor is reached at ~1E-5 V with current system
- At what liftoffs can we still accurately collect data?
- What is the relationship between intercept of long-time decay and liftoff?
- Can the 2 methods be combined to determine depth and remaining diameter of rebar?
- 1" Sample was examined, as it gave strongest response

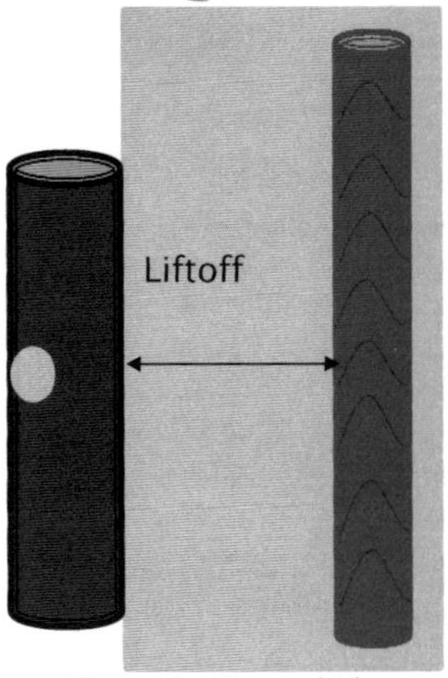

Figure 7 - General idea behind liftoff with rebar

Varying Liftoff for 1'' Rebar

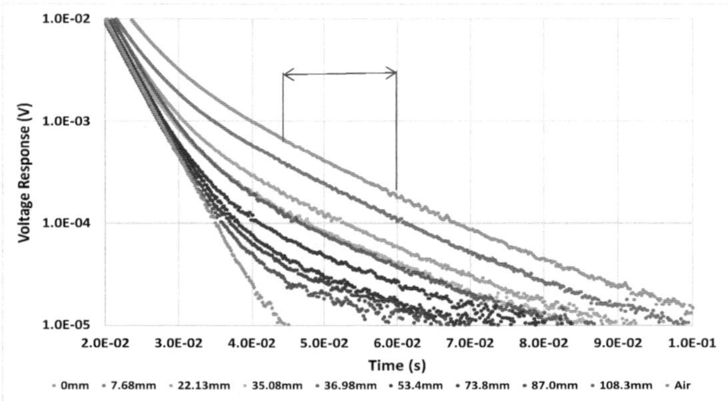

Figure 12 - Zoomed in area of the transient response on a semi-log graph for varying liftoff between solenoid probe and rebar. Response due to just air is also plotted.

- Slopes are all similar
- Voltage range of linear response in semi-log plot varies
- Slopes were at looked at over the range of 0.045s-0.06s

Relationship Between Intercept & Liftoff

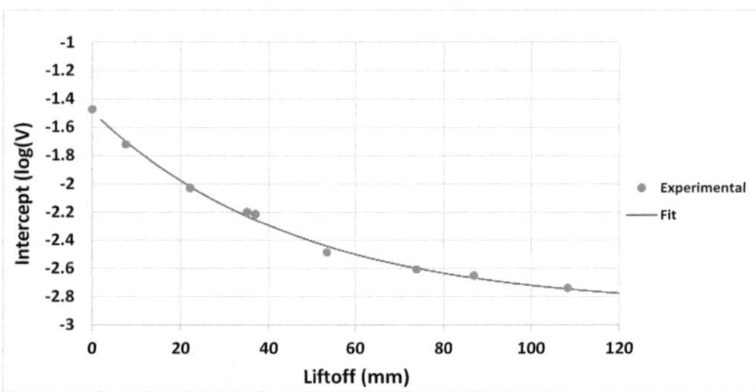

Figure 13 -Exponential decay relationship between liftoff from rebar and long time intercept

- As expected, slopes remain similar, while intercept values change.
- A discrete relationship between liftoff and intercept was found
- Exponential relationship present between intercept of the long time decay and the liftoff
- Decay term of 46 mm. The signal decreases by 39% every 46 mm.

$$b = 1.39 * e^{-LO/46} - 2.88$$

b - intercept value [log(V)]; LO - liftoff [mm]

Variance of Coupling Angle

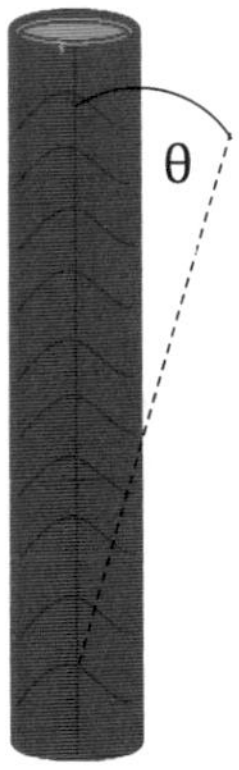

Figure 14 - Diagram of coupling angle

Coupling Angle

- Solenoid coil is being used to collect data
- Expected to have a stronger response when rod is inline with rebar (θ=0°)

Experimental Method

- 1" rebar was used
- Zero liftoff (in contact)
- Collected data at angles of 0°, 20°, 30°, 45° and 80°
- As θ increased, the slope of the long-term decay barely changed, while the response decreased

Coupling Angle Response

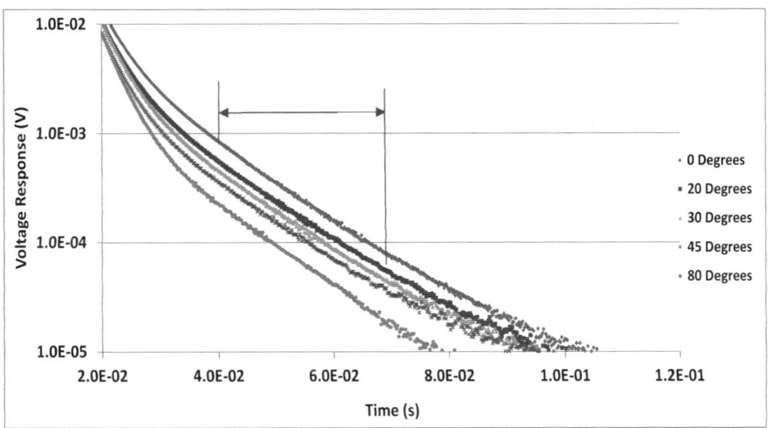

Figure 15- Transient response on a semi-log graph at varying angles between solenoid coil and rebar

- Long time decay section reached at ~0.03 s for all measurements
- Range up to 80° tested

Coupling Angle Results

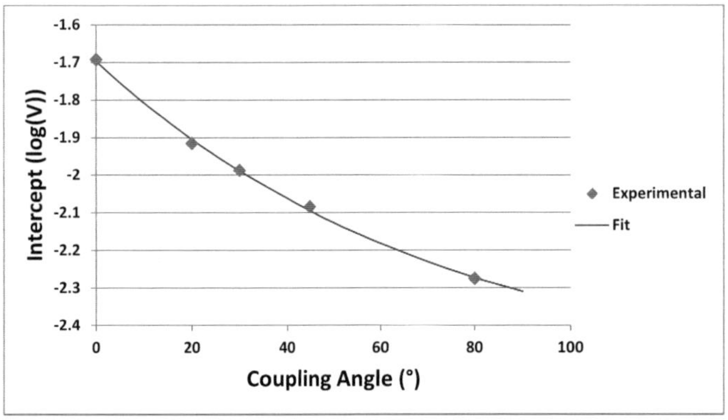

Fig 16 - Exponential decay relationship for coupling angle

$$b = 0.86e^{-\theta/71} - 2.56$$

Slopes remain similar, intercept decreases with increased θ

- Shows exponential decay relationship
- With a L/2*sinθ conversion, decay constants are similar

NDT in Canada 2019 | June 18-20 | Edmonton, AB

17

Rebar Junctions

- Potential issues arise when junctions are considered
- Due to nature of the size of rebar available for testing and lack of repeatable sizes 1'' & ¾'' rebar were used
- It is expected that superposition of the individual signals will occur

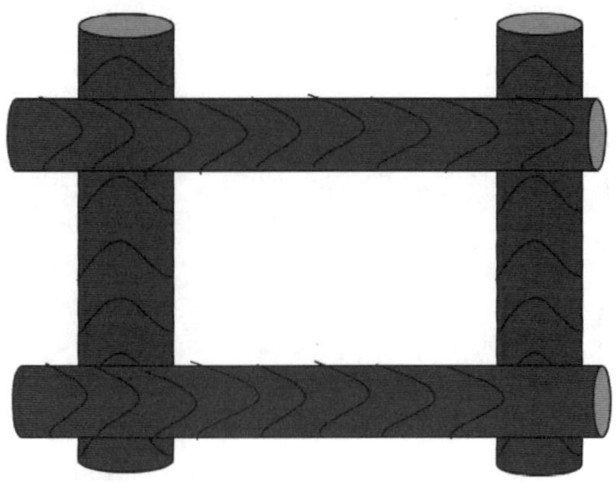

Figure 16- Crossed rebar area that will also affect signal

NDT in Canada 2019 | June 18-20 | Edmonton, AB

18

Junction Configuration Examined

- Example configuration:
  - 1" Rebar in front of ¾" Rebar
  - Probe is parallel to ¾" Rebar

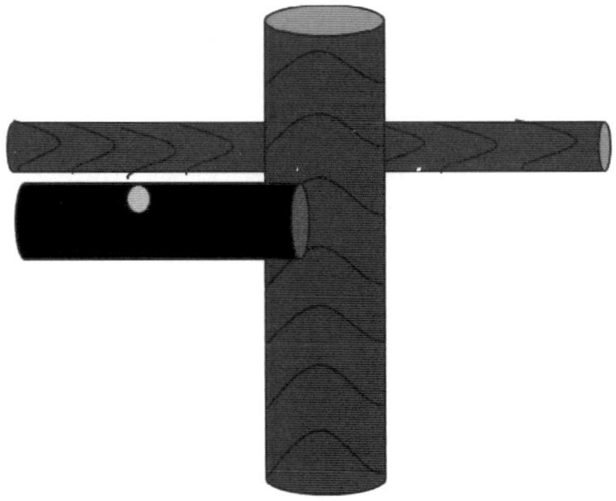

Figure 17- Measured configuration

Junction Response

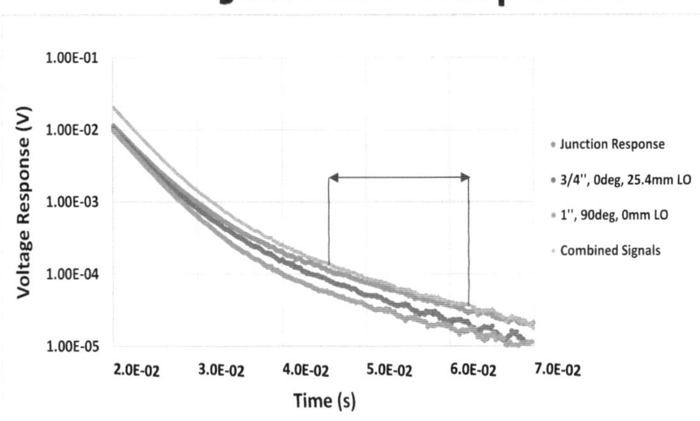

Figure 18- Transient response on a semi-log graph, while looking at example where 1'' rebar is in front of ¾ '' rebar and probe is aligned with the ¾'' rebar.

- Demonstrates superposition principle
- Signal response similar to addition of two individual signals
- Good agreement between superposed response (yellow) & experimental (blue) over sampling range

Conclusions

- Varying diameter of rebar changes slope of long-time decay in a semi-log plot. Inverse of the slope varies as radius squared, in agreement with theory
- For increasing liftoff, slopes remain similar, but intercept decays exponentially
- Varying coupling angle also changes transition voltage and response is similar to liftoff
- In the case of junctions signal response is dominated by the rebar that the probe is parallel to
- Effect of signal superposition is observed in this case

NDT in Canada 2019 | June 18-20 | Edmonton, AB 21

Future Work

- Optimize coil diameter for target rebar diameter
- Examine junctions with same sized rebar
- Perform field trials on rebar embedded in concrete

NDT in Canada 2019 | June 18-20 | Edmonton, AB 22

Acknowledgements

GROUPE MEQUALTECH Inc.

MONTRÉAL | QUÉBEC | BÉCANCOUR | SHERBROOKE

NSERC
CRSNG

Questions?

References

[1] J. D. Jackson, "Magnetostatics, Faraday's Law, Quasi-Static Fields," in *Classical Electrodynamics*, Hoboken: John Wiley and Sons, 1999, p. 17423.

[2] R. Nave, "Faraday's Law," *Hyper Physics*. [Online]. Available: http://hyperphysics.phy-astr.gsu.edu/hbase/electric/farlaw.html.

[3] B. Von Wedensky, "Concerning the Eddy Currents Generated by a Spontaneous Change of Magnetization (Translated - Jan. 2008)," *Ann. Phys.*, vol. 64, 1921.

Automatic Defect Detection for X-Ray inspection: Identifying defects with deep convolutional network

Roger BOOTO TOKIME[1,2], Xavier MALDAGUE[1], Luc PERRON[2]
[1]ECE, Université Laval, Québec, QC, Canada
[2]Lynx Inspection Inc., Québec, QC, Canada
E-mails: rogerbooto@gmail.com,
xavier.maldague@gel.ulaval.ca,
lperron@lynxinspection.com

Abstract

In this day and age, parts manufacturers are expected to meet the stringiest quality standards despite the increasing diversity and complexity of the parts they are making. Traditional inspection techniques are often inadequate or too expensive for mass production, so companies rely on a statistical approach to meet their quality objectives. To help reduce the costs associated with parts inspection and make 100% inspection possible, we propose a new method, based on artificial intelligence, that can perform automated defect recognition (ADR) in a high volume part production environment. In this paper, we will discuss the application of a state of the art deep convolutional network method that has been proven effective in the automated detection and identification of defects through semantic segmentation of radiographic imagery. The performance of a deep convolutional network, which is based on an encoder-decoder architecture called SegNet, are presented using real-life porosity samples in welded parts.

Keywords— Deep learning NDT, Automatic weld inspection, Automatic defect detection (ADR), X-Ray inspection, Non-destructive evaluation (NDE), X-Ray convolutional network, X-Ray Artificial Intelligence

1. Introduction

Since their discovery in 1895 by Wilhelm Conrad Röntgen[1, 2, 3], x-rays have been used extensively for evaluating and validating the quality of industrially manufactured products [4, 5, 6] in the context of non-destructive testing (NDT). Nowadays, several techniques and methods make it possible to guarantee a certain level of quality of the parts just manufactured and among these methods, there is the non-destructive evaluation method. Unlike destructive evaluation, non-destructive evaluation is the art of being able to inspect a manufactured part without physically destroying it. Several technologies make it possible to achieve such a feat [7, 8, 9], including technologies based on the exploitation of x-rays, the technology on which this article is based. Traditionally, x-ray (radiographic) assessment was done manually by an NDT technician through the illumination of a radiographic film. This way of doing things has many advantages, but its biggest drawback is that it is not scalable. Indeed, with the increasing number of parts machined each year, it would take a huge number of technicians to inspect the parts properly. As a result, manufacturers have fallen back on the digitization of radiographic films, also known as computed radiography (CR) and have developed processes for creating digital radiographic images (DR). Statistical methods have been developed to analyze a large quantity of parts by inferring an acceptance quality level value [10, 11] through the analysis of a representative sample set taken from multiple batches of manufactured parts. Doing so allows the inspection to be expandable, but does not guarantee a 100% inspection ratio. We have developed a new method based on artificial intelligence that is able to perform automated defect detection using radiographic images in the context of a high volume production environment while being sufficiently economical to allow 100% inspection. In the field of deep learning by computer vision, this is the equivalent of performing what is called semantic segmentation [12], in other words, assigning a class number to one or several pixels that corresponds to the type of content, in this case, defect (1) or no defect (0). This paper discusses the application of a deep convolutional network based on an encoder-decoder architecture called SegNet. The paper is broken down as follows: 1) A review of the literature on encoder-decoder architecture, 2) The decomposition of the database, 3) A description of the network used, 4) The results obtained and 5) A summary of the results obtained.

2. Literature review

In the field of digital imaging, several problems are solved by drawing on a set of techniques from several related disciplines. With such a foundation, it is then possible to design systems to solve complex problems. In this chapter, we will explore in detail how we can design systems based on techniques from related disciplines, such as Convolution and Signal Processing, Neural Network, Convolutional Neural Network and Encoder-Decoder.

2.1. Convolution and signal processing

In mathematics, convolution is an operation composed of two functions, ***f*** and ***g*** on the same infinite domain, consisting in finding the integral (or the sum if the domain is discrete) around a point in the domain. In the field of signal processing, functions ***f*** and ***g*** can be one-dimensional time signals or two-dimensional spatial signals but we will focus on two-dimensional signal processing for this article. In image processing, functions ***f*** and ***g*** correspond to the input image (***f***) and the convolution kernel (ω) is also called a convolution filter. As we can observe in Equation 1, we can extract an interesting property of convolution, which is the fact that the application of a particular filter can extract the same information in all input images. Later in this article we will see how this is used.

$$g(x,y) = \omega \circledast f(x,y) = \sum_{s=-a}^{a}\sum_{t=-b}^{b} \omega(s,t) f(x-s, y-t)$$

$\circledast$: convolution operator

$g(x,y)$: Filtred image (1)

$f(x,y)$: Original image

ω : Filter kernel

-a $\leq$ s $\leq$ a and -b $\leq$ t $\leq$ b : Every element of the filter

2.2. Neural network

In biology, animals with brains transmit and receive information to different parts of their bodies through a system of neurons. The information that circulates in the system is in the form of an electrical signal and the cell responsible for the transmission is the neuron. A neuron is composed of five main parts: dendrite, the cell body, the nucleus, the axon and the terminal axons. When a signal is emitted in the system, the neuron receiving the signal via the dendrites will be activated if the electrical disturbance is high enough [13] and then transmit a signal via the terminal axons. This property then allows us to model the neuron as a function receiving a signal from one or multiple sources (input signal) and the image of this function corresponds to the activation value of the neuron (output signal). This model is the one used by designers of artificial neural networks as shown in Figure 1. However, it is important to note that this model is a simplification of what is actually happening in biology, therefore, this model cannot be used as a substitute for studying the behaviour of biological neurons.

2.3. Convolutional neural network

In sections 2.1 and 2.2, we introduced two important properties in signal processing: convolution and artificial neurons. Based on these properties, we can then create a particular type of artificial neural network that is being called a convolutional neural network. Convolutional networks are neural networks with filters designed to extract specific type of information from the signal and use it as input signal for the activation function. This process allows a machine learning system designer to narrow down the information necessary to learn from the available general features available in order to solve a particular problem. Depending on the network design being used, the machine learning system can even learn by itself, in which case we are talking about an automatic learning system. As we have just seen, an automatic learning system based on a convolutional neural network architecture is a system capable of generating a filter bank to extract the information necessary to solve a given problem. Such learning can be done supervised, unsupervised[14] or by reinforcement[15]. In this article, we will describe a supervised learning system. Note that the convolutional network has the interesting ability to spatially reduce the signal once filtered. This allows the representation of spatial information in a smaller space, which essentially encodes the information. This principle is the basis of a network architecture called encoder-decoder that we will detail in the next section.

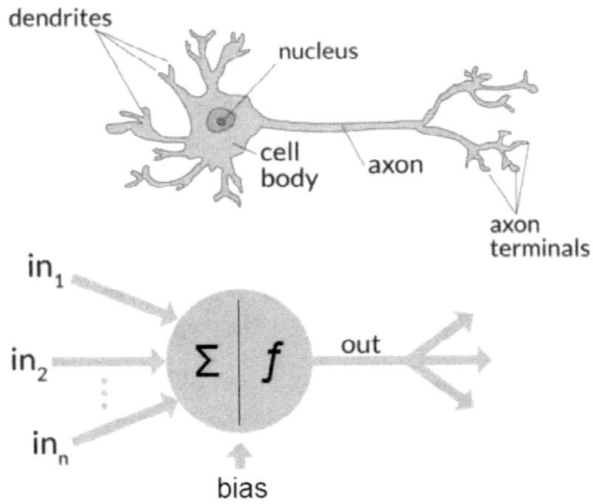

Figure 1: Artificial and biological neuron analogy

2.4. Encoder-decoder

An encoder is a network that takes an input signal and outputs a map of characteristics. These characteristic vectors contain the information, the features, that represent the input. The decoder is a network similar to the encoder, but unlike the encoder, it takes the characteristic vector of the encoder and gives the best possible correspondence with the intended actual input or output. Encoders are trained with decoders. A loss function is based on the calculation of the differences between the actual and reconstructed input. An optimizer will try to train both the encoder and the decoder to reduce this loss of reconstruction. Once trained, the encoder will provide a vector of characteristics for the input that can be used by the decoder to build the input with the characteristics that matter most to make the reconstructed input recognizable as the real input[16]. In our article we wish to perform a semantic segmentation of the porosity defects present in welding samples.

3. Database

For this experiment, we used a public database called GDXray [17]. This database contains several samples of radiographic images including images of welding with porosity defects. The database already contains segmented image samples, which is a good basis for training a small network. On the other hand, the quantity of images available is insufficient for a generalization of the problem. In order to generalize the problem, we decided to cut the large images that were about 1200 px in height and 4000 px in width into several smaller 256x256 images. This had the effect of increasing the amount of data available to the network. It also facilitated training on a graphics card, because a small fixed image size that can easily be stored in memory. In addition to generating small images from the original GDXray images, we have artificially generated several new images by performing basic transformations such as a 342° rotation at 18° intervals and generating noisy images by adding Gaussian noise to each generated image. We also performed x, y and diagonal translations and created negative type images of each image. Finally, we ended up with a database composed of more than 20 thousand 256x256 grayscale and binary images from the original 720 distinct images available in GDXray based on the following calculations:

$$720 * (r + n + i + tx + ty + txy) = 23760$$

$$720 * (19 + 1 + 1 + 3 + 3 + 3) = 23760$$

Where **r** is the number of rotations performed, **n** is the number of noisy images generated, **i** is the number of inverse images generated and **tx**, **ty**, **txy** are the number of translations performed on the x, y axes and the diagonal. In our experiments, we separated our data so that we used 90% of our images as training data, and 10% as test data that our network will never have seen. In addition, we performed a cross validation of our training data by separating 75% for training and 25% for validation. However, we did measure the reliability of our predictions using a metric called F1. This metric is used to verify the reliability and generalizability of our classification model, as it takes into consideration precision and recall metrics. Precision is a measure to calculate the ability of our system to predict pixels belonging to both classes in the right regions and recall allows us to calculate the sensitivity of our system when predicting true positives. Thus, the score given by the F1 metric gives a good idea of the prediction and classification capabilities of our model. During our experiments, we managed to obtain an F1 score of **80%** for our SegNet model and this is a very promising result.

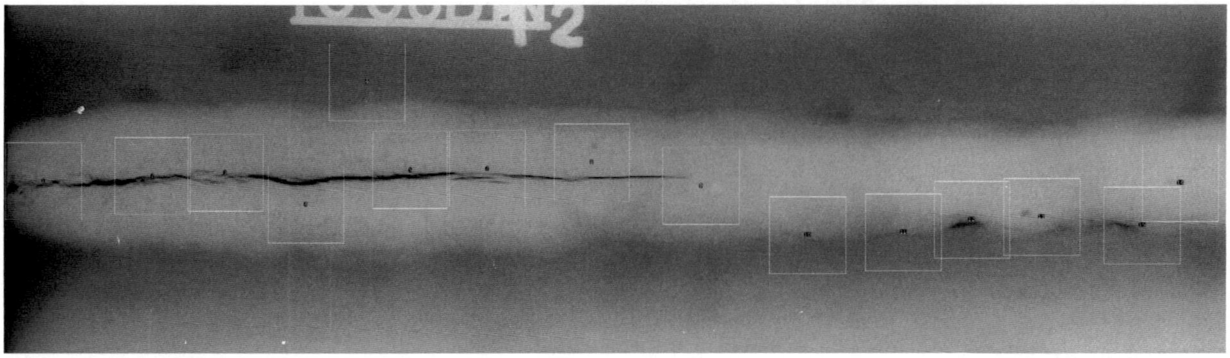

(a) Select a zone from the original image

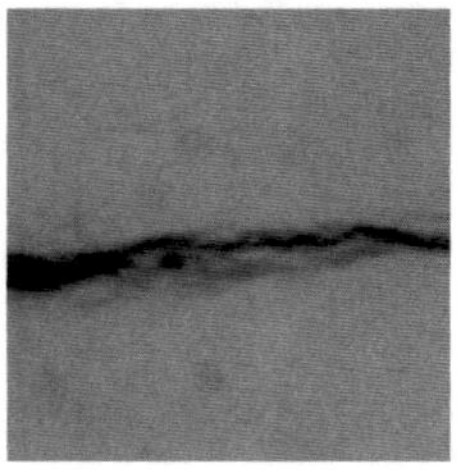

(b) Cropped 256x256 input image

(c) Associated image ground truth

Figure 2: Example of how we created our database

4. Network description

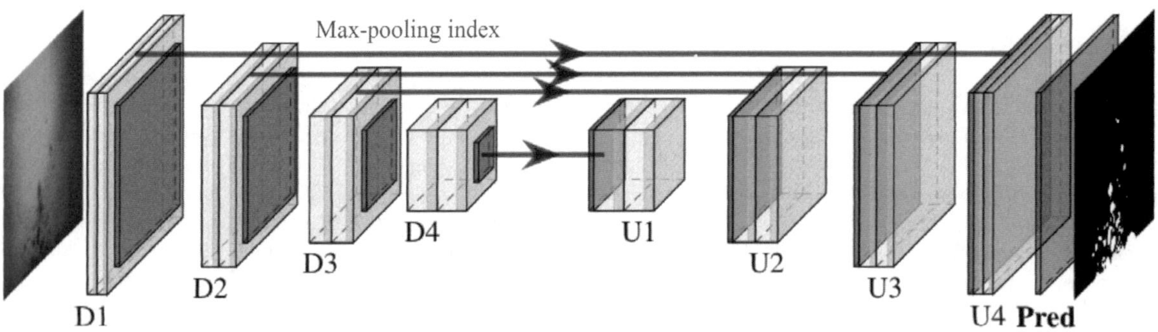

Figure 3: Our SegNet architecture diagram

4.1. Encoder

In Figure 3, we can see the two sections of our encoder-decoder architecture, the encoder being on the left while the decoder is on the right. The encoder consists of 4 convolution blocks (D1 - D4) and 4 pooling layers. Convolution blocks perform the following operations: convolution, batch normalization and application of an activation function. We will now detail the usefulness of each of the operations and, if applicable, explain our choices. Let's start with convolution. As mentioned before, convolution is a mathematical operation consisting of summing the products as shown in equation 1. To convolve an image to a filter, we drag the filter onto the image and depending on the type of dragging, also called stride, the filtered image may or may not be the same size as the input image. Let us now move on to the normalization of batches. Batch normalization speeds up learning when we have features with different value scales, for example, from 0 - 1 and 0 - 1000. This not only reduces the scale of values, but also preserves the importance of each characteristic[18]. As can be seen in Figure 4, the convolution blocks are composed of 6 layers with layers 3 and 6 being activation layers whose activation functions are exponential linear unit (ELU) and scaled exponential linear unit (SeLU)[19, 20] respectively.

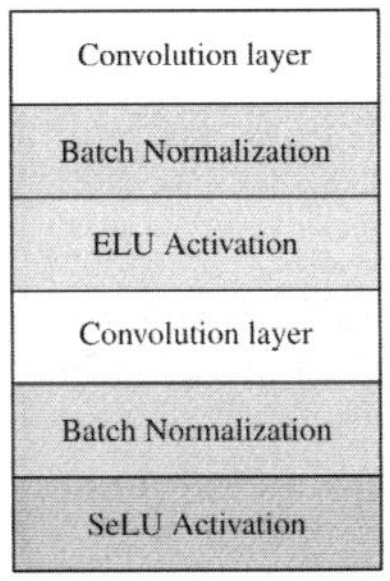

Figure 4: Encoder blocks: Convolution block operations for D1 to D4

The choice of these activation functions is based on the following properties of each function; 1) They keep the simplicity and speed of calculation of the rectified linear unit (ReLU) activation function, which is the reference activation function in most state of the art deep learning models, when the values are greater than zero. 2) They treat values near or below zero in two different ways; as indicated in their name, exponentially and exponentially scaled. As a result, the network is in continuous learning mode, because unlike ReLU, the ELU and SeLU functions are unlikely to disable entire layers of the network by propagating zero values in the network. This phenomenon is known as the dying ReLU[21, 22]. The last layer of each encoder block is a pooling layer that consists of generating a feature map at each resolution level. As shown in Figure 5, this operation, in our case, reduces the size of the image by a factor of two each time it is applied. This operation allows to keep the pixels representing the elements that best represent the image. To do this, we keep the largest pixel value in a kernel of any size and the position of this pixel which allows us to have a spatial representation of the pixels of interest. As a result, the network learns to encode not only the essential information of the image, but also its position in space. This approach makes it easier to reconstruct the information performed by the decoder, the architecture of which we will detail in the next paragraph.

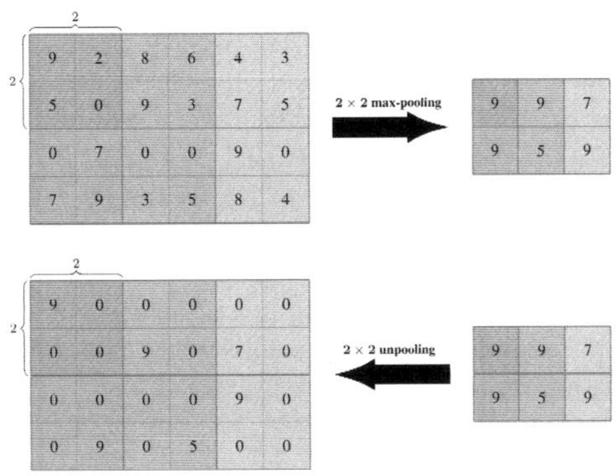

Figure 5: Pooling operation example

4.2. Decoder

The decoder consists of 4 convolution blocks (U1 - U4) and 4 unpooling layers. Convolution blocks perform the same operations as D1 - D4, but are organised a bit differently as described in Figure 6. The blocks U1 and U2 have one more block of convolution, batch normalization and activation, because we unintendedly pasted an extra block to the decoder section during one of our experiments and obtained great results. The blocks U3 and U4 follows the same convention in terms of operations as Dx except that the last layer of U4 is the prediction layer, which means that the activation function will not be ELU or SeLU, but the sigmoid function, in our case. At the end, we compare our prediction with the ground truth image by using the Dice loss function [23].

5. Results

In Figure 7, we can observe the prediction results of our model on a radiographic image. Since our training images have a size of 256x256, in order to cover the entire surface of the test image, we generated a mask in which we manually delimited the area where the weld is located. Then we used a technique called "sliding window" that allows us to make pixel-by-pixel predictions in

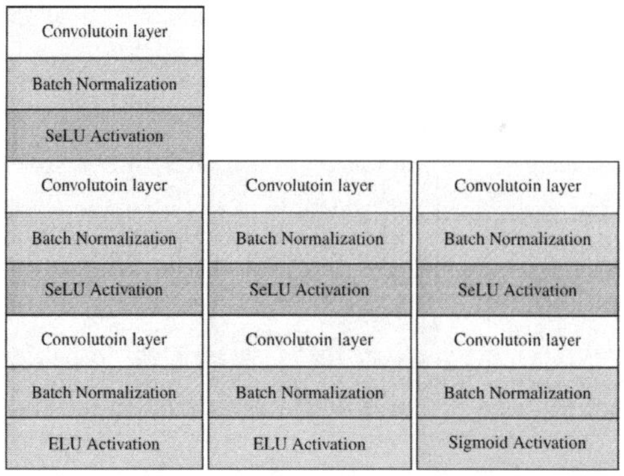

Figure 6: Decoder blocks: Convolution block operations for U1 and U2 (left), U3 (middle) and U4 (right)

the selected area. The results of this manipulation can be seen in Figure 7.n. As explained in sections 2.4 and 4 of this article, an encoder-decoder network architecture can reconstruct an input signal into an output signal by minimizing the differences between predictions and the ground truth. In our case, the output values will be between 0 (no defect) and 1 (defect) so we can interpret these values as a probability value that a given pixel represents an area containing a defect or not. In Figure 7, image **a** represents the manually selected area in which we will predict the location of defects. The images **b** to **e** represent a close-up view of the yellow outlined areas in the original image **a**. The images **f** to **i** represent the predictions made by our network. The representation chosen to show our results is a heat map in which the dark blue represents the pixels where the network does not predict any defect and in red the pixels where the network predicts a strong representation of a defect. The images **j** to **m** represent the ground truth images associated with the framed areas in the original image **a**.

Our experimental results show that our deep machine learning approach to performing semantic segmentation can be applied to detect porosity defects in radiographic images representing a welded area. However, we can observe that a section to the right of our test images was not predicted. This is due to the fact that our images were not a multiple of 256 in width. Therefore, this section is not considered when applying the sliding window. The problem can easily be solved by ensuring that the acquired images are multiples of the size of the cropped images, but this is not the most practical solution in a production environment.

We measured the reliability of our predictions using a metric called F1. This metric is used to verify the reliability and generalizability of our classification model, as it takes precision into consideration and recall metrics as expressed in equation 2. Precision is a measure to calculate the ability of our system to predict pixels belonging to both classes in the right regions and recall allows us to calculate the sensitivity of our system when predicting true positives. Thus, the score given by the F1 metric gives a good idea of the prediction and classification capabilities of our model. During our experiments, we managed to obtain an F1 score of **80%** for our SegNet model which is a very promising result.

$$Precision = \frac{truePositive}{truePositive + falsePositive}$$

$$Recall = \frac{truePositive}{truePositive + falseNegative} \tag{2}$$

$$F1 = 2 * \frac{precision * recall}{precision + recall}$$

6. Conclusion

In summary, we have demonstrated that deep learning through semantic segmentation can detect porosity defects in a weld. To do this, we implemented a deep convolutional network based on an encoder-decoder architecture called SegNet. The network has been adapted to our needs as described in Figure 3 and the results can be seen in Figure 7. In Figure 7, we can clearly see in a visual way that our approach is able to detect porosity defects in welds and in section 5, we have explained how we have been able to evaluate the quality of our results using statistical measurements that illustrate that our model is able to predict and classify pixels of a radiographic image representing a weld with an **80%** efficiency. Based on these promising results, our future work will be based

Figure 7: Prediction results after applying sliding window

on the application of a network model that classifies the different types of porosities, in other words, we want to be able to label the different types of defects that we can predict. In addition, we will explore the application of smaller network models, which could accelerate training time.

References

[1] A. B. Reed, "The history of radiation use in medicine," *Journal of Vascular Surgery*, vol. 53, no. 1, pp. 3S–5S, 2011.

[2] R. Halmshaw, "The discovery of x-rays and the early history of industrial radiography," *INSIGHT*, vol. 37, no. 9, pp. 669–671, 1995.

[3] R. I. Frankel, "Centennial of röntgen's discovery of x-rays," *West J Med*, vol. 164, pp. 497–501, Jun 1996. PMC1303625[pmcid].

[4] N. Brierley, C. Bellon, and B. Lazaro Toralles, "Optimized multi-shot imaging inspection design," *Proc Math Phys Eng Sci*, vol. 474, pp. 20170319–20170319, Aug 2018. 30220862[pmid].

[5] A. Fischer, T. Lasser, M. Schrapp, J. Stephan, and P. B. Noël, "Object specific trajectory optimization for industrial x-ray computed tomography," *Scientific Reports*, vol. 6, pp. 19135 EP –, Jan 2016. Article.

[6] X. Xiao, A. Ferro, T. Ma, C. Y. Han, X. Zhou, and W. Wee, "Adaptive reference image set selection in automated x-ray inspection," *Journal of Electrical and Computer Engineering*, vol. 2014, p. 7, 2014.

[7] P. Zhu, Y. Cheng, P. Banerjee, A. Tamburrino, and Y. Deng, "A novel machine learning model for eddy current testing with uncertainty," *NDT & E International*, vol. 101, pp. 104 – 112, 2019.

[8] C. I.-C. X. P. V. M. N. P. Avdelidis, T.-H. Gan, "Infrared thermography as a nondestructive tool for materials characterisation and assessment," 2011.

[9] J. García-Martín, J. Gomez-Gil, and E. Vázquez-Sánchez, "Non-destructive techniques based on eddy current testing," *Sensors (Basel, Switzerland)*, vol. 11, pp. 2525–65, 12 2011.

[10] E. Corporation, "Quality control procedures qcp 15.1."

[11] J. A. Melchore, "Sound practices for consistent human visual inspection," *AAPS PharmSciTech*, vol. 12, pp. 215–221, Jan 2011. 21203872[pmid].

[12] R. B. Tokime, H. Elassady, and M. A. Akhloufi, "Identifying the cells' nuclei using deep learning," *2018 IEEE Life Sciences Conference (LSC)*, pp. 61–64, 2018.

[13] Z. S. e. a. Lodish H, Berk A, *Molecular Cell Biology. 4th edition. New York: W. H. Freeman.* 2000. https://www.ncbi.nlm.nih.gov/books/NBK21535/.

[14] A. Jung, "A gentle introduction to supervised machine learning," *CoRR*, vol. abs/1805.05052, 2018.

[15] Y. Li, "Deep reinforcement learning," *CoRR*, vol. abs/1810.06339, 2018.

[16] K. A. Siddiqui, 2019. https://www.quora.com/What-is-an-Encoder-Decoder-in-Deep-Learning.

[17] V. Z. U. M. G.-L. I. Z. I. L. H. C. M. "Mery, D.; Riffo, "Gdxray: The database of x-ray images for nondestructive testing.," 2015. 34.4:1-12.

[18] S. Ioffe and C. Szegedy, "Batch normalization: Accelerating deep network training by reducing internal covariate shift," *CoRR*, vol. abs/1502.03167, 2015.

[19] G. Klambauer, T. Unterthiner, A. Mayr, and S. Hochreiter, "Self-normalizing neural networks," *CoRR*, vol. abs/1706.02515, 2017.

[20] D. Pedamonti, "Comparison of non-linear activation functions for deep neural networks on MNIST classification task," *CoRR*, vol. abs/1804.02763, 2018.

[21] L. Lu, Y. Shin, Y. Su, and G. Karniadakis, "Dying relu and initialization: Theory and numerical examples," 03 2019.

[22] A. F. Agarap, "Deep learning using rectified linear units (relu)," *CoRR*, vol. abs/1803.08375, 2018.

[23] C. H. Sudre, W. Li, T. Vercauteren, S. Ourselin, and M. J. Cardoso, "Generalised dice overlap as a deep learning loss function for highly unbalanced segmentations," *CoRR*, vol. abs/1707.03237, 2017.

DETECTION OF CORROSION UNDER INSULATION (CuI) USING ADVANCED PIPE INSPECTION TECHNOLOGY

CUI

1

Ankit Vajpayee, P.Eng. C.Eng.
CINDE,
Edmonton, Canada

CINDE
CANADIAN INSTITUTE FOR NON-DESTRUCTIVE EVALUATION

NDT in Canada
NDTC 2019
Inform - Engage - Advance
June 18 - 20
River Cree Resort & Casino
Edmonton, Alberta

Russell NDE Systems Inc. are from Alberta, Canada

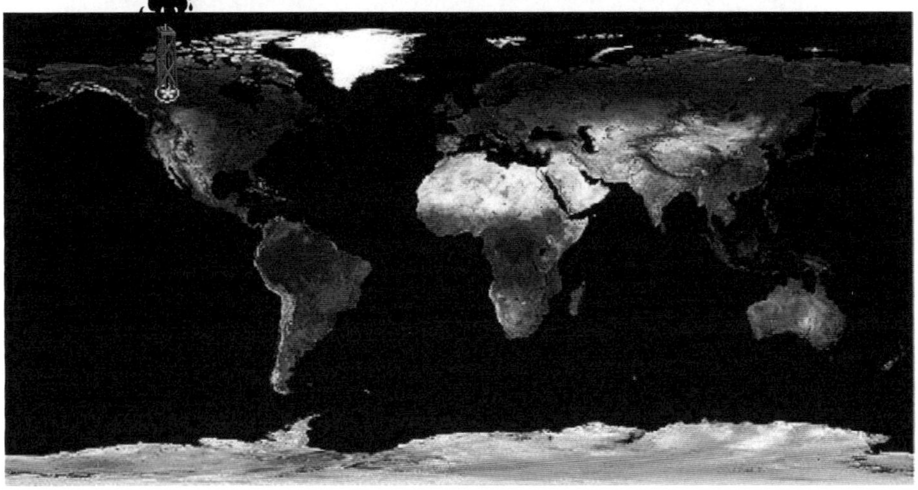

CINDE
CANADIAN INSTITUTE FOR NON-DESTRUCTIVE EVALUATION

NDT in Canada
NDTC 2019
Inform - Engage - Advance
June 18 - 20
River Cree Resort & Casino
Edmonton, Alberta

Graphically:

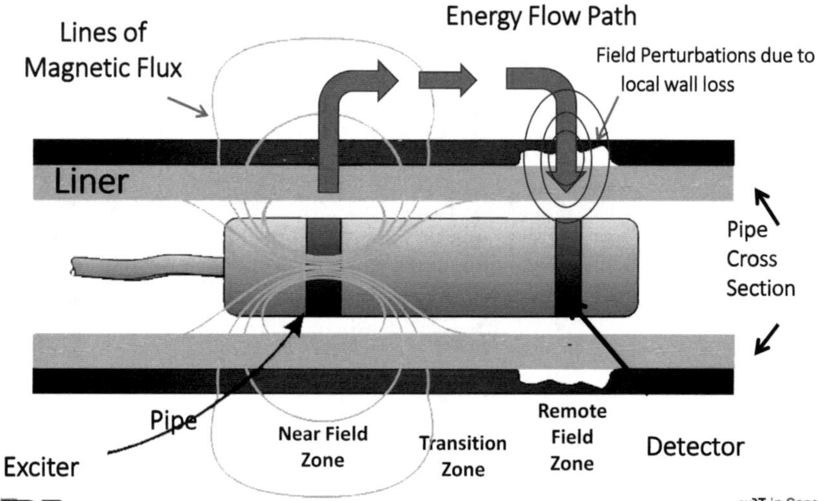

CINDE CANADIAN INSTITUTE FOR NON-DESTRUCTIVE EVALUATION

NDT in Canada NDTiC 2019 Inform - Engage - Advance

FEA Results

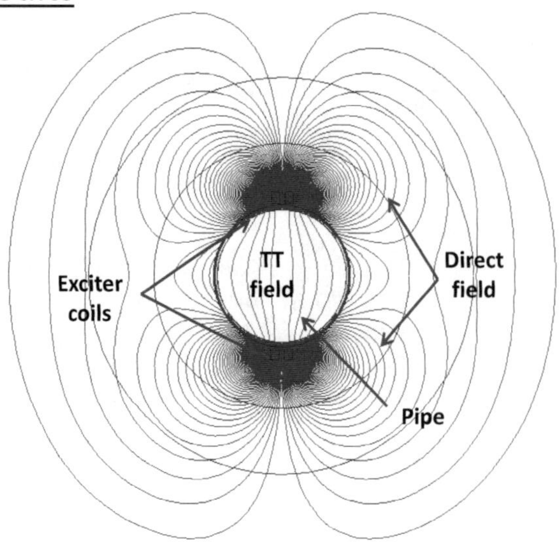

Magnetic field pattern for a Bracelet TT probe on an steel pipe

CINDE CANADIAN INSTITUTE FOR NON-DESTRUCTIVE EVALUATION

NDT in Canada NDTiC 2019 Inform - Engage - Advance

CUI & CUF detection

- Any inspection technique must cover a large area for fast scanning
- Must be sensitive to both large area general thinning and small local pitted areas
- Must be suitable for piping and tank walls
- Electro-magnetic probes offer a viable option

CINDE CANADIAN INSTITUTE FOR NON-DESTRUCTIVE EVALUATION

NDT in Canada NDTC 2019 Inform - Engage - Advance June 18 - 20 River Cree Resort & Casino Edmonton, Alberta

The Challenge

- Relying solely on visual searches for sources of moisture ingress into the insulation system *is largely ineffective.*
- Moreover, there are not always externally-visible signs of CUI presence.
- This makes the use of random, isolated searches for *CUI wholly inefficient.*

CINDE CANADIAN INSTITUTE FOR NON-DESTRUCTIVE EVALUATION

NDT in Canada NDTC 2019 Inform - Engage - Advance June 18 - 20 River Cree Resort & Casino Edmonton, Alberta

A Needle in a Haystack

- Although the mechanism (the chemical reaction) resulting in CUI is well understood and the root cause of the problems are known, ***finding CUI is not that easy.***
- Once moisture gets into the insulation, it is ***very hard to predict where it will start pooling and initiate CUI.***

CINDE
CANADIAN INSTITUTE FOR NON-DESTRUCTIVE EVALUATION

Electro-magnetic solutions for CUI detection

Modified RFT

(Remote Field Technology)

CINDE
CANADIAN INSTITUTE FOR NON-DESTRUCTIVE EVALUATION

NDT in Canada
NDTC 2019
Inform - Engage - Advance

Butane Sphere Leg. Inspection through Fire Proofing

CINDE
CANADIAN INSTITUTE FOR NON-DESTRUCTIVE EVALUATION

NDT in Canada
NDTiC 2019
Inform - Engage - Advance
June 18 - 20 River Cree Resort & Casino Edmonton, Alberta

Standard 16-ch Bracelet™ Probe

Pole scan system

NDT in Canada
NDTiC 2019
Inform - Engage - Advance
June 18 - 20 River Cree Resort & Casino Edmonton, Alberta

Standard 16-ch Bracelet™ probe

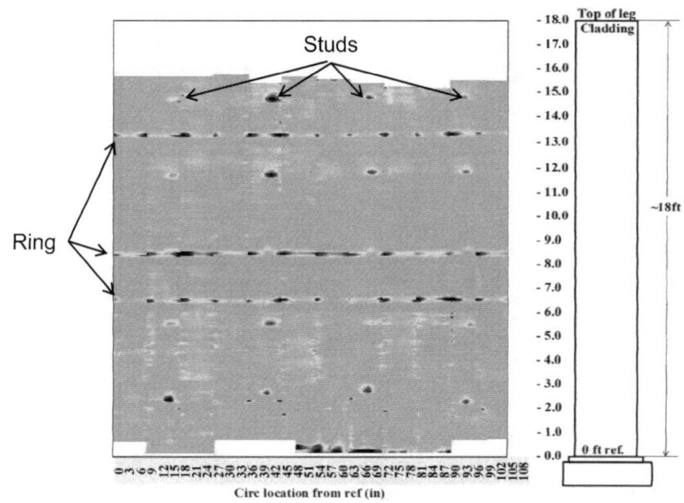

Typical CUF Bracelet probe colour map

RFT for CUI detection on insulated tanks & vessels

CINDE
CANADIAN INSTITUTE FOR NON-DESTRUCTIVE EVALUATION

NDT in Canada
NDTiC 2019
Inform - Engage - Advance
June 18 - 20
River Cree Resort & Casino
Edmonton, Alberta

Bracelet Probe™

- Bracelet Probe design means that it covers a wide circumferential extent of piping as it scans axially.
  - Effective area coverage
  - Can cover entire pipe circumference in series of scans;
  - each scan covers 10" of circumferential width.

    Not limited to a single tangent view or the need to perform grid scans
- Axial speeds of up to 12 feet per minute.
- Probe performance relies on a number of operational variables such as:
  - Test frequency (air gap, sample thickness)
  - Insulation thickness
- This not only dictates scan speed but also the threshold of detection.

CINDE
CANADIAN INSTITUTE FOR NON-DESTRUCTIVE EVALUATION

Color map presentation - Bracelet for CUI detection

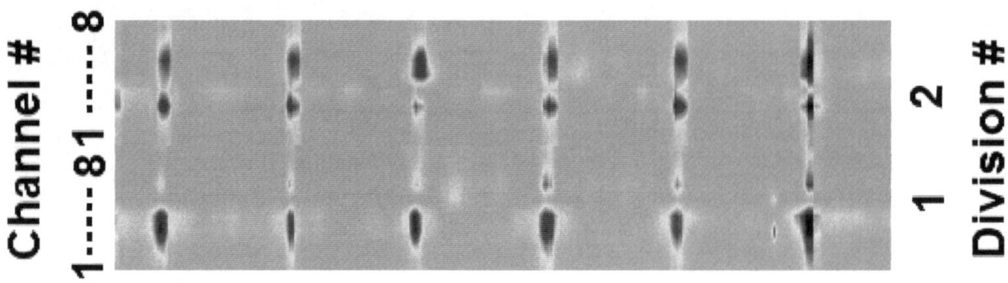

NDT in Canada
NDTC 2019
Inform - Engage - Advance

CINDE
CANADIAN INSTITUTE FOR NON-DESTRUCTIVE EVALUATION

NDT in Canada
NDTiC 2019
Inform - Engage - Advance
June 18 - 20
River Cree Resort & Casino
Edmonton, Alberta

Applications

- Applications Include:
  - Insulated facility and cross-country piping.
  - Insulated vessels and storage tanks
  - Sphere legs – with fireproofing.

CINDE
CANADIAN INSTITUTE FOR NON-DESTRUCTIVE EVALUATION

Bracelet Probe on Tank Shell, Siberia

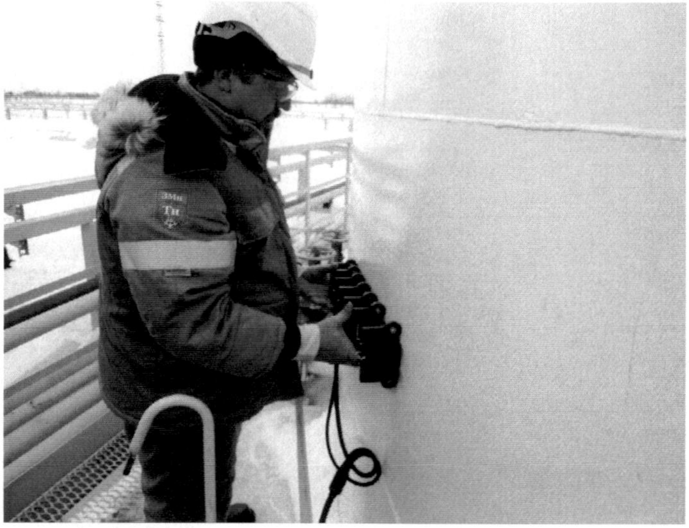

CINDE
CANADIAN INSTITUTE FOR NON-DESTRUCTIVE EVALUATION

NDT in Canada
NDTiC 2019
Inform - Engage - Advance
June 18 - 20
River Cree Resort & Casino
Edmonton, Alberta

Bracelet Probe on Plant Piping, Colombia

CINDE
CANADIAN INSTITUTE FOR NON-DESTRUCTIVE EVALUATION

NDT in Canada
NDTiC 2019
Inform - Engage - Advance
June 18 - 20
River Cree Resort & Casino
Edmonton, Alberta

Bracelet Probe on Plant Piping, Malaysia

CINDE
CANADIAN INSTITUTE FOR NON-DESTRUCTIVE EVALUATION

NDT in Canada
NDTiC 2019
Inform - Engage - Advance
June 18 - 20
River Cree Resort & Casino
Edmonton, Alberta

Bracelet Probe on Tank Shell, Saudi Arabia

CINDE
CANADIAN INSTITUTE FOR NON-DESTRUCTIVE EVALUATION

NDT in Canada
NDTiC 2019
Inform - Engage - Advance
June 18 - 20
River Cree Resort & Casino
Edmonton, Alberta

Bracelet Probe on Sphere Leg, Texas

CINDE
CANADIAN INSTITUTE FOR NON-DESTRUCTIVE EVALUATION

NDT in Canada
NDTiC 2019
Inform - Engage - Advance
June 18 - 20
River Cree Resort & Casino
Edmonton, Alberta

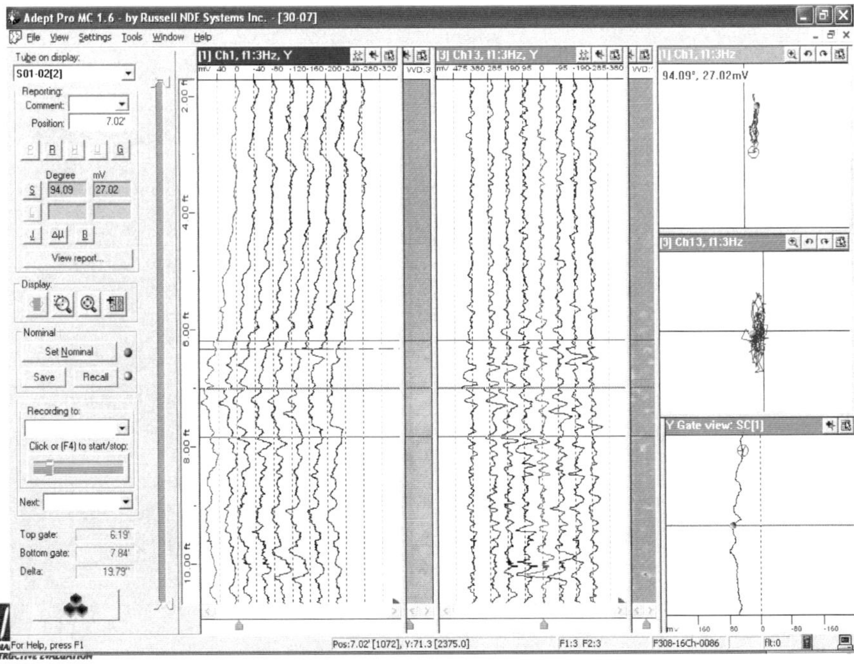

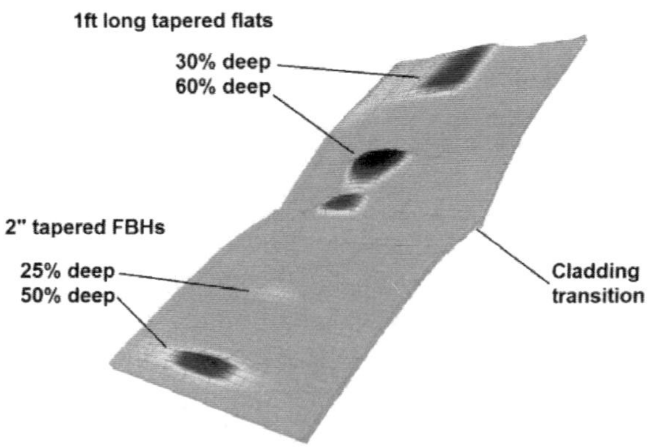

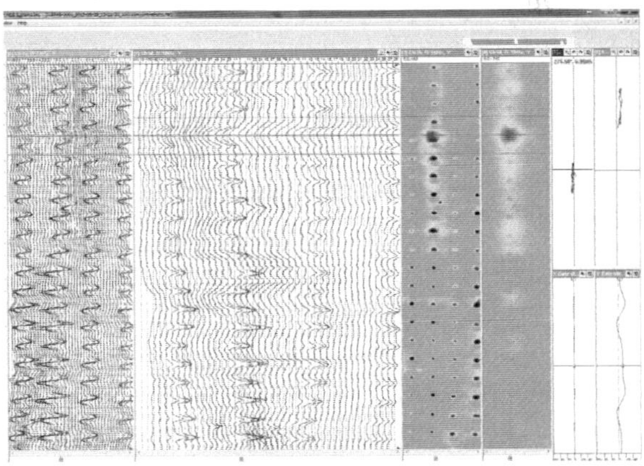

Easy to understand, graphical representation

CINDE CANADIAN INSTITUTE FOR NON-DESTRUCTIVE EVALUATION

NDT in Canada NDTC 2019 Inform - Engage - Advance June 18 - 20 River Cree Resort & Casino Edmonton, Alberta

Calibration for localized pits, example

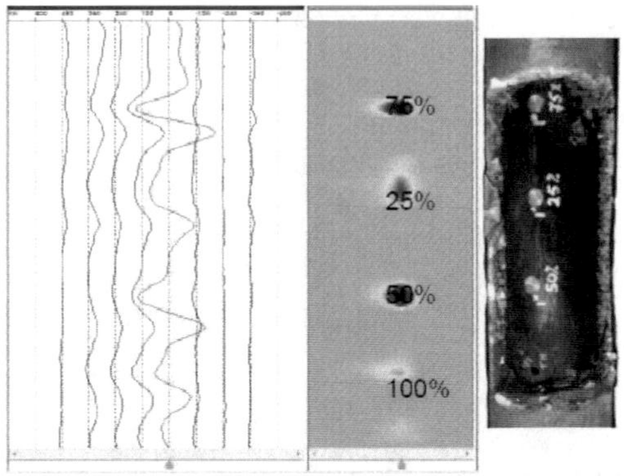

Ø1" 25%, 50%, 75% and Ø1/2" 100% deep OD FBHs on 0.200" wall Ø6" CS pipe with 1" thick foam insulation and outside aluminum cladding (removed in photo)

CINDE
CANADIAN INSTITUTE FOR NON-DESTRUCTIVE EVALUATION

NDT in Canada
NDTC 2019
Inform - Engage - Advance
June 16 - 20
River Cree Resort & Casino
Edmonton, Alberta

Calibration ½" Plate (3.5" Insulation 1/8" Aluminum clad)

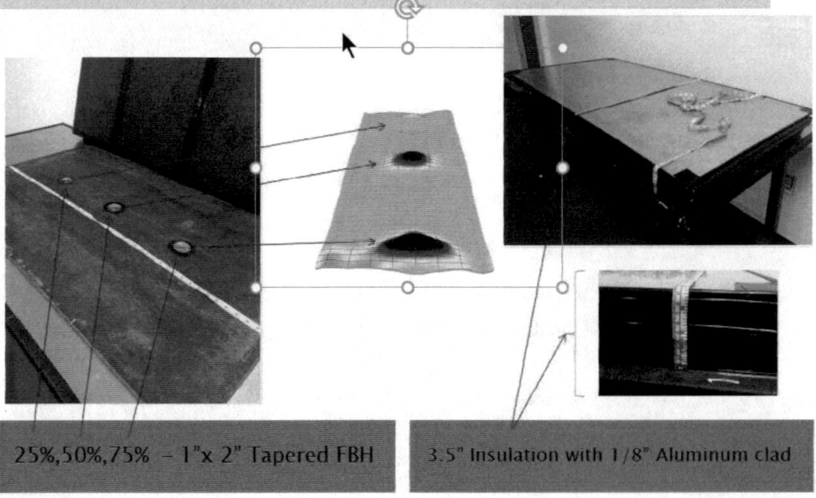

25%,50%,75% - 1"x 2" Tapered FBH

3.5" Insulation with 1/8" Aluminum clad

CINDE
CANADIAN INSTITUTE FOR NON-DESTRUCTIVE EVALUATION

NDT in Canada
NDTC 2019
Inform - Engage - Advance
June 16 - 20
River Cree Resort & Casino
Edmonton, Alberta

Industry Trials with Real World CUI

We have participated in several blind trials for Industry-supported groups who have gathered real life samples

- Application has undergone extensive client-specific blind and field trials, in addition to having participated in industry-sponsored trials

Results of blind trials showing 90% POD for damage exceeding 25% wall loss

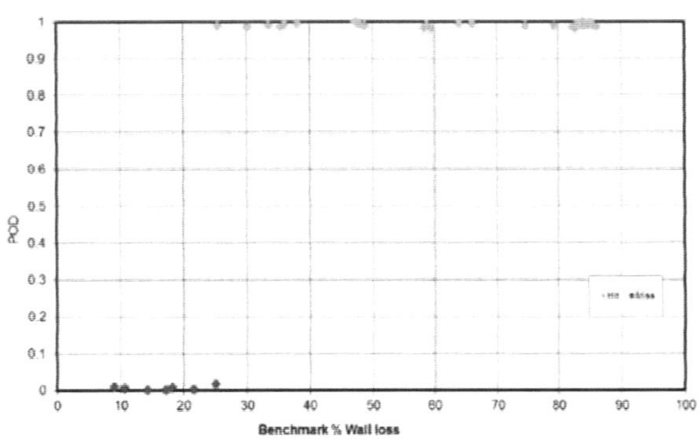

CINDE
CANADIAN INSTITUTE FOR NON-DESTRUCTIVE EVALUATION

NDT in Canada
NDTC 2019
Inform - Engage - Advance
June 18 - 20 River Cree Resort & Casino Edmonton, Alberta

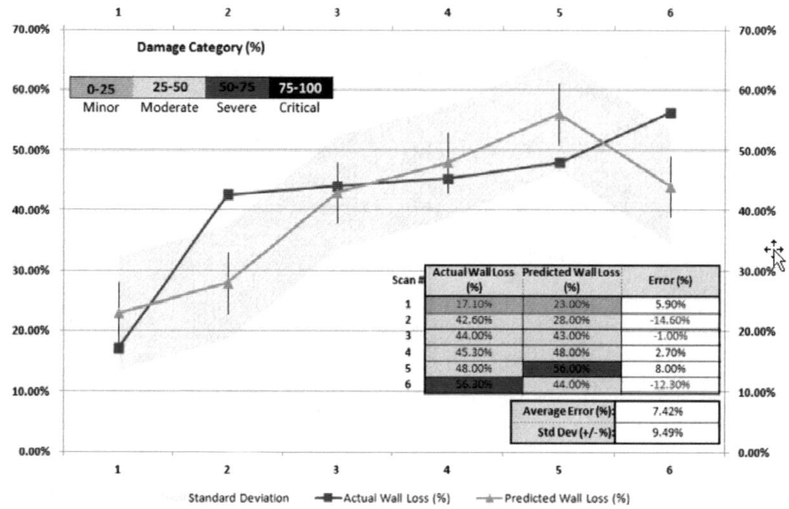

Scan #	Actual Wall Loss (%)	Predicted Wall Loss (%)	Error (%)
1	17.10%	23.00%	5.90%
2	42.60%	28.00%	-14.60%
3	44.00%	43.00%	-1.00%
4	45.30%	48.00%	2.70%
5	48.00%	56.00%	8.00%
6	56.30%	44.00%	-12.30%

Average Error (%):	7.42%
Std Dev (+/-%):	9.49%

CINDE
CANADIAN INSTITUTE FOR NON-DESTRUCTIVE EVALUATION

NDT in Canada
NDTiC 2019
Inform - Engage - Advance
June 18 - 20
River Cree Resort & Casino
Edmonton, Alberta

The RFT Technology

- Concept is simple: large area scanning at reasonable speed
- 16 sensors spread across 10" scan path, giving relatively high resolution to small areas of CUI on outside of pipe or vessels under 2"-4" of insulation
- Aluminum, stainless and cloth covers are acceptable
- Results are semi-quantitative within +/- 15% categories

CINDE
CANADIAN INSTITUTE FOR NON-DESTRUCTIVE EVALUATION

NDT in Canada
NDTiC 2019
Inform - Engage - Advance
June 18 - 20
River Cree Resort & Casino
Edmonton, Alberta

Benefits:

- A major benefit of the system is in the data it provides to the end user.
- Due to the nature of the application, a CUI location is ***fully mapped with length and width, in addition to an estimate for wall loss.***
- These are metrics that can be fed into a fitness for service calculation and used to articulate a response to the find.
  - ***A response based on sound engineering judgment, with input providing a whole picture of the damage being assessed.***

CINDE
CANADIAN INSTITUTE FOR NON-DESTRUCTIVE EVALUATION

NDT in Canada
NDTiC 2019
Inform - Engage - Advance
June 18 - 20 River Cree Resort & Casino Edmonton, Alberta

Final thoughts

- Like any NDT application, ***its use cannot follow a "one size fits all" approach.***
- Requires clear understanding of the challenges, and suitability for use as part of a detailed tank inspection program.
- Together with fit-for-purpose procedures, proper training, and ongoing system enhancement ***this application becomes a valuable tool in any CUI Program.***

*****GOOD DECISIONS START WITH GOOD INFORMATION*****

CINDE
CANADIAN INSTITUTE FOR NON-DESTRUCTIVE EVALUATION

NDT in Canada
34
NDTiC 2019
Inform - Engage - Advance
June 18 - 20 River Cree Resort & Casino Edmonton, Alberta

High-Resolution, Non-Destructive Imaging with Accelerator-Based Neutron Sources

KATIE RITTENHOUSE| DIRECTOR OF SALES| PHOENIX, LLC

CINDE
CANADIAN INSTITUTE FOR NON-DESTRUCTIVE EVALUATION

NDT in Canada
NDTiC 2019
Inform - Engage - Advance
June 18 - 20
River Cree Resort & Casino
Edmonton, Alberta

What is neutron imaging?

Neutron imaging is an NDT method that characterizes the internal structures of industrial components that are otherwise hidden from other NDT methods.

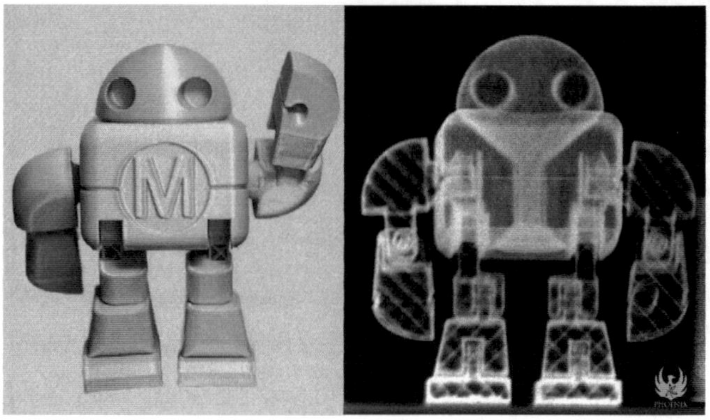

Why Neutron Imaging for NDT?

- Neutrons pass easily through high density metals and provide detailed information about internal low-density materials, cracks and voids.
- This property is extremely important for components with thick outer shells and internal structures such as engine turbine blades, munitions and spacecraft components.

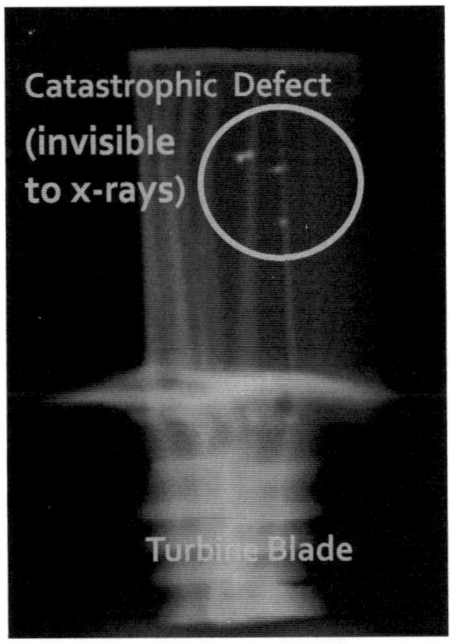

NDT in Canada 2019 | June 18-20 | Edmonton, AB

Physics of Neutron Imaging

Because neutrons are inherently electrically neutral, they do not interact with the atomic electrons of elements in the same way that photons or electrons do. Cross section determined by nucleus composition, not material density

- Easily penetrate dense metal
- Absorbed or stopped readily by low density and hydrogen-rich materials. Water and plastics are great neutron shields!
- Because Gadolinium and Boron have a uniquely high cross section for neutron attenuation they can be used as contrast enhancers

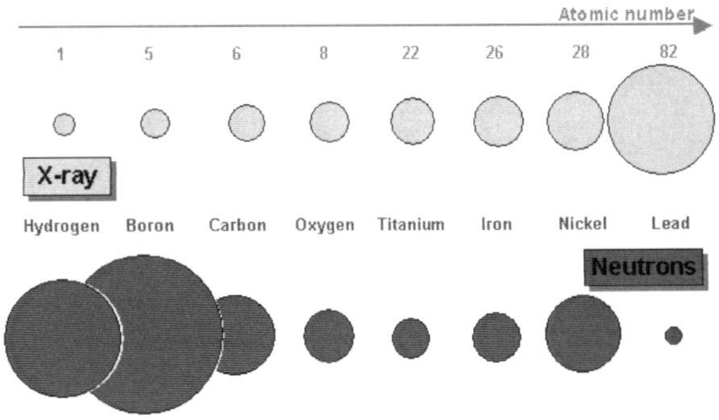

NDT in Canada 2019 | June 18-20 | Edmonton, AB

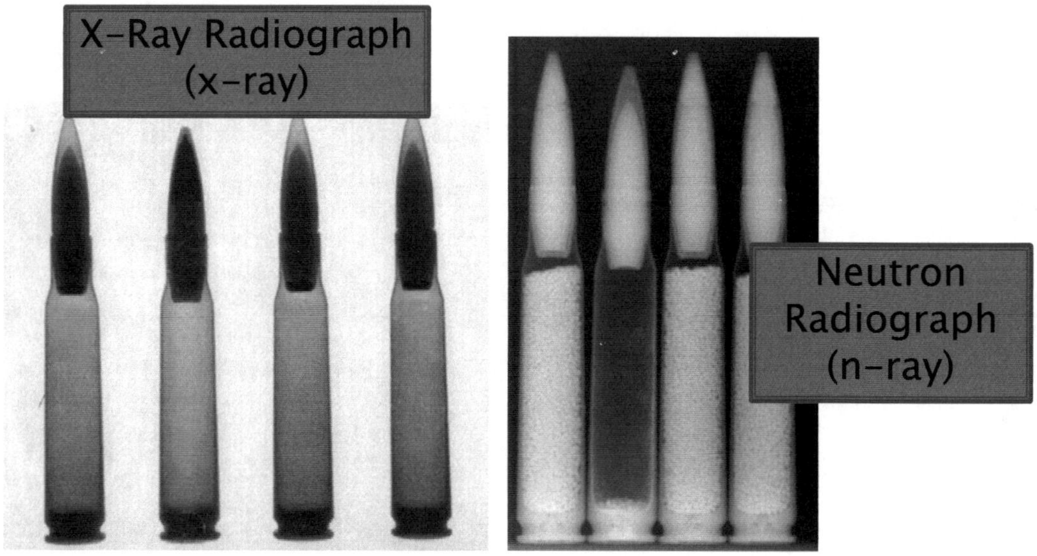

NDT in Canada 2019 | June 18-20 | Edmonton, AB

nRay Applications: Composites

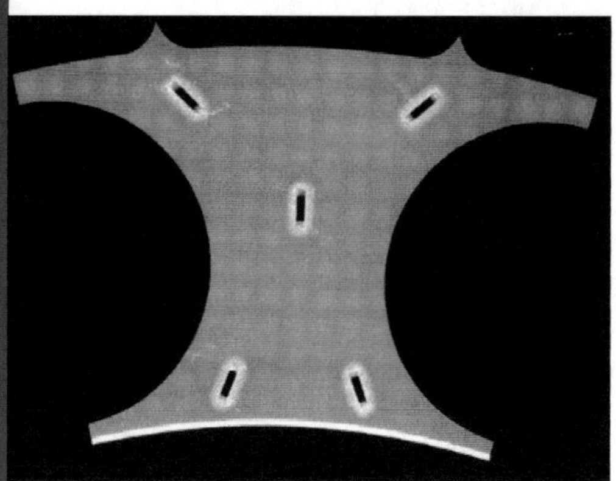

Neutron Image

Tooling ports visible - uneven distribution of BN in ceramic matrix composites (CMC)

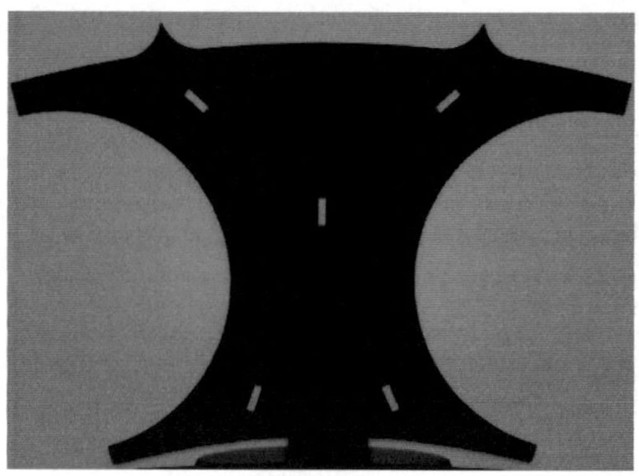

X-Ray

NDT in Canada 2019 | June 18-20 | Edmonton, AB

nRay Applications: O-Ring Identification

NDT in Canada 2019 | June 18-20 | Edmonton, AB

Unique Properties: Gd Penetrant

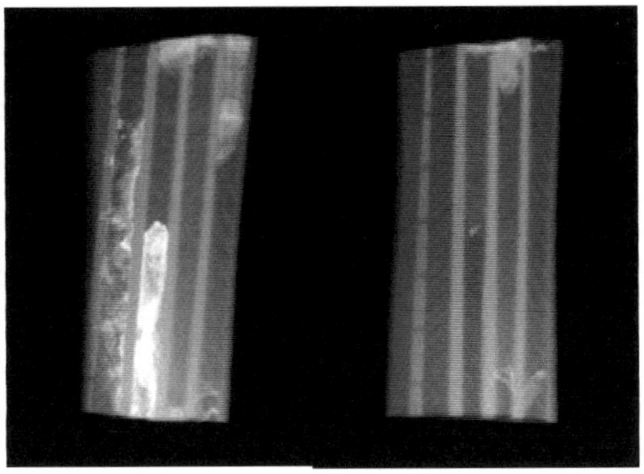

- Gadolinium has a very high cross section for absorption for thermal neutrons, meaning it reacts strongly to neutron radiation.
- The penetrant process causes foreign matter or gaps to appear bright on radiographic images, standing out in stark contrast to their surroundings.

NDT in Canada 2019 | June 18-20 | Edmonton, AB

Neutron Imaging History

- The technique has been used for decades on critical, high cost of failure aerospace and defense components.
- Until now, specialized research reactors were the only practical sources for neutron imaging
  - Availability is limited and shrinking
  - Significant logistical challenges, especially with energetic materials

NDT in Canada 2019 | June 18-20 | Edmonton, AB

Phoenix Neutron Imaging Center

- First operation expected fall of 2019
- Capable of accepting energetic materials
- ATF Compliant & Permitted
- Compliant with DoD Ammunition and Explosives Safety and Security Manuals (4145.26 & 5100.76)
- Phoenix accelerator-based neutron generation system

Phoenix Neutron Imaging Center

NDT in Canada 2019 | June 18-20 | Edmonton, AB

Quality and Throughput

Calculations	Nuclear Reactor (current baseline)	Phoenix High Yield System
Regulatory Burden	High	Minimal
L/D Ratio	105	105
Thermal Neutron Flux @ Image Plane ($n/cm^2/s$)	3.00E+06	3.20E+05
Neutron : Gamma Ratio	>1E6	>1E6
Cadmium Ratio	>5	>4
Annual Film Capacity (# Exposures)	>20,000	>25,000
Annual CR Capacity (# Exposures)	n/a	>35,000

NDT in Canada 2019 | June 18-20 | Edmonton, AB

Future of Neutron Imaging

- Neutron radiography is and will continue to be a crucial part of ensuring the safety and reliability of critical components.
  - For example: canopy separation charges, ejection seats, flight termination charges
- Reactorless imaging systems will enable use for new applications:
  - For example: artillery shells, body armor, detonating fuzes, shape charges, grenades, etc.
  - Fast neutron imaging for larger components
- Advanced capabilities will become more accessible
  - Fast neutron imaging for large components
  - Fast and thermal neutron CT
  - Neutron/X-Ray data fusion for composite images

Thank you!

PHOENIX
TRANSFORMING NUCLEAR TECHNOLOGY

Katie.Rittenhouse@PhoenixWI.com
www.PhoenixWI.com

NDT in Canada 2019 | June 18-20 | Edmonton, AB

Sensitivity Response of Total Focusing Method (TFM) for Weld Inspection Versus Other Techniques

Jeremy CARIGNAN [1] Marie-Pierre DESPAUX [2], François LACHANCE [3], Philippe RIOUX [4]

[1] Centre de métallurgie du Québec, Trois-Rivières, Québec, Canada

[1] Phone: (819) 376-8707 x5026; courriel: jeremy.carignan@cegeptr.qc.ca

[2 3 4] Sonatest, Quebec City, Québec, Canada

[2 3 4] Phones: (418) 686-6222; e-mail: despauxm@sonatest.com, lachancef@sonatest.com, riouxp@sonatest.com

Abstract

Total Focusing Method (TFM) is an ultrasonic-based technique enhanced with the post-processing algorithm. TFM generates an image virtually focused at every pixel with increased probability of detection for certain applications specific scenarios where enhanced defect sizing and characterisation is critical. Although TFM is greatly promoted, there are some concerns, especially for weld inspection. This article evaluates the amplitude variation towards specific flaw orientation and compares it with other techniques, such as phased array ultrasonic testing (PAUT) and radiographic testing (RT).

Keywords: NDT (non-destructive testing or evaluation), Weld Inspection, Total Focusing Method (TFM), Full Matrix Capture (FMC), Phased Array Ultrasonic Testing (PAUT), Radiographic Testing (RT).

1 Introduction

This study was inspired by the doctoral thesis of Madame Kombossé Sy from the Université Paris-Saclay [1]. In her studies, Ms Kombossé claims that some defect orientations would respond better to certain propagation modes while using the Total Focusing Method (TFM) algorithm. In her thesis, Ms Kombossé conducts her conclusion mostly on CIVA simulations based on artificial alike flaws. The objective of this case study is to validate this claim with a real weld sample, having real flaws alike as well as adding the phased array ultrasonic testing (PAUT) sensitivity comparison. Other techniques such as radiography and metallography have been used along as a complementary tool for further verification.

2 FMC/TFM Glossary

2.1 Full Matrix Capture

Full matrix capture (FMC) is an acquisition technique where each element of a phased-array transducer is individually pulsed and the echoes are received from each individual element, including the transmitter. It generates a matrix signal with raw A-scan information. In fact, the FMC collects many data at once. This data can be further analysed in the post-acquisition phase.

2.2 Total Focusing Method

Total focusing method (TFM) is the post-acquisition phase of an FMC inspection. It processes the raw A-scan signals with the help of algorithms to display the FMC data in a clear visual format [1]. Every computed pixel of a TFM image is the summation of all the signals passing through this particular pixel.

2.3 Propagation Modes

The TFM uses different wave propagation modes based on the flaw orientation, as seen in Figure 1. Some propagation modes are suitable depending on the flaw orientation.

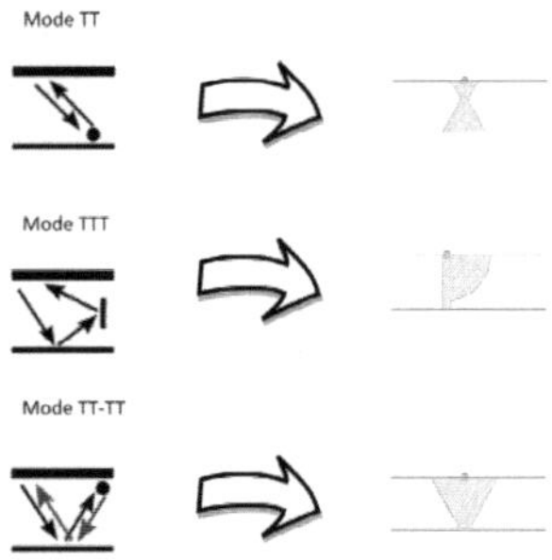

Figure 1 - Recommended propagation mode depending on the flaw orientation in the weld

3 Methodology

A 19 mm thick carbon steel plate with an unknown weld geometry and unknown flaws are used for the inspection, as per Figure 2. The groove angle was approximated at 45° and for the purpose of this study, PAUT, FMC/TFM, radiography Testing (RT) and metallography processes will be used to detect the flaws.

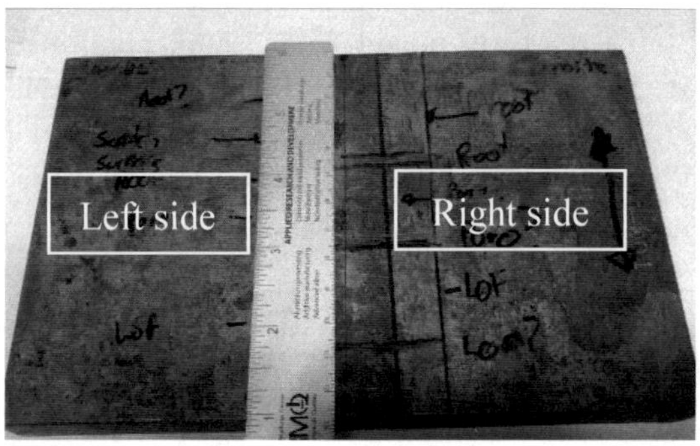

Figure 2 – The inspected plate prior to destructive testing

3.1 Ultrasonic Methodology

The Sonatest VEO+ platform along with 5 MHz DAAH transducer (model D1A) was used to record the PAUT and FMC data. Velocity, wedge delay, sensitivity and TCG calibrations were performed in order to compare the amplitude difference between flaws. Furthermore, the PAUT and TFM acquisitions were performed at the same time with the same probes and PA unit. Hence, the only parameters that could influence the results were the flaw orientation and the amplitude reference levels.

3.2 Radiography Methodology

The initial exposures were carried out with analogue film and then adapted to other techniques. The three expositions (0, +45° and -45°) were performed for each technique to detect possible lack of fusion and to determine the discontinuity heights.

Table 1 - Analogue film parameters

Distance between the source of the film	914 mm
Energy	200 kV
Exposure	2700 mA•sec.
Film brand	Carestream Industrex MX125
Pb intensity screen	0.127 mm front and 0.254 mm back
Filtration added to the X-ray tube	None
Optical density	2.5

3.2.1 Computed Radiography Technique

Both the exposure and the energy level were kept the same compared to the analogue technique. However, the laser power and the photomultiplier (PMT) were adjusted in combination with a Carestream Flex HR plate. A copper screen was used between the back of the imaging plate and the intensifying screen to prevent any lead fluorescence that could degrade the signal-to-noise ratio (SNR) of the image. A SNR of 130 was attained in the weld area and a contrast-to-noise ratio (CNR) of more of 3.5 was obtained.

3.2.2 Digital Radiography Technique

The main objective was to achieve similar SNR and CNR values to those of the computed radiography (CR) exposures. A 14-bit RayzorX Pro digital detector array (DDA) from Vidisco was used. The energy was raised to 240 kV, additional filtration was added to the tube window and the DDA was put farther away from the floor to reduce the scattered radiation and reduce the noise to achieve the acceptable SNR and CNR.

3.3 Metallography Methodology

Metallography were made regarding certain defects found with PAUT and TFM. All samples were grinded with silicon carbine paper submerged in water from 180 to 1200 grits [2]. Later different polycrystalline diamond solutions were used for polishing. The final step was accomplished with 0.05 µm colloidal alumina [2]. The etching reagent, Nital 2%, was applied for approximately 15 seconds [2].

4 Results

4.1 Radiography, Ultrasound & Metallography Overview of the Sample

Figure 3 illustrates the results obtained from all the non-destructive techniques. The first image is an End View: i.e. a cross section view from the left side of the weld. The second one is a radiography film of the same weld, and the other End View is from the right side of the weld.

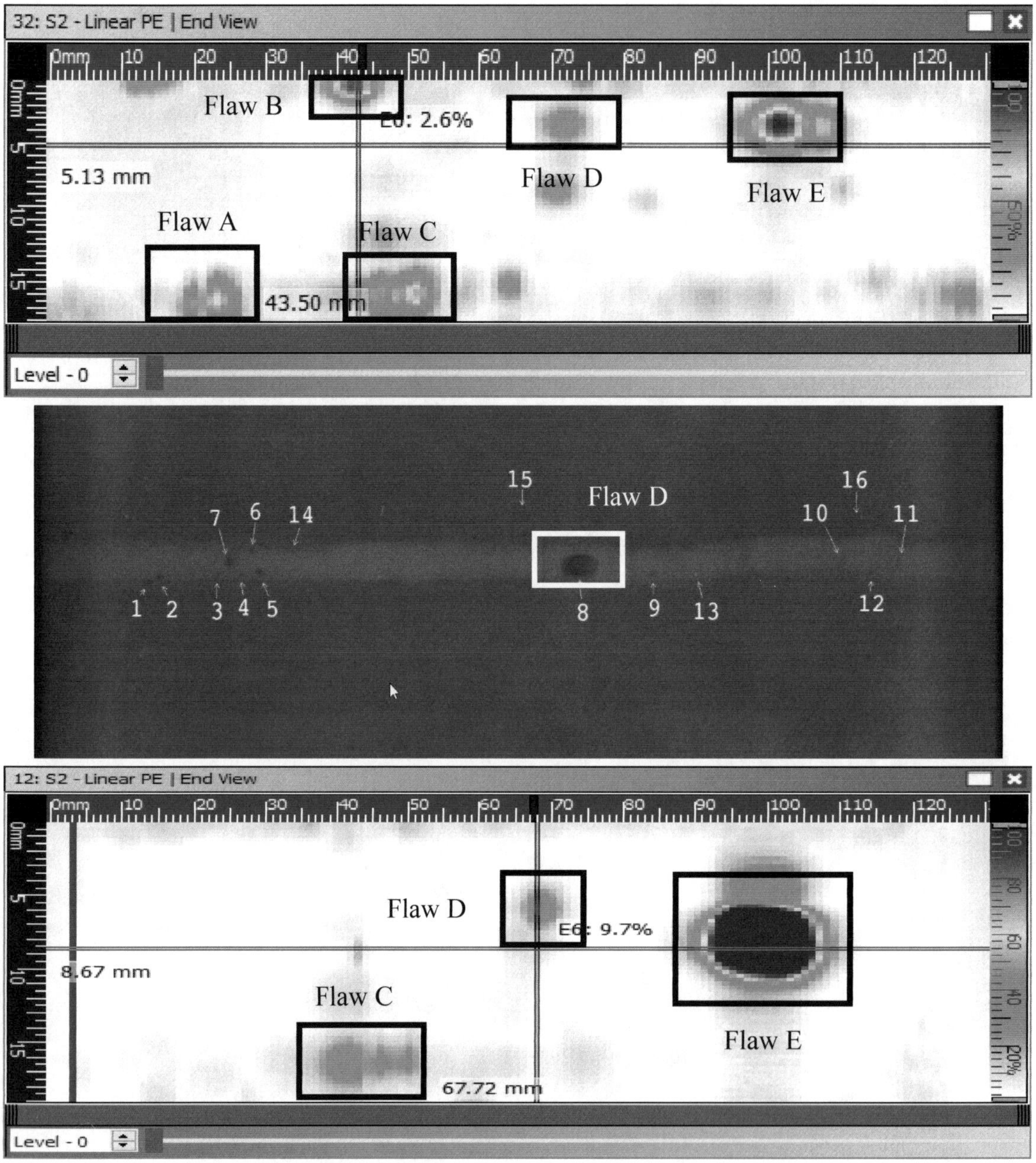

Figure 3 – The juxtaposed figures provide a better overview of the discontinuities in the plate.

In general RT is more sensitive in comparison to PAUT when dealing with volumetric flaws such as porosity and inclusion. On the other hand, PAUT is more sensitive for planar flaws such as crack or lack of fusion. All the numbers written on the X-ray except #13, which is an inclusion, correspond to porosities. Meanwhile the PAUT detected planar discontinuities were not visible in radiography. The same flaws have been detected by both PA and FMC/TFM test methods.

Metallography, since it is a destructive test, can eliminate any doubts about the defect's nature. The lack of fusion (flaw A) is visible at the root level, as seen in Figure 4, where there is a drastic change in microstructure at this location. Moreover, some porosities are visible at the root. Figure 5 shows two cracks landing to the surface. The big black spot, of the Figure 6, located in the middle of the weld, is the only porosity detected by PAUT and TFM. The fine line at the bevel of the Figure 7 is a lack of fusion.

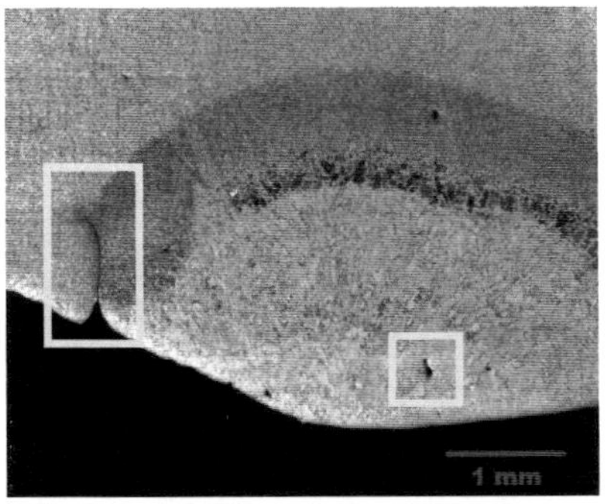

Figure 4 – Metallography of the flaw A

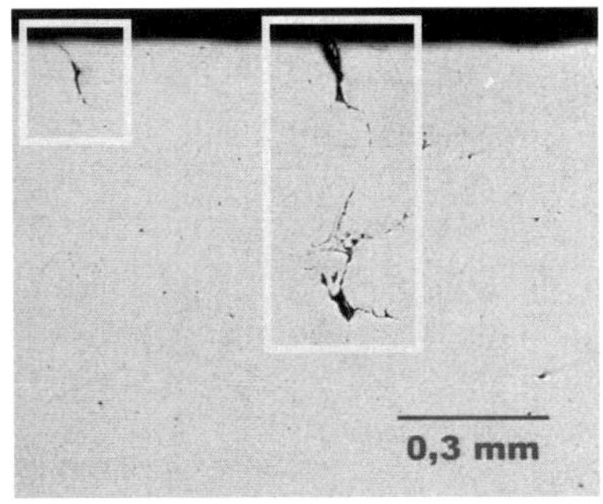

Figure 5 – Metallography of the flaw B

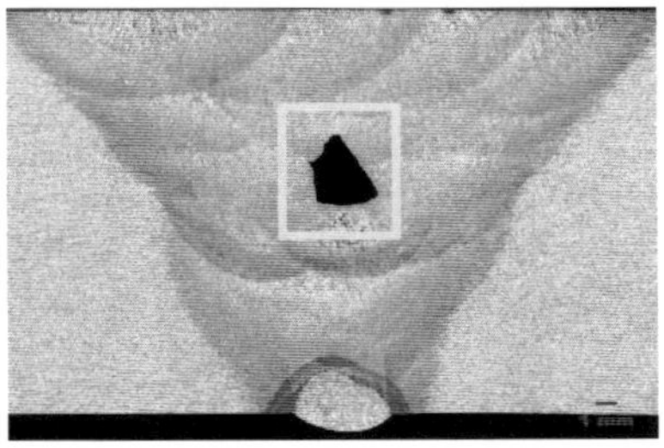

Figure 6 – Metallography of the flaw D

Figure 7 – Metallography of the flaw E

4.2 PAUT & TFM Differences

The major advantages of the TFM images compared to the PAUT scan are the improved resolution and the fact that the acoustic energy is virtually focused everywhere inside the weld volume. Such benefits are only achieved when the image is processed using the proper propagation mode towards the flaw orientation. Section 4.3 emphasises the importance of selecting the appropriate propagation mode according to the expected flaw orientation. Table 2 illustrates, in more details, the defect detected by ultrasonic techniques. For TFM pictures, only best propagation modes are chosen, the location and the acoustic signature of the flaws are included in this table.

Table 2 – PAUT and TFM defect characterisation comparison

Flaw A		**Flaw B**	
PAUT	TFM	PAUT	TFM
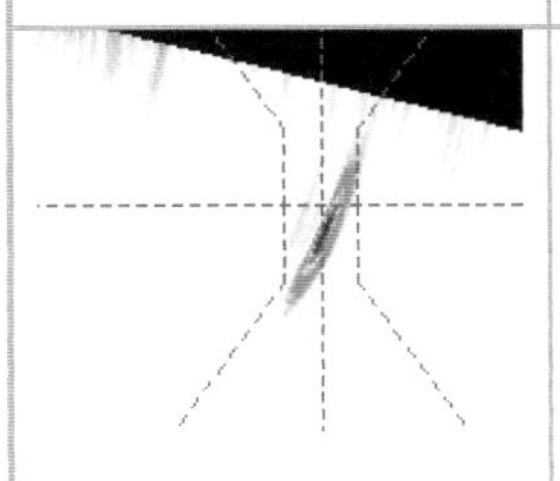	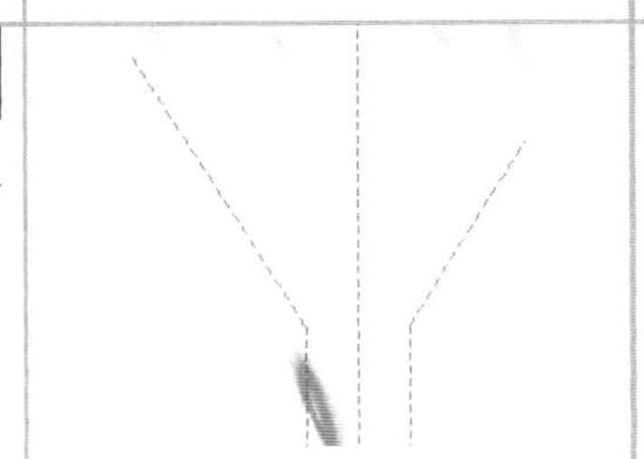	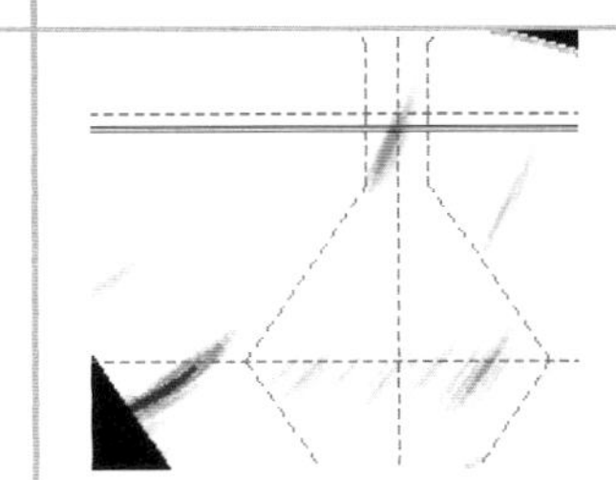	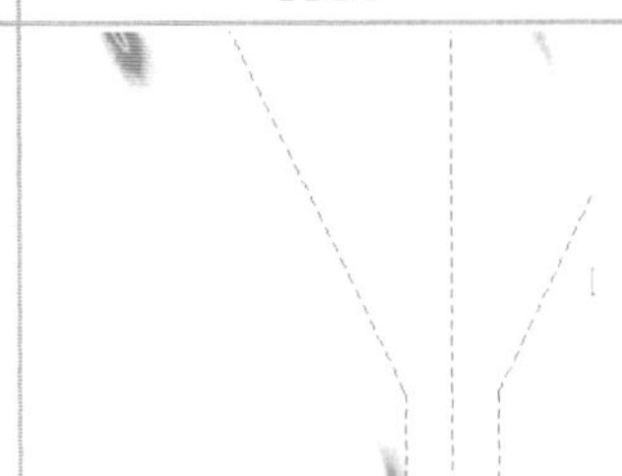
Type of defect: Lack of fusion near the root		Type of defect: Crack	
Location: 20 mm based on the scan axis		Location: 42 mm based on the scan axis	
Acoustic information: Strong amplitude response at the bevel or at the root		Acoustic information: Planar defect with distinguished amplitude response	
Is it detectable with: ☐RT ☒Metallography (see Figure 4)		Is it detectable with: ☐RT ☒Metallography (see Figure 5)	

Flaw C		Flaw D	
PAUT	TFM	PAUT	TFM
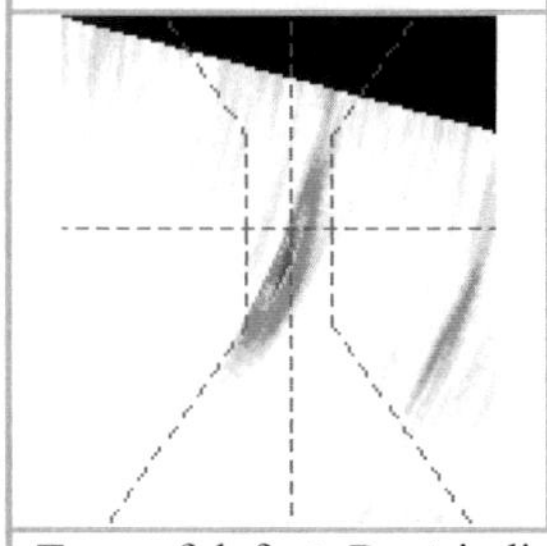	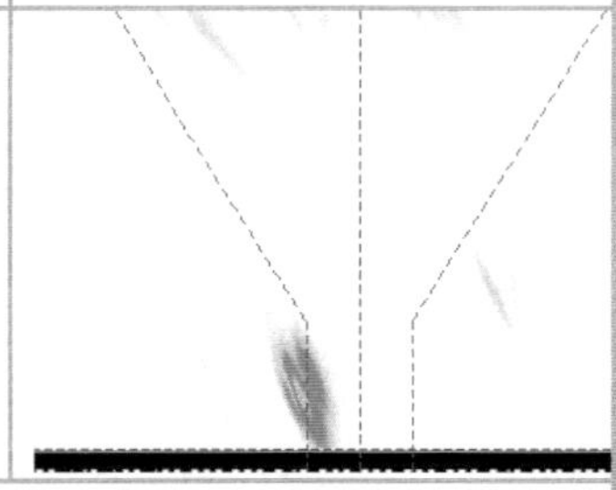	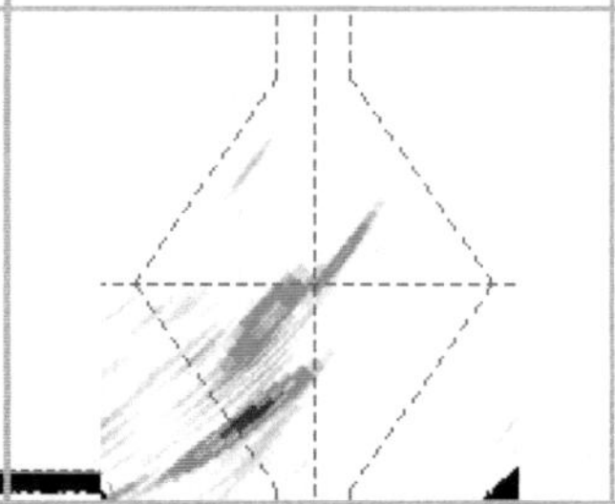	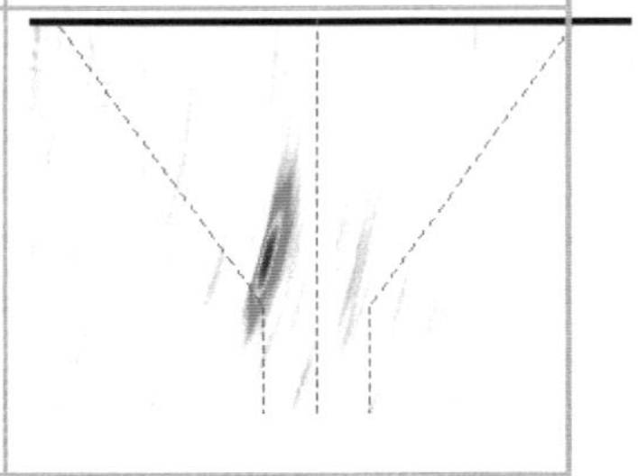
Type of defect: Root indication		Type of defect: Porosity	
Location: 50 mm based on the scan axis		Location: 70 mm based on the scan axis	
Acoustic information: Signal at the root		Acoustic information: Echo cluster in the weld volume	
Is it detectable with: ☐RT ☒Metallography		Is it detectable with: ☒RT (see Figure 3, #8 of the radiography film) ☒Metallography (see Figure 6)	
Flaw E			
PAUT		TFM	
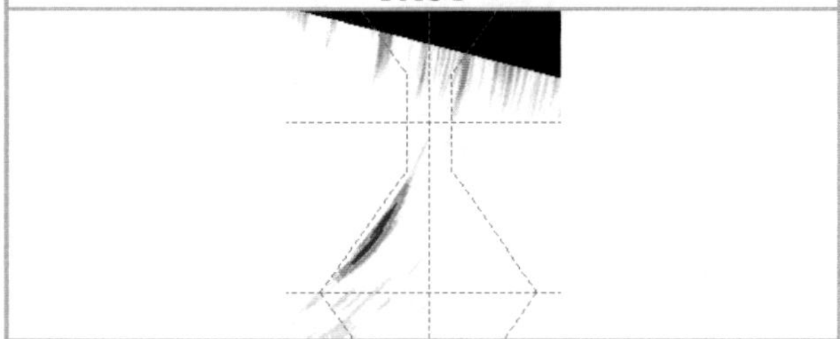		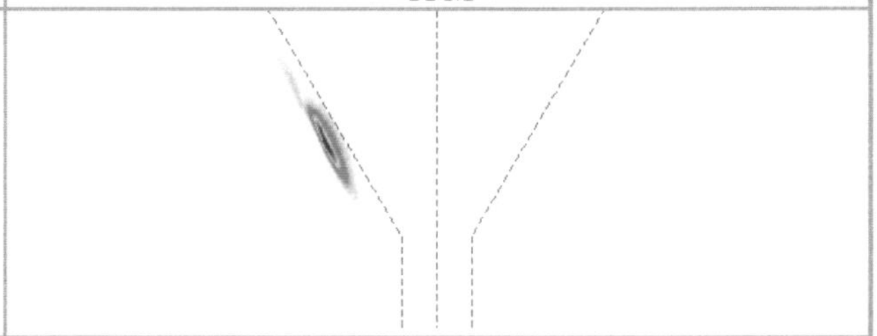	
	Type of defect: Lack of fusion		
	Location: 100 mm based on the scan axis		
	Acoustic information: Strong amplitude response at the bevel		
	Is it detectable with: ☐RT ☒Metallography (see Figure 7)		

Apart from the improvement in resolution, the physical representation of the amplitude is the main difference between PAUT and TFM. The image reconstruction algorithm of the TFM makes the amplitude comparison difficult. However, both amplitude representations are still meant to represent the sound energy level. However, they are not in the same reference any more. Therefore, the discontinuity in the scan with the biggest amplitude for a given propagation mode would be considered as a reference for the incoming comparison and the same process applies to PAUT. The relative reference is then converted to a decibel value in order to correlate the flaw orientation directly with the suitable ultrasonic technique. The following section is highlighting the influence of the flaw orientation towards the chosen propagation mode algorithm.

4.3 Amplitude Consistency Towards Flaw Orientation

Table 3 and Table 4 present the amplitude of the 5 defects inside the test sample for each propagation modes (TT, TTT, TT-TT) and PAUT. An automatic tool has been used to extract the maximum amplitude value for each flaw. The green boxes represent the flaw having the maximum amplitude response of the recorded weld for a given mode. The green flaw was used as a reference when it came to the decibel difference value computation towards the other flaw.

Table 2 – Amplitude analysis of the acquisition of the left side of the weld

Mode	Acquisition of the left side of the weld									
	Flaw A Lack of fusion		Flaw B Crack		Flaw C Porosities		Flaw D Porosities		Flaw E Lack of fusion	
	Amplitude (%)	dB diff.	Amplitude (%)	dB diff.	Amplitude (%)	dB diff.	Amplitude (%)	dB diff.	Amplitude (%)	dB diff.
PAUT	100.8	-6.0	177.8	-1.0	102.0	-5.8	112.0	-5.0	200.0	0.0
TT	100.0	0.0	47.0	-6.6	80.6	-1.9	84.9	-1.4	67.4	-3.4
TTT	100.0	0.0	10.0	-20.0	85.7	-1.3	21.5	-13.4	27.4	-11.2
TT-TT	60.4	-4.4	84.9	-1.4	68.1	-3.3	39.7	-8.0	100.0	0.0

The left ultrasonic acquisition included all the discontinuities detected in the weld sample. The PAUT detected all the defects within 6 dB difference compared to the reference level. All the individual TFM modes exceeded this decibel difference for at least one defect compared to the reference level. As for example for the TTT mode, the crack with an amplitude of 10% have approximately 20 dB difference and at this level the flaw is considered undetected. Another example is the TT-TT mode, where the porosity has 8 dB difference compared to the mode reference. In such cases the flaw could easily be left undetected.

Table 3 – Amplitude analysis of the acquisition of the right side of the weld

Mode	Acquisition of the right side of the weld									
	Flaw A Lack of fusion		Flaw B Crack		Flaw C Porosities		Flaw D Porosities		Flaw E Lack of fusion	
	Amplitude (%)	dB diff.	Amplitude (%)	dB diff.	Amplitude (%)	dB diff.	Amplitude (%)	dB diff.	Amplitude (%)	dB diff
PAUT	---	---	---	---	97.1	-4.4	83.3	-5.8	161.9	0.0
TT	---	---	---	---	34.1	-9.3	100.0	0,0	76.4	-2.3
TTT	---	---	---	---	45.1	-6.9	31.7	-10.0	100.0	0.0
TT-TT	---	---	---	---	9.1	-20.8	9.9	-20.1	100.0	0.0

For the right side, only three flaws were detectable. For the TT-TT propagation mode, the lack of fusion has a strong amplitude compared to other flaws. This is due to the sensitivity of this method to lack of fusion orientation inside the test piece. Such specific behaviour emphasises the importance of selecting the appropriate reflector reference type and orientation prior to inspecting a weld with only one propagation mode.

Table 4 shows the average decibel variants compared to the reference, according to a given mode or technique. Such variations for the individual propagation mode usually occurs when the propagation mode and the flaw orientations are not compatible with each other. This table highlight the fact that PAUT acquisition technique is less sensitive to defect orientations compared to individual propagation modes such as TFM technique.

Table 4 – Means of the decibel deviation for left and right acquisitions

Mode	Mean dB deviation
PA	-4.78
TT	-4.58
TTT	-9.96
TT-TT	-12.37

Moreover, Table 4 highlights the unlikeliness of qualifying a weld qualification sample, based on ASME section V, Article IV, Mandatory Appendix IX when using only one propagation mode (3). Indeed, according to this standard, in order to qualify a test piece a notch is required on every bevel as well as a minimum of one vertical surface notch.

5 Conclusion

This paper identifies and compares a combination of techniques for weld inspection by based on the approach proposes in the thesis of Ms Kombossé. This was achieved by utilising the simulation results in an experimental setup, based on a real FMC/TFM acquisition testing a physical weld sample with verified defects. Moreover, the PAUT and TFM comparisons support the fact that efficient resolution and flaw sizing are achieved better utilising the TFM approach compared to the PAUT technique with the appropriate propagation mode.

The sensitivity towards to the flaw orientation indicates the possible amplitude variations for different propagation modes. In some application scenarios for a given combination of TFM modes on a defined flaw orientation, the decibel difference could reach up to 20 dB. Considering such amplitude change, it is doubtful that a single propagation mode could validate the integrity of a weld qualification sample. Therefore, utilising all three TFM propagation modes would improve the resolution and sizing capabilities of the inspection and minimising the risk of negligent due to defect orientations.

References

1. **Sy, Kombossé.** *Étude du développement de méthodes de caractérisation de défauts basée sur les reconstructions ultrasonores TFM.* Paris : Université Paris-Saclay, 2018.

2. **Association technique de traitement thermique.** *Métalographie et techniques d'analyse.* Paris : Dunod, 2004.

3. **American Society of Mechanical Engineers.** Section V, Article IV, Mandatory Appendix IX.

A Review of The Benefits of PWI/TFM for Weld Inspection

CINDE
NDT in Canada Conference 2019

CINDE
CANADIAN INSTITUTE FOR NON-DESTRUCTIVE EVALUATION

NDT in Canada
NDTiC 2019
Inform - Engage - Advance
June 18 - 20
River Cree Resort & Casino
Edmonton, Alberta

Table of Contents

NDT in Canada 2019 | June 18-20 | Edmonton, Alberta

Next Generation Ultrasonic Inspection

Historical Constraints

Historical Constraints

- **Too Slow!**
  - The volume of data (i.e. A-Scans) generated by FMC/TFM is much larger than PAUT (n vs. n^2) = greater processing time = slower inspection speeds
- **Too Much Data!**
  - The volume of data (i.e. A-Scans) generated by FMC/TFM is much larger (n vs. n^2) = much larger file size
- **Lack of Sensitivity/Penetration**
  - With a single element transmitting, sensitivity and penetration may be reduced

Next Generation Ultrasonic Inspection

A Review of PWI/TFM and Associated Benefits

5

NDT in Canada 2019 | June 18-20 | Edmonton, Alberta

Benefits of PWI (PWI/TFM)

- PWI acquisition with TFM image processing = PWI/TFM
- Represents an optimization of both FMC data set acquisition and TFM image post-processing
- 5x-100x faster than traditional FMC/TFM
- Provides greater sensitivity
- Very good penetration/SNR for attenuative materials (Ti, Inconel, etc.)
- Smaller file sizes
- Generally, an "up-correlation" in inspection frequency; e.g. austenitic welds inspected with 5MHz
- Reduced technique development time

NDT in Canada 2019 | June 18-20 | Edmonton, Alberta

PWI (PWI/TFM) Explained

- PWI/TFM utilizes a larger transmitting aperture than FMC/TFM, but with the proper algorithmic image processing, the data set can still be used to create a TFM image

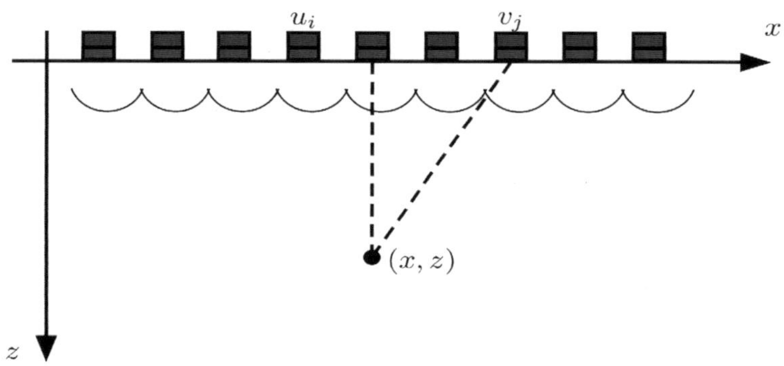

PWI (PWI/TFM) Explained

- As seen in the image below, several elements are transmitting at once, but the image reconstruction still considers each elemental transmit/receive pair, as in "standard" FMC/TFM

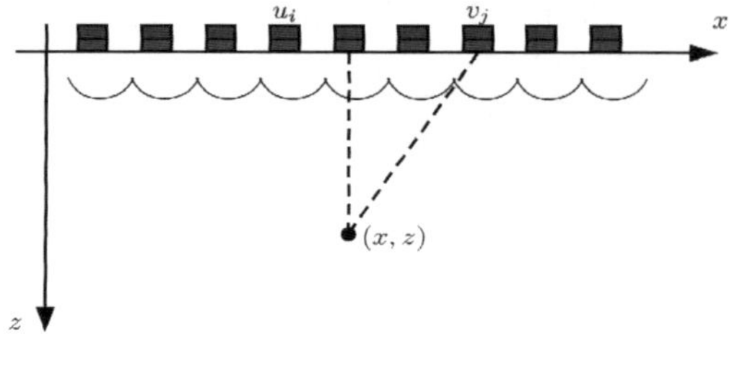

PWI Inspection Speed

Inspection Speed (Example)

- Inspection speed (TPAC historical)
  - For 300mm metal path with 128 Element Probe x 2mm Pitch
    - 2.5MHz = ~2"/s
    - 5MHz = ~1"/s
- With recent new software developments by TPAC, inspection speeds of >12"/s are achievable

NDT in Canada 2019 | June 18-20 | Edmonton, Alberta

PWI Inspection Speed - Don't Blink!

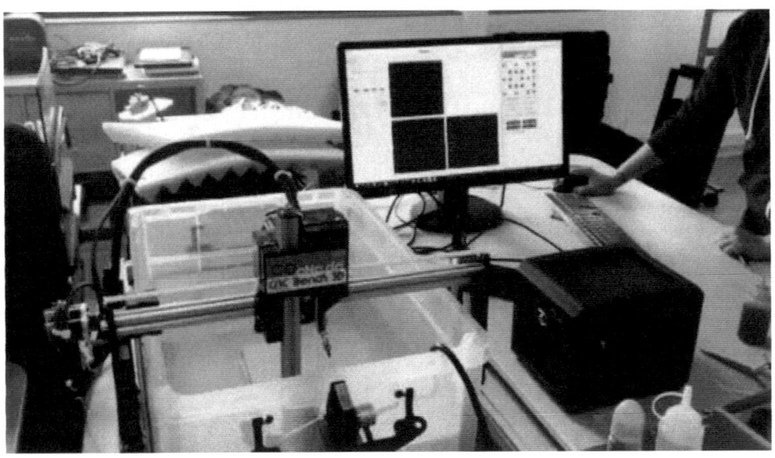

NDT in Canada 2019 | June 18-20 | Edmonton, Alberta

PWI/TFM Sensitivity

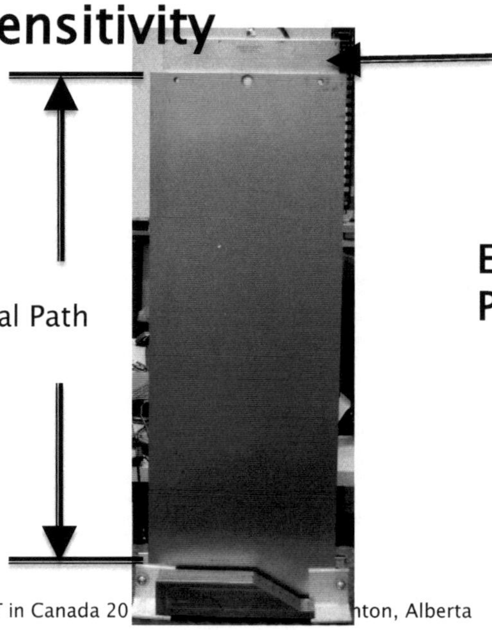

Extended Metal Path Inspection

PWI Sensitivity

Extended Metal Path Inspection

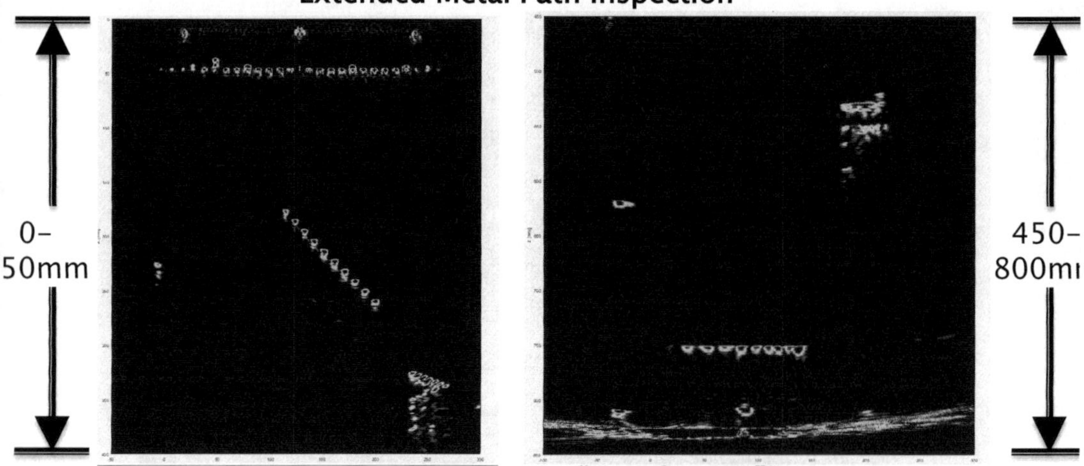

One Setup, No "Focal Laws"!

Review of Wave Modes

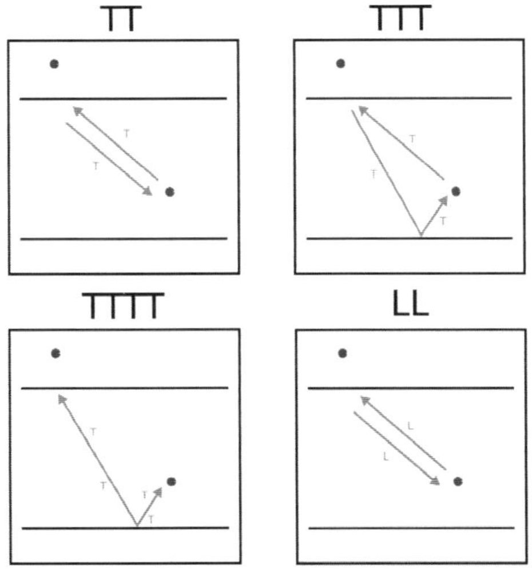

Review of Wave Modes

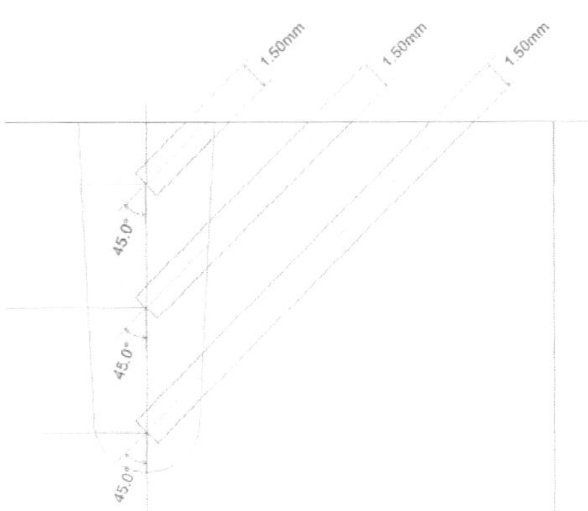

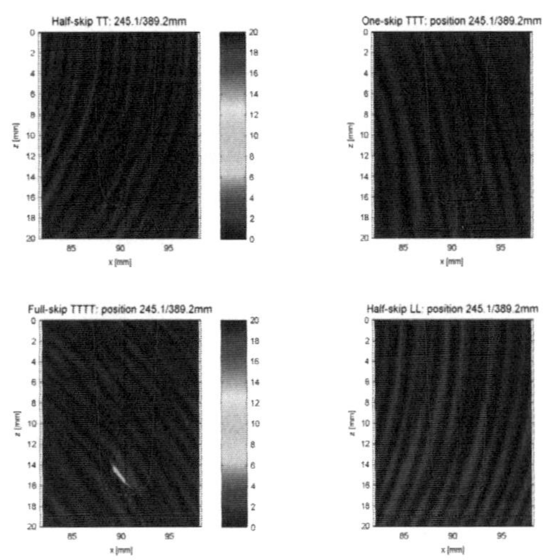

NDT in Canada 2019 | June 18-20 | Edmonton, Alberta

Multi-Modal PWI Acquisition

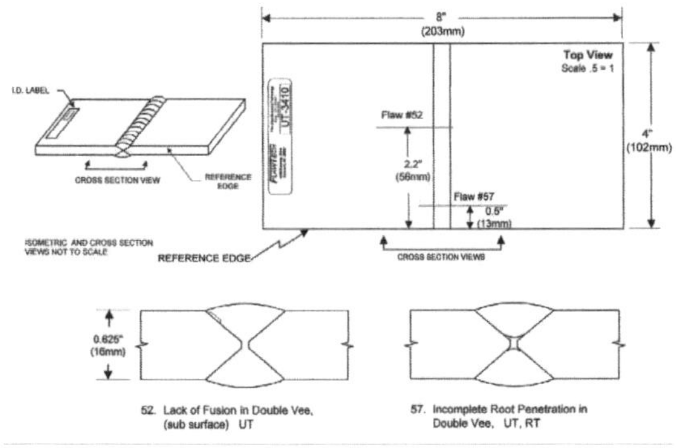

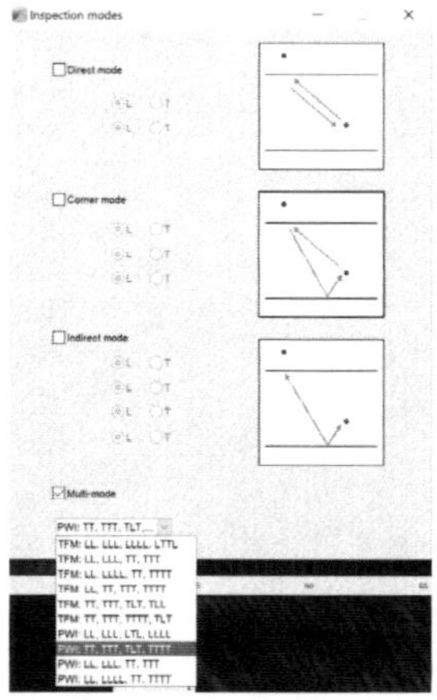

Mutiple wave mode acquisition available in TPAC PWI/TFM

NDT in Canada 2019 | June 18-20 | Edmonton, Alberta

Multi-Modal PWI Acquisition

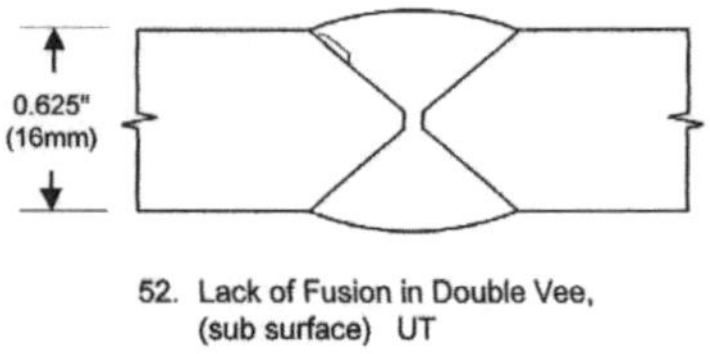

52. Lack of Fusion in Double Vee, (sub surface) UT

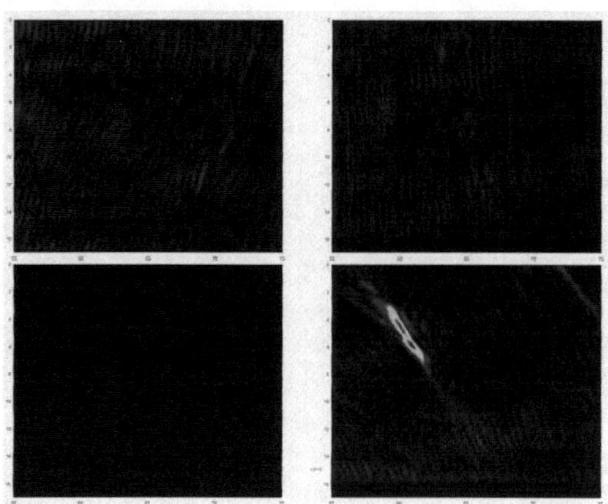

"Modal Merge" also available in TPAC analysis

NDT in Canada 2019 | June 18-20 | Edmonton, Alberta

Multi-Modal PWI Acquisition

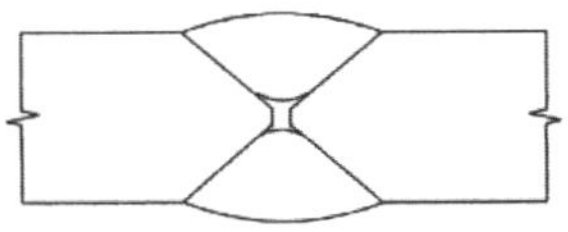

57. Incomplete Root Penetration in Double Vee, UT, RT

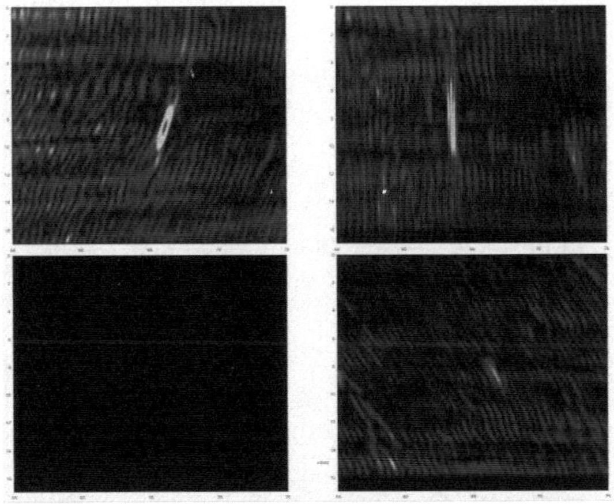

"Modal Merge" also available in TPAC analysis

Next Generation Ultrasonic Inspection

PWI/TFM Inspection Examples

Carbon Steel Plate

- Part
  - ~2" T
  - Manufactured Slag Inclusion
- Probe
  - 5MHz
  - 64 Element
  - No wedge
  - Centered on weld
  - TT Mode

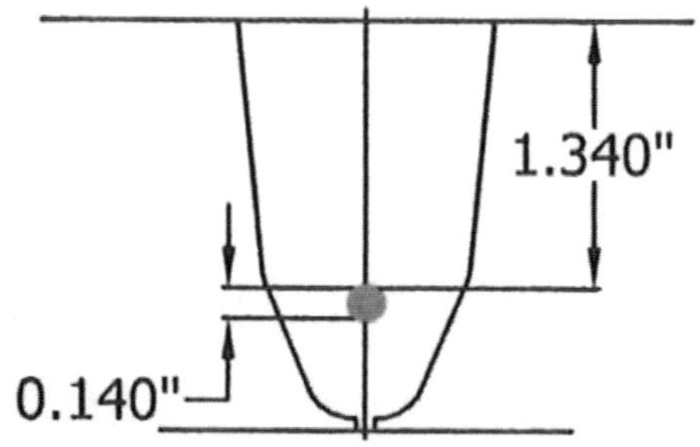

NDT in Canada 2018 | June 19-21 | Halifax, NS

Carbon Steel Plate

- PWI acquisition with TFM image reprocessing
- 1M TFM reconstrucion points
  - 1000 x 1000

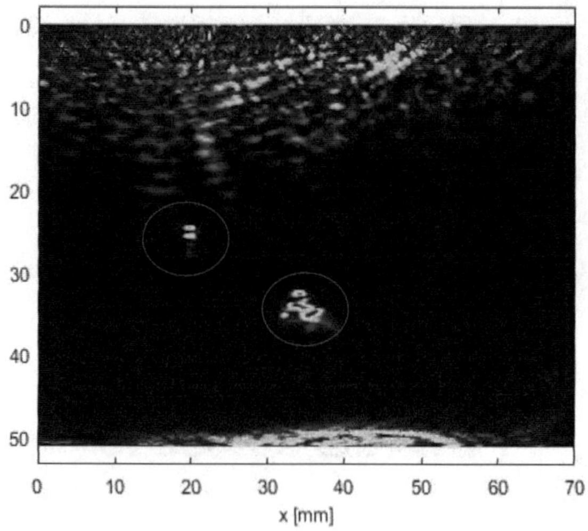

Carbon Steel Plate

- PWI acquisition with TFM image reprocessing
- 1M TFM reconstrucion points
  - 1000 x 1000

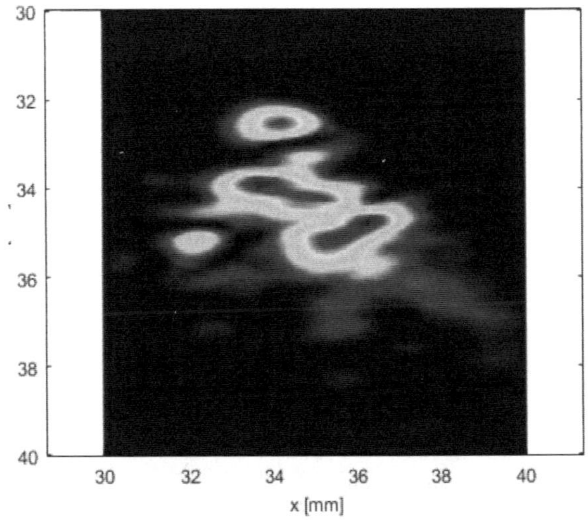

4" Thick Plate

- Part
  - Carbon Steel
  - 4" T DV
  - Inclusion, I.P., Crack
- Probe
  - 5MHz
  - 64 Element
  - TT Mode

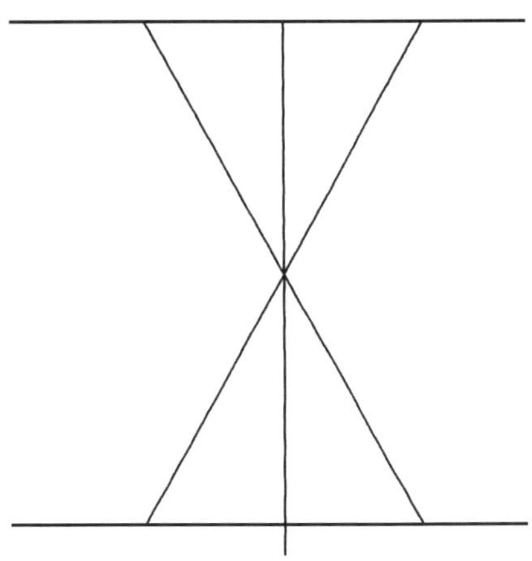

4” Thick Plate

- PWI acquisition with TFM image reprocessing
- 1M TFM reconstrucion points (1000 x 1000)

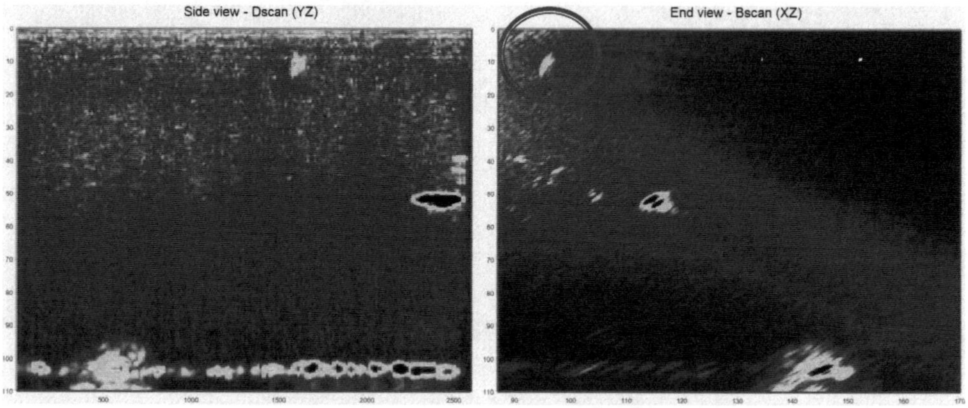

NDT in Canada 2019 | June 18-20 | Edmonton, Alberta

CRA (Inconel 59) Welded SDH Block

- Part
  - Carbon Steel Pipe with Alloy 59 Weld
  - ~1.75" T
  - Ø0.125" SDH at ¼, ½, and ¾ T
- Probe
  - 5MHz
  - 64 Element
  - LL Mode

NDT in Canada 2019 | June 18-20 | Edmonton, Alberta

CRA (Inconel 59) Welded SDH Block

- PWI acquisition with TFM image reprocessing
- ~250K TFM reconstrucion points
  - 500 x 500

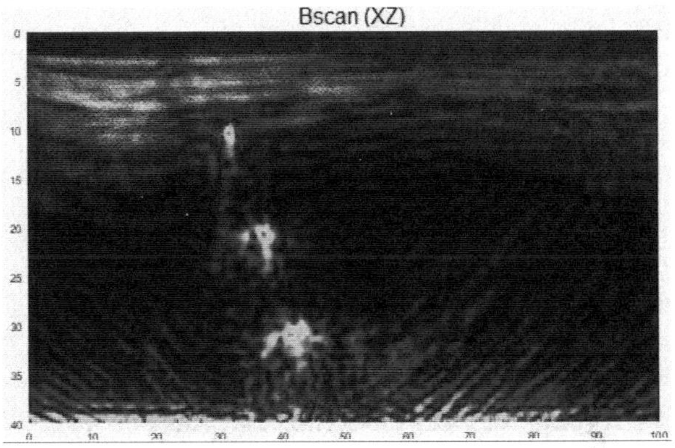

NDT in Canada 2019 | June 18-20 | Edmonton, Alberta

Dissimilar Metal SDH Block

- Part
  - SS > Inc Weld > CS
  - ~1.25" T
  - SDH
- Probe
  - 5MHz
  - 64 Element
  - LL Mode

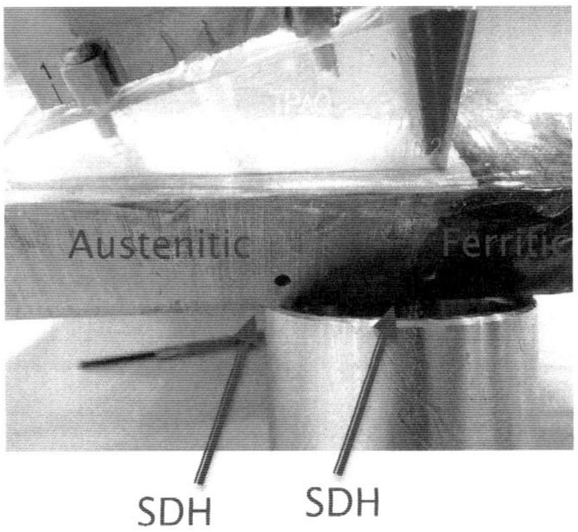

NDT in Canada 2019 | June 18-20 | Edmonton, Alberta

Dissimilar Metal SDH Block

- PWI acquisition with TFM image reprocessing
- ~250K TFM reconstrucion points
  - 500 x 500

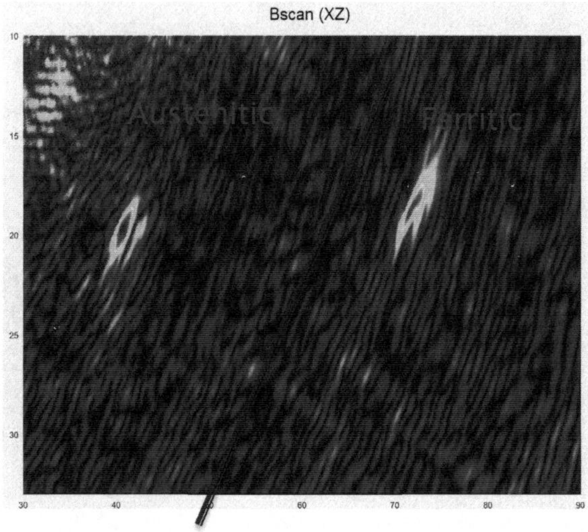

NDT in Canada 2019 | June 18-20 | Edmonton, Alberta

Conclusion

- PWI/TFM offers:
  - Faster acquisition speeds
    - TPAC can offer >12"/s
  - Superior resolution of TFM imaging
    - Better characterization and sizing
  - Higher frequency/higher resolution inspection of attenuative materials compared to PAUT
  - Reduced technique development time
    - Being focused "everywhere" reduces the need for determining focal laws/using multiple groups
  - TPAC can offer inspection with multiple wave modes in one acquisition
  - Greater (full) coverage with one line scan (each side)

NDT in Canada 2019 | June 18-20 | Edmonton, Alberta

Thank You!

Questions or Comments?

NDT in Canada 2019 | June 18-20 | Edmonton, Alberta

Development of Radiographic Image Processing Algorithms at CNL

Krassimir STOEV [1]

[1] Canadian Nuclear Laboratories, Chalk River, Ontario K0J 1J0, Canada
Phone: +1-613-584-3311 extension 43970, Fax: +1-613-584-8110; e-mail: Krassimir.Stoev@CNL.ca

Abstract
Image processing software is an important tool during analysis and interpretation of digital radiography images. CNL is currently in the process of developing software, which will be used for implementation and testing of new image processing algorithms. The current state of the CNL image processing software is discussed. Basic functionality is described. Examples of currently implemented advanced image processing techniques are presented. Details of different bilateral filtering/smoothing algorithms are compared.

Keywords: radiography, image processing

1. Introduction

Digital radiography (DR) is a fast and reliable volumetric inspection and diagnostics method which can be used for testing of a variety of objects made from different materials. DR offers several advantages in comparison to classical film radiography, such as improved detector dynamic range (i.e., a wider range of material thickness can be radiographed simultaneously), improved productivity (due to higher sensitivity of the digital detectors significantly reducing exposure times), improved safety due to lower exposure levels, reduced costs of consumables, etc. As a result, DR allows for integration and automation of the data collection, storage, archiving and transmission processes, and offers improved information flow and data management. Image processing is very important to digital radiography, and allows for better interpretation of radiographs. CNL has developed a prototype image processing software, which is constantly being upgraded and improved. This software is used not only for processing of radiographic images, but also as a test bed for development and evaluation of new image processing algorithms [1].

2. Description of the software and implemented algorithms

In 2010, CNL started development of an image processing software (called CNL Image Processing Tool, or IPT), which was intended to be used for implementation and testing of new image processing algorithms. At the current stage, all basic editing and image manipulation functions are implemented [1], and work is underway on the development of advanced image processing algorithms and their implementation and testing. The main window of the software has three regions: menus, control buttons, and display area (see Figure 1). The IPT software is designed to work with 16-bit TIFF images, but can read and save other formats of images. The following basic image processing options are available in the menus and in the control buttons area: Selection Tool, Zoom (1% to 1000%, fit width, height, stretch, selection), Magnifying Glass Tool, Invert Image, Rotate, Flip Horizontally or Vertically, Crop Selection, Multiply, Intensity Filtering (i.e., Median, Average, Gaussian, Min/Max), Spatial Filters (Gradient, Laplacian, Sobel, Prewitt, Shift/Difference, Line Segment), Crop, Copy, Paste, Invert, Despeckle, Set or Stretch Brightness / Contrast / Gamma / Histogram Contrast, Line-Profile Tool, Histogram Display, Equalize Image Histogram, Unsharp Mask, Smoothing, Sharpen, Emboss. All of these options are basic image manipulation options, and can be used for initial assessment of the radiographic images.

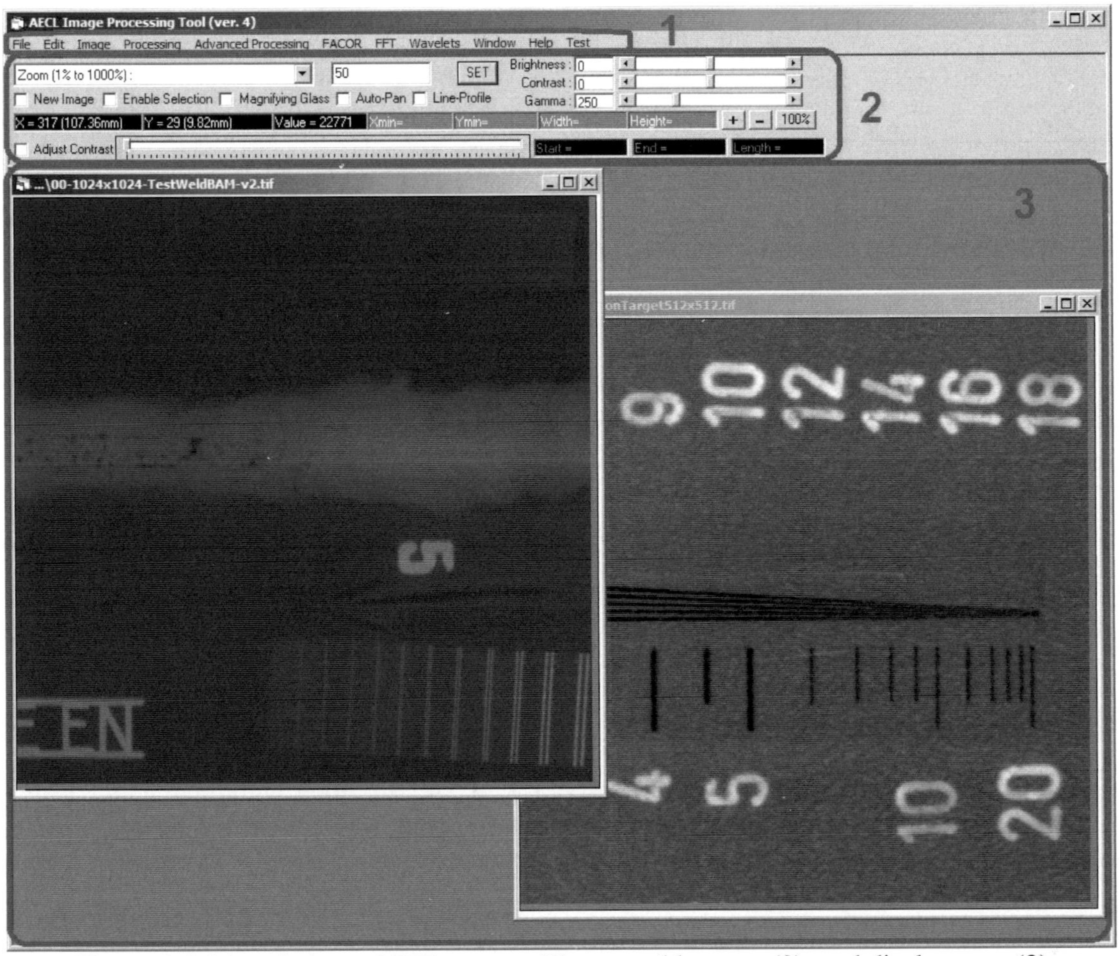

Figure 1. Main window of IPT: menus (1), control buttons (2), and display area (3)

After implementing the basic image editing, manipulation and processing options, work was concentrated on the development of advanced image processing algorithms and their implementation. Several types of algorithms were developed and tested:

- Methods for image noise suppression (such as smoothing and differential operators based on fitting with orthogonal polynomials, bilateral smoothing filters, etc.).
- Methods for image contrast enhancement (such as local contrast enhancement techniques based on local signal variance, adaptive unsharp mask, adaptive contrast enhancement, FACOR filter, etc.).
- Methods for image grey-scale manipulation (such as pixel-based mathematical transformations, continuous linear unrestricted image intensity adjustment, non-linear continuous intensity remapping, global and local non-linear histogram equalization, adaptive histogram equalization (AHE), contrast-limited AHE (CLAHE), etc.).
- Methods for statistical image analysis (such as 16-bit line-profile, horizontal and vertical integrated line profile, image and area statistics, algorithm for removing outlier values, etc.).
- Methods based on two-dimensional Fast Fourier Transform (FFT)
- Methods based on two-dimensional Discrete Wavelet Transform (DWT)

Some of these methods will be discussed in detail in the following sections.

2.1 Noise suppression methods

Noise suppression, also called smoothing, is usually the first step applied during the image processing of radiographic images. The aim of this step is to reduce the random noise in the image without introducing systematic or non-linear deviations from the original image, which could lead to introduction of image processing artefacts in the final processed image. The following advanced smoothing algorithms were developed and implemented: methods for correction of outlier pixels, fully selectable 3x3, 5x5, and 7x7 convolution filters, polynomial 2D smoothing, fuzzy median smoothing, bilateral 2D smoothing, and adaptive local smoothing. Two-dimensional bilateral smoothing was identified as the most promising noise suppression method, because it takes into account both the closeness of the pixels (i.e., the filter includes coefficients that fall off with distance) and similarity of the pixels (i.e., the filter includes coefficients that decay with dissimilarity). The most important property of the bilateral smoothing filters is that they preserve edges, while other smoothing filters smear edges. The originally proposed bilateral 2D smoothing [2] used Gaussian functions for both spatial-domain coefficients and for grey-level coefficients. A generalized formula for bilateral filtering was proposed:

$$S_1(i,j) = \sum_{k=-M}^{M} \sum_{m=-M}^{M} A(i+k, j+m) \cdot Y(i+k, j+m)$$

where ***Y(i, j)*** represents the original image, ***S(i, j)*** represents the smoothed image, ***A(i+k, j+m)*** are the weights/coefficients of the bilateral 2D smoothing operator. In the formulae below, $\boldsymbol{\mu}$ and $\boldsymbol{\sigma}$ are the average and the standard deviation over pre-defined number of pixels, $\boldsymbol{\alpha_1}$ and $\boldsymbol{\alpha_2}$ are the multiplication coefficients, and ***M*** and ***N*** ***(N<M)*** are the filter width/region and a subset of the filter region. Several functions for bilateral 2D smoothing operator were implemented and tested:

$$A(i+k, j+m) = e^{-\alpha_1 \cdot \left(\frac{|\mu_N(i,j)-\mu_M(i,j)|}{\sigma_M(i,j)+\sigma_N(i,j)}\right) \cdot (k^2+m^2)} \cdot e^{-\alpha_2 \frac{|Y(i+k,j+m)-\mu_N(i,j)|}{\sigma_N(i,j)} \cdot \frac{|Y(i+k,j+m)-Y(i,j)|}{|\mu_N(i,j)-\mu_M(i,j)|}} \tag{1}$$

$$A(i+k, j+m) = e^{-\alpha_1 \cdot \left(\frac{\sigma_M(i,j)}{\sigma_N(i,j)}\right) \cdot \left(\frac{|\mu_N(i,j)-\mu_M(i,j)|}{\sigma_N(i,j)+\sigma_M(i,j)}\right) \cdot (k^2+m^2)} \cdot e^{-\alpha_2 \cdot \frac{|Y(i+k,j+m)-Y(i,j)|}{|\mu_N(i,j)-\mu_M(i,j)|}} \tag{2}$$

$$A(i+k, j+m) = e^{-\alpha_1 \cdot \left(\frac{\sigma_M(i,j)}{\sigma_N(i,j)}\right) \cdot \left(\frac{|\mu_N(i,j)-\mu_M(i,j)|}{\sigma_N(i,j)+\sigma_M(i,j)}\right) \cdot (k^2+m^2)} \cdot e^{-\alpha_2 \cdot \frac{|Y(i+k,j+m)-Y(i,j)|}{|Y(i+k,j+m)-\mu_N(i,j)|}} \tag{3}$$

$$A(i+k, j+m) = e^{-\alpha_1 \cdot \left(\frac{\sigma_M(i,j)}{\sigma_N(i,j)}\right) \cdot \left(\frac{|\mu_N(i,j)-\mu_M(i,j)|}{\sigma_N(i,j)+\sigma_M(i,j)}\right) \cdot (k^2+m^2)} \cdot e^{-\alpha_2 \cdot \frac{|\mu_N(i,j)-\mu_M(i,j)|}{\sigma_M(i,j)} \cdot \frac{|Y(i+k,j+m)-Y(i,j)|}{|Y(i+k,j+m)-\mu_N(i,j)|}} \tag{4}$$

$$A(i+k, j+m) = e^{-\frac{|\mu_N(i,j)-\mu_M(i,j)|}{\sigma_M(i,j)} \cdot \frac{|Y(i+k,j+m)-\mu_N(i,j)|}{\sigma_N(i,j)} \cdot \frac{(k^2+m^2)}{\sigma_2}} \tag{5}$$

$$A(i+k, j+m) = e^{-\frac{|Y(i+k,j+m)-\mu_N(i,j)|}{\sigma_N(i,j).\sigma_M(i,j)} \cdot \frac{(k^2+m^2)}{\sigma_2}} \tag{6}$$

$$A(i+k, j+m) = e^{-\left(\alpha_1 \cdot \frac{|Y(i+k,j+m)-Y(i,j)|}{\sigma_M^2(i,j)} + \frac{(k^2+m^2)}{.\sigma_2^2}\right)} \tag{7}$$

$$A(i+k, j+m) = \frac{1}{\left(1+\frac{[Y(i+k,j+m)-Y(i,j)]^2}{\sigma_1^2}\right)} . \frac{1}{\left(1+\frac{(k^2+m^2)}{\sigma_2^2}\right)} \tag{8}$$

$$A(i+k,j+m)=\frac{1}{\left(1+\frac{[Y(i+k,j+m)-Y(i,j)]^2}{\sigma_1^2}\right)}+\frac{1}{\left(1+\frac{(k^2+m^2)}{\sigma_2^2}\right)} \tag{9}$$

$$A(i+k,j+m)=\frac{1}{\left(1+\frac{|Y(i+k,j+m)-Y(i,j)|}{\sigma_N(i,j)+\sigma_M(i,j)}\right)}\cdot\frac{1}{\left(1+\frac{(k^2+m^2)}{\sigma_2^2}\right)} \tag{10}$$

$$A(i+k,j+m)=\frac{1}{\left(1+\alpha_1\cdot\frac{[Y(i+k,j+m)-\mu_N(i,j)]^2}{\sigma_N(i,j).\sigma_M(i,j)}\right)}\cdot\frac{1}{\left(1+\frac{(k^2+m^2)}{\sigma_2}\right)} \tag{11}$$

$$A(i+k,j+m)=\frac{1}{\left(1+\frac{[Y(i+k,j+m)-\mu_N(i,j)]^2}{\sigma_1^2}\right)}+\frac{1}{\left(1+\frac{(k^2+m^2)}{\sigma_2^2}\right)} \tag{12}$$

All of the above functions differ from the original formula for the coefficients for bilateral smoothing, see Equation (13).

$$A(i+k,j+m)=e^{-\left(\frac{|Y(i+k,j+m)-Y(i,j)|}{\sigma_1}+\frac{(k^2+m^2)}{\sigma_2^2}\right)} \tag{13}$$

The original image and two processed images are shown in Figure 2 and their operators noted in the caption. Results from the comparison of different implementations of the bilateral 2D smoothing operators are shown in Table 1. The area shown in red in Figure 2(a) was used to calculate the signal-to-noise ratio (SNR). The presented images are in 8-bit JPG format, while the original were bigger images (512x512 pixels) in 16-bit TIF format, so it is not possible to correctly evaluate the spatial resolution from the images presented herein. As can be seen from Table 1, only Equations (2), (3) and (10) preserve the spatial resolution as per the original image, with Equations (2) and (10) achieving the same improvements in SNR as a 7x7 Gaussian filter. The original bilateral smoothing [2] and Equation (8) achieve better SNR and better spatial resolution than polynomial smoothing. Equations (9) and (12) achieve better SNR and better spatial resolution than polynomial 5x5 averaging. It should be noted that the results presented in Table 1 are for specific values of the parameters for each filter, and different selection of parameters will lead to different results. Optimization of the parameters of the proposed bilateral 2D smoothing operators was outside the scope of the current research.

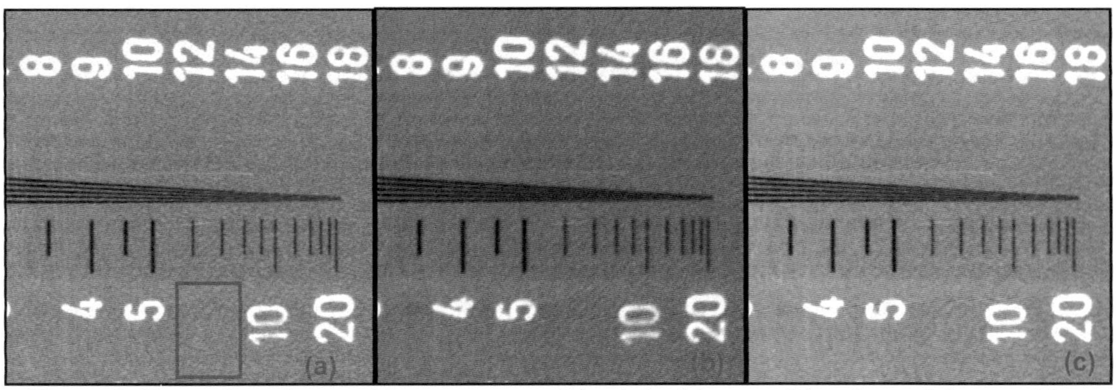

Figure 2. Examples of smoothing operators: (a) original image, (b) polynomial smoothing, and (c) bilateral smoothing using Eq. (9).

Table 1. Comparison of SNR and LP (Line-Pairs) of different smoothing algorithms.

Image	SNR	Spatial Resolution [LP/mm]	Filter Parameters
Original	17.4	9	
Selectable Filter 5x5 (Average)	30.7	4.5	
Selectable Filter 7x7 (Gaussian)	25.1	7	
Polynomial Smoothing	27.1	6	F1, 2, 2
Original Bilateral Smoothing	29.3	8	3, 3, 0.05
New Bilateral Smoothing Eq. (1)	23.7	8	1, 0.1, 3, 0.1
New Bilateral Smoothing Eq. (2)	25.1	9	1, 0.1, 3, 0.4
New Bilateral Smoothing Eq. (3)	22.6	9	1, 0.2, 3, 1
New Bilateral Smoothing Eq. (4)	19.0	8	1, 0.2, 3, 0.5
New Bilateral Smoothing Eq. (5)	19.0	7	1, 3, 1
New Bilateral Smoothing Eq. (6)	18.5	8	1, 3, 0.001
New Bilateral Smoothing Eq. (7)	21.8	8	1, 2, 1
New Bilateral Smoothing Eq. (8)	27.3	8	3000, 3, 1
New Bilateral Smoothing Eq. (9)	33.6	6	3000, 3, 1
New Bilateral Smoothing Eq. (10)	25.1	9	1, 3, 2
New Bilateral Smoothing Eq. (11)	22.6	8	1, 0.1, 3, 3
New Bilateral Smoothing Eq. (12)	34.5	5.5	1, 3000, 3, 1

Additionally, formulae and an algorithm for a 2D smoothed derivatives calculation was developed. Results from applying this algorithm are presented in Figure 3. The smoothed derivatives operation combines noise suppression with visibility enhancement, and can be useful for simultaneous observation of objects with different thickness in the radiographs. It should be noted that the original image looks darker in comparison to the case when viewed on professional radiography monitor, due to the limited contrast for images in PDF and Microsoft Office files. Nevertheless, the improvements demonstrated in this presentation cannot be achieved with simple intensity and brightness adjustments of the original image.

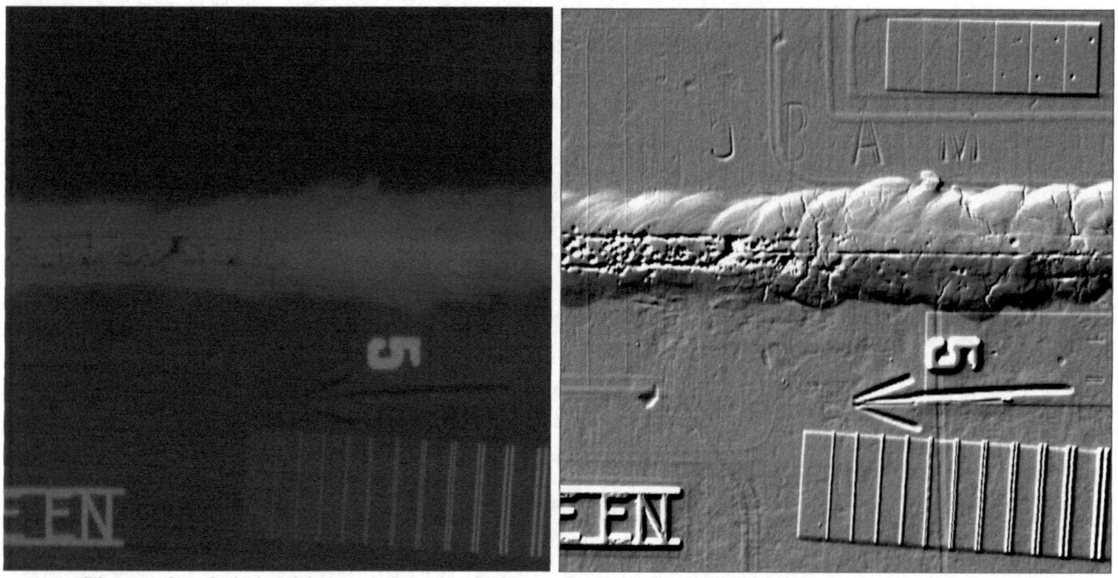

Figure 3. Original image (on the left) and smoothed 2D derivative (on the right).

2.2 Contrast enhancement methods

Contrast enhancement methods are used for improvement of the visibility of details in digital radiographs. Contrast enhancement allows for simultaneous visualization and observation of defects and internal parts which could not be viewed simultaneously in the original image, for example details and/or defects located in parts of the object which have a significant difference in thickness. The wide range of thickness variation will lead to a very wide range in grey-scales in the radiographic image. The human eye can distinguish only about 32 different grey-levels, so it will be impossible to observe internal details in the radiographic image without application of appropriate image enhancement techniques. It is important to visualise and observe simultaneously the whole thickness range recorded in the radiographs in order to perform proper diagnostics and interpretation. This requires strong image enhancement algorithms and advanced visualization methods. Several different types of algorithms for contrast enhancement were developed and implemented:

- Sharpening filters (17 sharpening filters with selectable parameters were implemented),
- Local contrast enhancement algorithms,
- Adaptive unsharp mask algorithms,
- Adaptive contrast enhancement,
- Contrast enhancement by digital equalization (CEDE).

Examples of image contrast enhancement methods are presented in Figure 4.

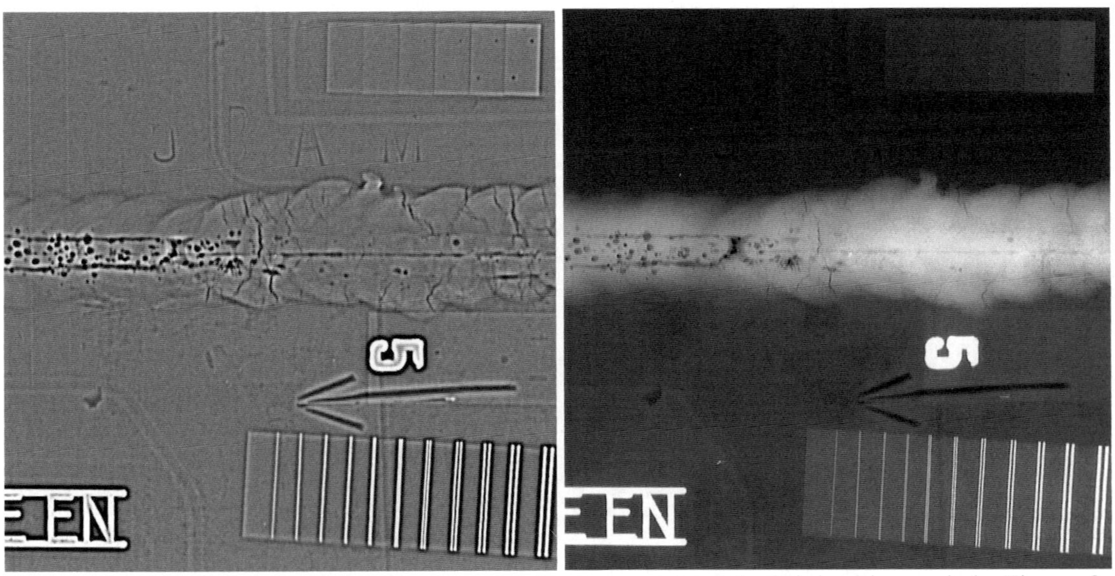

Figure 4. CEDE (left), and local contrast enhancement (right). Original image is in Figure 3.

2.3 Grey-scale manipulation methods

Most of these algorithms are based on global or localized histogram equalization, or on other advanced methods like differential hysteresis filtering. Some of the developed and implemented algorithms are:

- Sigmoidal intensity adjustment with multiple types of functions,
- Auto sigmoidal intensity adjustment,
- Auto gamma histogram equalization,
- Local histogram equalization,
- Contrast limiting adaptive histogram equalization (CLAHE),

- Random sampling local histogram equalization (RSLHE),
- Local intensity stretching with bi-linear interpolation (LISBLI),
- Local intensity rank filtering.

Several examples of images processed with grey-scale manipulation methods are presented in Figure 5.

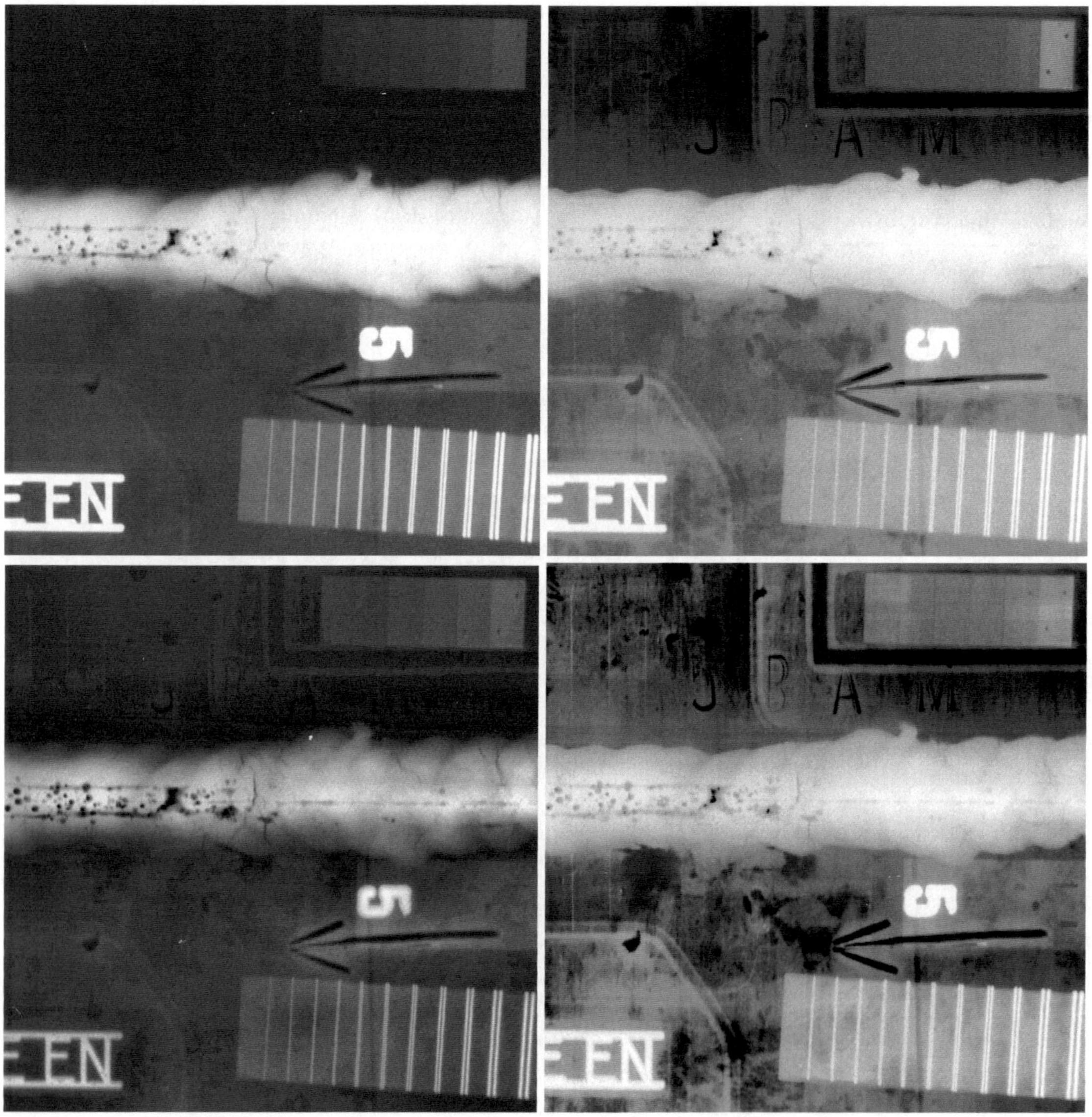

Figure 5. Auto sigmoidal intensity adjustment (left top), auto gamma histogram equalization (right top), CLAHE (left bottom), and RSLHE (right bottom). Original image is in Figure 3.

2.4 Statistical image analysis methods

Image processing methods, based on calculation of different local statistical parameters, were developed and implemented. Some of the implemented algorithms are based on the following local statistical quantities: signal variance, entropy, adaptive histogram entropy, signal-to-noise ratio, contrast-to-noise ratio, etc. Examples of statistical methods are shown in Figure 6.

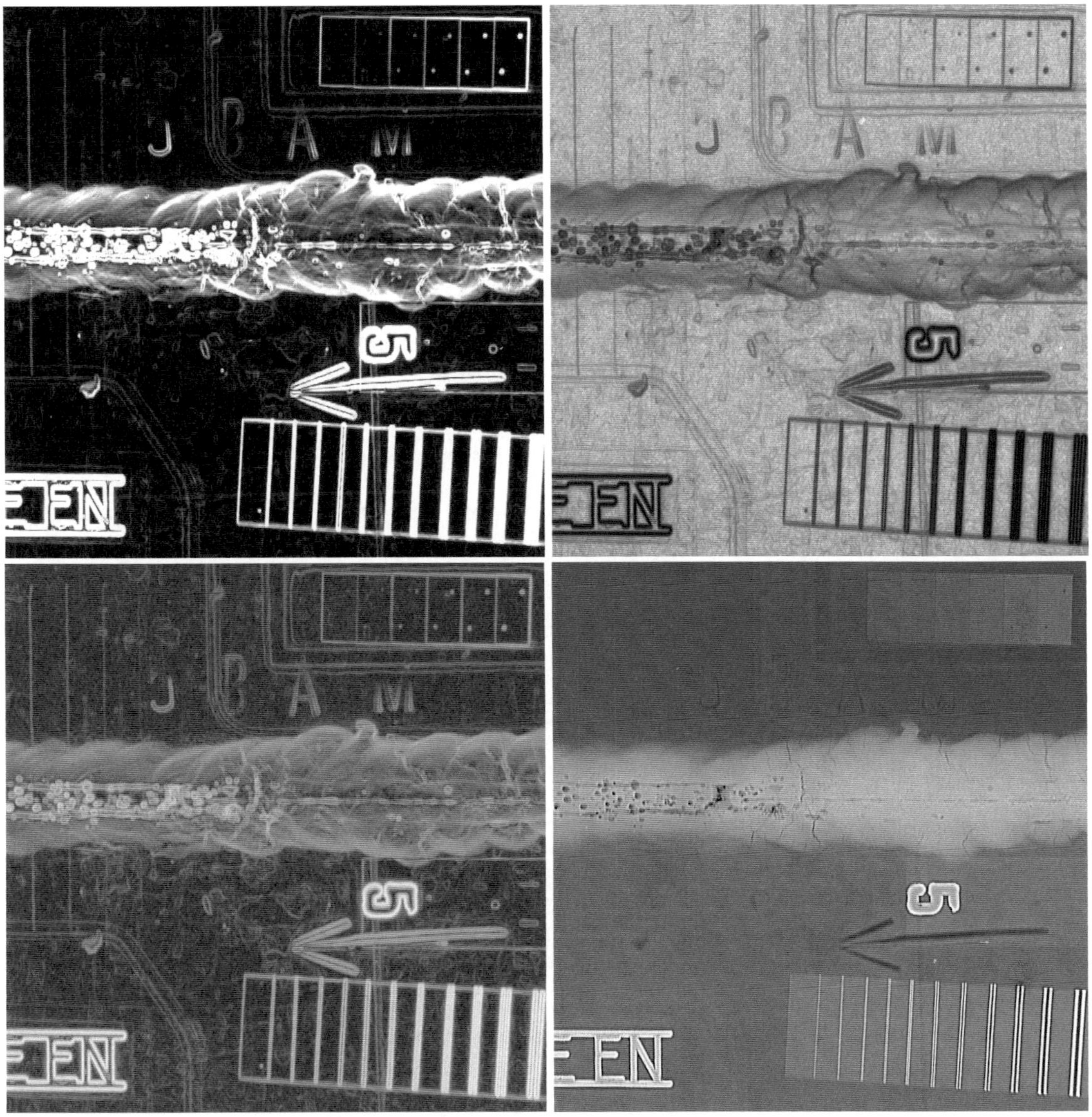

Figure 6. Image processing methods based on local signal variance (left top), local SNR (right top), logarithmic signal variance (left bottom), and local entropy (right bottom). Original image is in Figure 3.

2.5 FFT based methods

The Fourier transform is widely used in different fields of science. It is based on two basic functions: sine and cosine. These functions have infinite range in the space domain, but they are completely localized (δ-functions) in the frequency domain. Because of the choice of the basic functions, Fourier analysis is very well suited for studying periodic signals. In other words, Fourier analysis is not local in space, but is local in frequency. The most widely used digital implementation of the Fourier analysis is the Fast Fourier Transform (FFT), which allows for fast transformation/computation when the number of the points is 2^N, where N is an integer number > 3. The signal filtering is done in the frequency domain, usually by applying a bandpass filter, which can lead to simultaneous reduction of the random noise and enhancement of details visibility. An example of FFT-based image processing is presented in Figure 7.

2.6 DWT based methods

The wavelet transform maps the input data into a new space, the basic functions of which are quite localized both in the space domain and in the frequency domain. The term "wavelet" means "small wave" and represents a localized wave-like function. Wavelets are localized in frequency as well as space, i.e. their rate of variation is restricted in both domains, which make them very suitable for analysing fast-changing (in time or space) signals. While the Fourier analysis is unique (i.e., there are only two basic functions: sine and cosine), the wavelet analysis is not unique, and there are many possible sets of wavelets which one can use. The trade-off between different wavelet sets is between their compactness versus their smoothness. Another advantage of the wavelet transform is that it is less computationally complex (i.e., it is faster than the Fast Fourier Transform). The discrete wavelet transform (DWT) is implemented by means of filter banks. The input signal has been split into frequency bands, by means of application of low-pass and high-pass filters. The DWT is usually implemented using a pyramidal algorithm, and at the end we have two sets of coefficients: details coefficients for all decomposition levels, and the approximation coefficients for the last level.

Two types of filtering of the input image were developed using DWT: a low-pass filter (or convolution with a smoothing function) and a high-pass filter (or convolution with a differentiating function). The following DWT image smoothing operators were implemented in the software: hard threshold, soft threshold, affine threshold, Garrote threshold, and smoothly clipped absolute deviations (SCAD) threshold. The following DWT image enhancement operators were implemented in the software: phase only reconstruction, piecewise linear function, gamma function, log function, Stahl Function, and gamma-sigmoidal function. In addition, several combined smoothing and enhancement DWT operators were implemented, based on sigmoidal, power, and exp-gamma functions. All of the above described operators use the same filter function for all details coefficients of all decomposition levels and for the approximation coefficients for the last level. Separately, two multi-scale enhancement algorithms were developed and implemented, based on application of a different filter/enhancement function for details coefficients of each different decomposition level. An example of DWT image processing is presented in Figure 7.

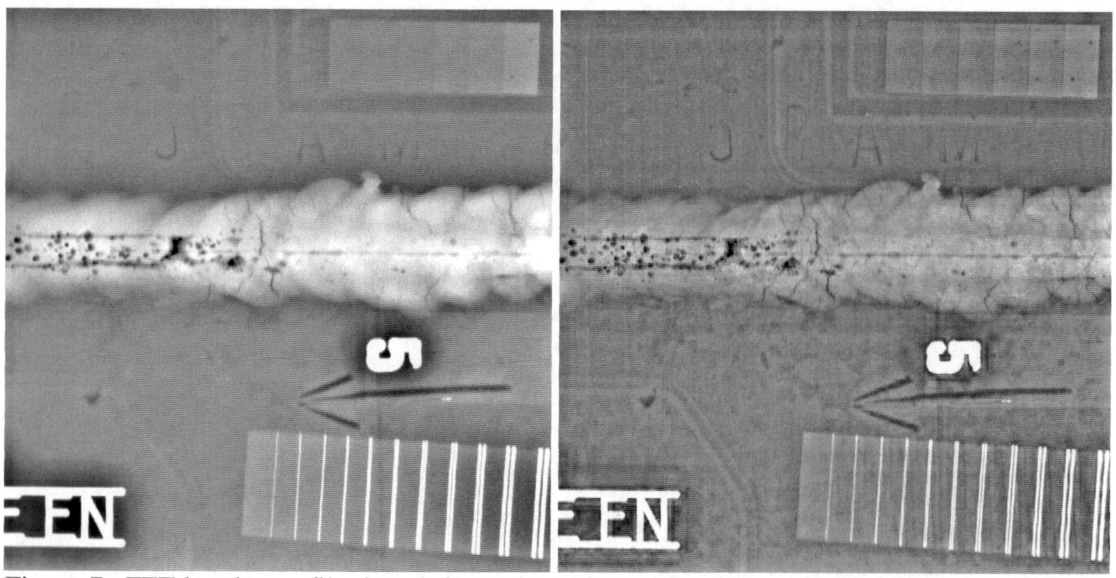

Figure 7. FFT band-pass filtering (left), and multi-level DWT based image processing (right). Original image is in Figure 3.

3. Conclusions

Digital radiography is significantly faster and offers improved detection latitude and better contrast, sensitivity and SNR in comparison to classical film radiography, which could lead to financial advantage during inspections. Advanced image processing can bring noticeable improvement in the visibility and detection of details in the radiographic images. CNL has developed prototype image processing software, to be used for implementation and testing of new image processing algorithms. At the current stage, all basic editing and image manipulation functions are implemented, together with some advanced image processing algorithms, which belong to three groups: noise suppression, contrast enhancement, and grey-scale manipulation for visibility improvement. Examples of the developed and implemented algorithms include polynomial and adaptive smoothing, statistical image evaluation tools, adaptive contrast enhancement, non-linear histogram adjustment algorithms, FFT and DWT image enhancement methods, etc. The developed software platforms can be used for further research work in the field of radiography image processing, and for the development and testing of new image processing algorithms. Development of advanced image processing methods would be beneficial not only to digital radiography, but to all other image-producing NDT methods, such as visual inspection (digital cameras are used on a regular basis for collecting data during visual inspections, and can greatly enhance the productivity of such methods as liquid penetrant and magnetic particles testing), ultrasound imaging, etc.

It should be noted that image processing can introduce artefacts into the images, and in this respect, the image processing software should be considered only as an aid during the interpretation and flaw-sizing of the digital radiographic images. Any indication detected in the enhanced radiograph should be confirmed with simple intensity stretching (this is equivalent to changing the intensity of the radiographic film viewer), as per the ASME code requirements [3]. Only indications that are confirmed in this way should be reported.

It would be useful for the radiographic community to generate several standardized digital radiography images (for example, images of welds and castings), which can be used to formalize and standardize the requirements and methods for testing radiographic image processing methods and software. The path forward is to organize a round-robin test for comparing capabilities of currently available image processing programs and for evaluating new image processing algorithms. Quantification of the capabilities of the image processing algorithms is needed in several areas, such as noise suppression, SNR improvement, minimum observable flaw or Image-Quality-Indicators feature in enhanced images, and evaluation of artefacts introduced by the image processing methods.

References

1. K. Stoev, "Comparison of Radiographic Image Processing Algorithms", Proceedings of 6th European Workshop on Structural Health Monitoring, v.2, p. 1639 (P29), July 2-6, 2012, Dresden, Germany (http://www.ecphm2012.com/Portals/98/BB/p29.pdf).
2. C. Tomasi and R. Manduchi, "Bilateral Filtering for Gray and Color Images", Proceedings of the 1998 IEEE International Conference on Computer Vision, Bombay, India, 1998.
3. ASME Boiler and Pressure Vessel Code, Section V "Nondestructive Examination", Article 2 "Radiographic Examination", Mandatory Appendix IV "Interpretation, Evaluation, and Disposition of Radiographic and Radioscopic Examination Test Results Produced by the Digital Image Acquisition and Display Process", 2010 edition.

New ASME FMC Rules

2019 ASME BPVC

CINDE
CANADIAN INSTITUTE FOR NON-DESTRUCTIVE EVALUATION

NDT in Canada
NDTiC 2019
Inform - Engage - Advance

June 18 - 20
River Cree Resort & Casino
Edmonton, Alberta

2019 ASME BPVC, Sec. V, Article 4. July 1, 2019

- *Mandatory Appendix XI Full Matrix Capture (FMC)*
- *Nonmandatory Appendix F Examination of Welds Using Full Matrix Capture (FMC)*

NDT in Canada 2019 | June 18-20 | Edmonton, AB

What's different from PAUT?

- Fracture mechanics only (MA IX)
- Cal Block
- *NO* Scanner Block
- Amplitude Fidelity Rule
- Essential Variables
- Saving Data, or not...

NDT in Canada 2019 | June 18-20 | Edmonton, AB

The FMC Cal Block and Managing Modes

SDH - resolution

Through Wall Slot - Exam volume coverage

≤11% Notch - Sizing verification

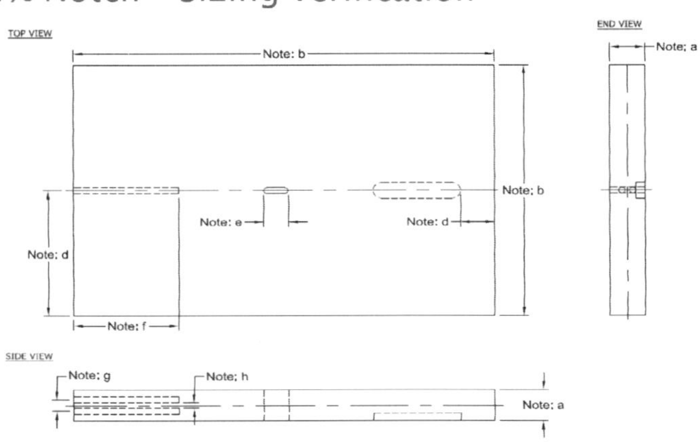

NDT in Canada 2019 | June 18-20 | Edmonton, AB

The Through Wall Slot

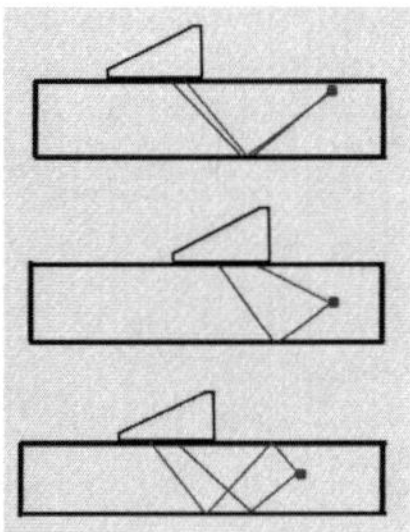

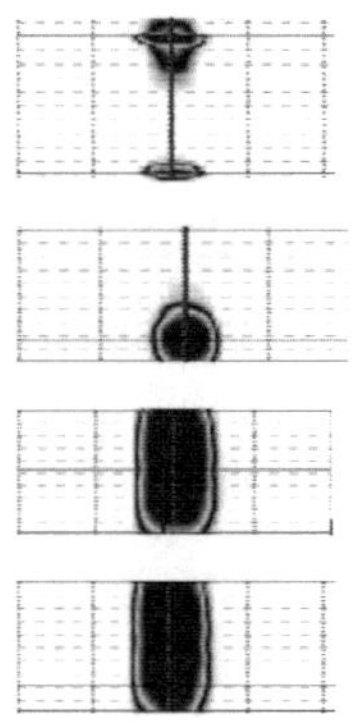

History 2013 to present

- *Applus shows up at Code Week*
- *OPG using a system for corrosion*
- *2015ish ASME splits WGUT to AUT*
- *Nov 2016 TG FMC splits and begins*
- *May 2017 FMC elevated to SWG*
- *Dec 2018 we hit the finish line*
- *July 1 2019 - Publication*

NDT in Canada 2019 | June 18-20 | Edmonton, AB

< 2 Years - How?

- *Interim web meetings*
- *Global collaboration (IIW, BINDT)*
- *Reported directly to SC V*
- *Very Talented and Dedicated group*
- *Presentations at every meeting*

NDT in Canada 2019 | June 18-20 | Edmonton, AB

Thank you!

any

Questions?

NDT in Canada 2019 | June 18-20 | Edmonton, AB

NRC·CNRC

NRC.CANADA.CA

Qualification and Process Control of Computed Radiography Systems

Muzibur Khan, Mike Brothers, Trent Gillis
Structures & Materials Performance Laboratory
Aerospace Research Centre

June 2019

National Research Council Canada — Conseil national de recherches Canada

Canada

Outline of Presentation

- ❑ Film vs Digital Radiography
- ❑ Introduction of Computed Radiography (CR)
- ❑ Qualification & Process Control of CR (ASTM E2445)
- ❑ CR for High Resolution Applications
- ❑ Summary

2

Objectives of the Study

- ❑ **Qualification of CR systems in accordance with ASTM E2445 standards**
- ❑ **Investigation of EPS concepts for aluminum alloy**
  - ❖ Baseline EPS performance of CR system/Imaging Plates
  - ❖ Estimation of minimum, range, linearity of exposure
- ❑ **Developing process control of CR imaging applications**
  - ❖ To ensuring a high degree of repeatability of inspection results
- ❑ **Investigation of CR for high resolution applications**

3

Background: Film vs Digital Radiography

- ❑ **Film for NDT applications is gradually diminishing (time, cost, hazardous chemical, lack of digital advantage)**
- ❑ **Digital Radiography (DR/CR) allow faster/easier image acquisition**
- ❑ **DR/CR Widespread application still poses significant challenges**
  - ❖ Initial cost of system, Steep learning curve
  - ❖ Lack of procedure to choose parameters
  - ❖ Lack of demonstrated system performance
- ❑ **Performance assessment required (If CR/DR can effectively provide equal or better performance than the existing film-based technology)**

4

Computed Radiography (CR)

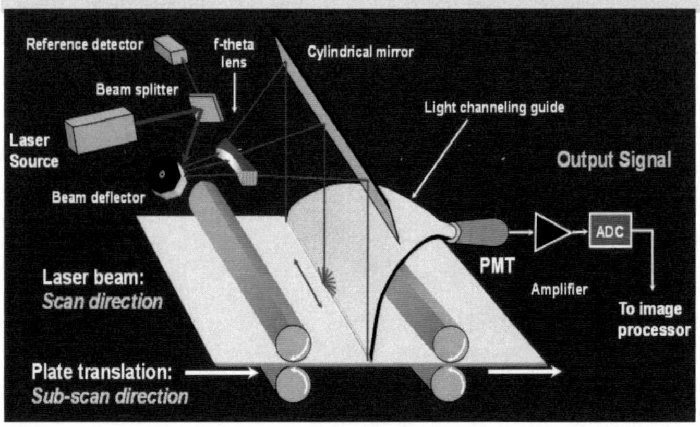

Ref: www.ndt-ed.org

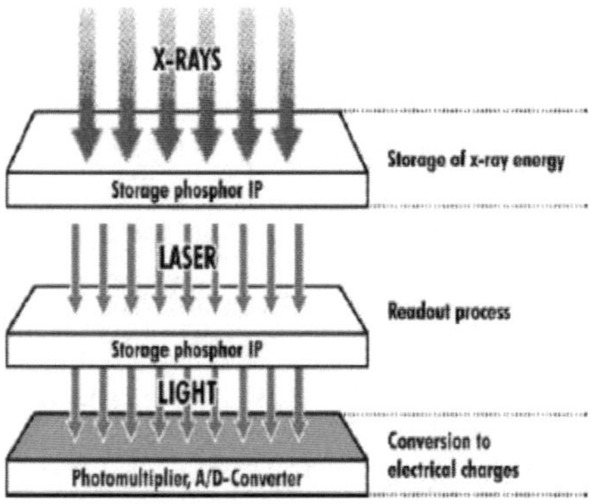

Ref: Korner, M., "Advances in Digital Radiography"

- ❑ CR uses a reusable phosphor imaging plate and allows faster image acquisition digitally, improved capabilities for analysis and digital storage

5

Contrast in CR

- ❑ **In general CR has better "<u>latitude</u>" than film**
- ❑ **CR can cover a wide range of thicknesses/densities in a single exposure**

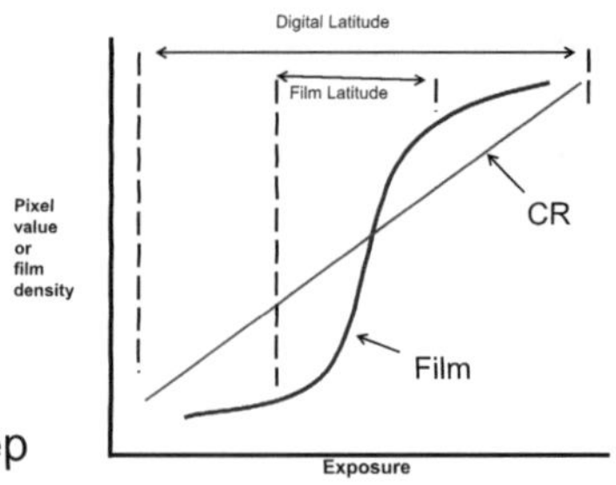

- ❑ **CR has better <u>contrast sensitivity</u> (CS)**

i.e. Capability to "window/level" to visualize a contrast step

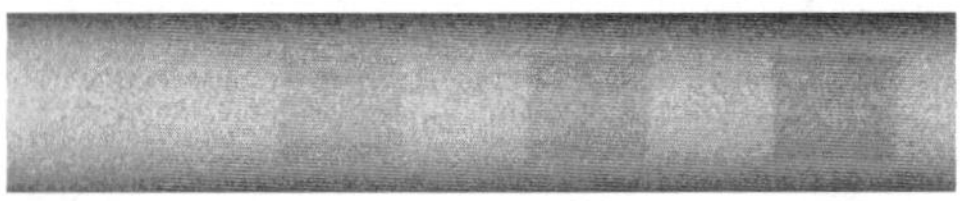

Ref: Digital and Film Radiography Comparison - Ken LaCivita

Primary Metrics for Image Quality in CR

- **Primary metrics for establishing the performance are**
  - Signal-to-Noise Ratio (SNR)
  - Contrast-to-Noise Ratio (CNR)
  - Basic Spatial Resolution (SR_b)
  - Higher the SNR, CNR or smaller the SR_b, better the quality
  - In film radiography (RT) these are evaluated indirectly by image quality indicators (IQI)
  - No direct relationship between radiographic sensitivity and flaw size

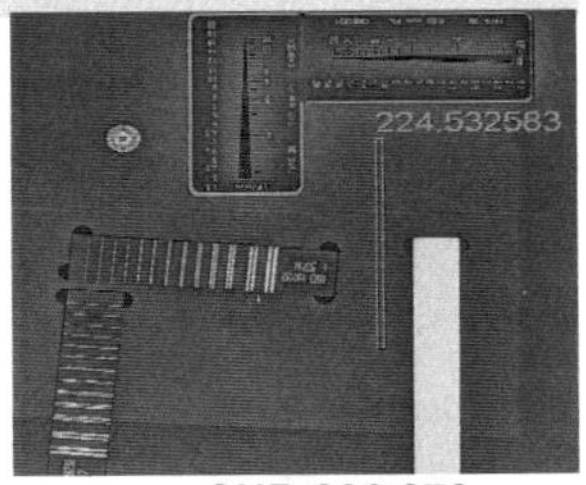

SNR 226.273

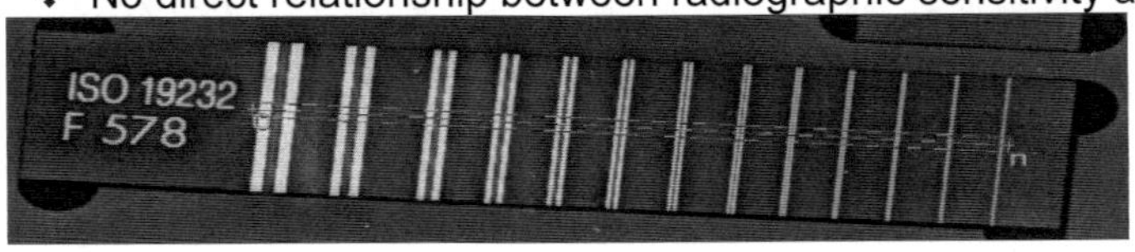

(SR_b)

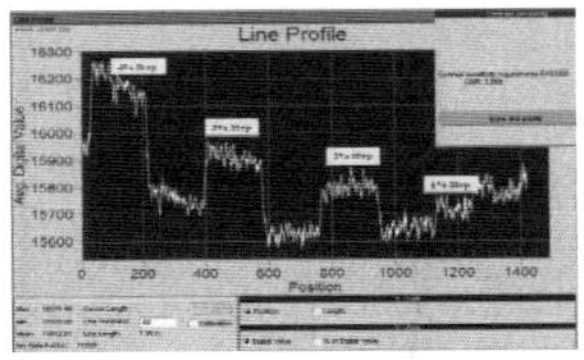

(CNR) < 2

7

NRC Computed Radiography

- Computed radiographic system (SR 70 µm, HR 35 µm)
  - Table-top flatbed scanner with contactless plate transport, inline erasure
  - 16 Bit, Scan width 14", Throughput 90 plates/hr (SR), 28 plates/hr (HR)
- CR image acquisition/analysis software (DICONDE compliant
- HR & medium speed (IPS) and extreme HR and slow speed imaging plates (IPU)
- BAM certified highest system Class IP 1 / 80 according to ISO 16371-1 and ASTM E2446

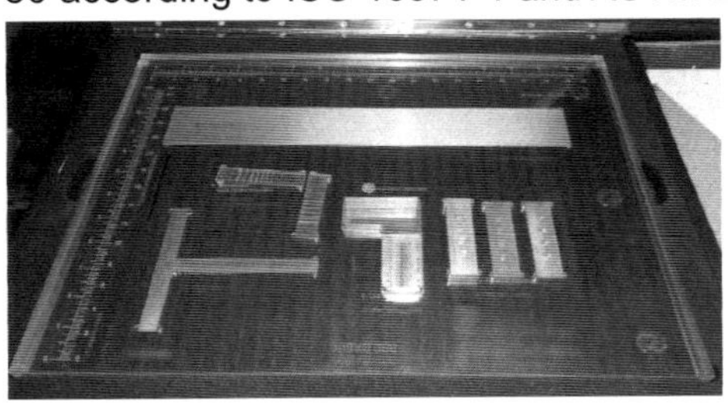

8

ASTM E 2445M-14 (Phantom)
(Practice for Qualification and Long Term Stability of CR Systems)

- ❑ Spatial Resolution and Unsharpness – Converging Line Pair
- ❑ Contrast Sensitivity (Contrast-to-Noise Ratio CNR)
- ❑ Spatial Resolution and Unsharpness – Duplex Wire
- ❑ Geometric Distortion
- ❑ Laser Beam Function
- ❑ Blooming or flare
- ❑ Slippage
- ❑ Shading
- ❑ IP Linearity
- ❑ Signal-to-Noise Ratio (SNR)
- ❑ EPS (Either SNR or EPS)

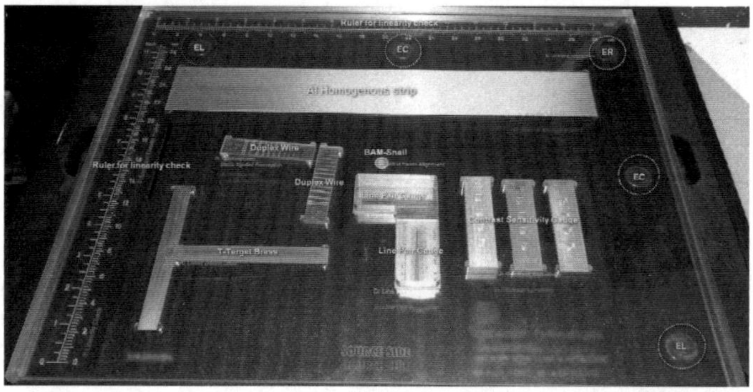

Basic Spatial Resolution (SR_B) (Duplex Wire Gauge)

- ❑ SR_B - ability to clearly see abrupt changes in an object, resolve/separate smaller features
- ❑ Film has better SR_B (Film grain ~0.6-0.8µm vs CR IP grain ~ 3-5µm, scanning res., others)
- ❑ Duplex Wire IQI in accordance with ASTM E2597
- ❑ Resolution is determined by the first unresolved wire pair (modulation depth <20%)
- ❑ Profile line requires parallel to gauge and positon/width of line have effect on BSR
- ❑ Best result – IPU (blue plates) and lower scanning sampling rate (scanning time /

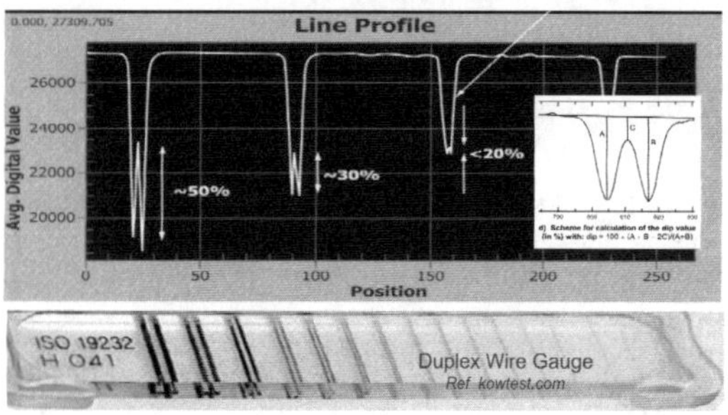

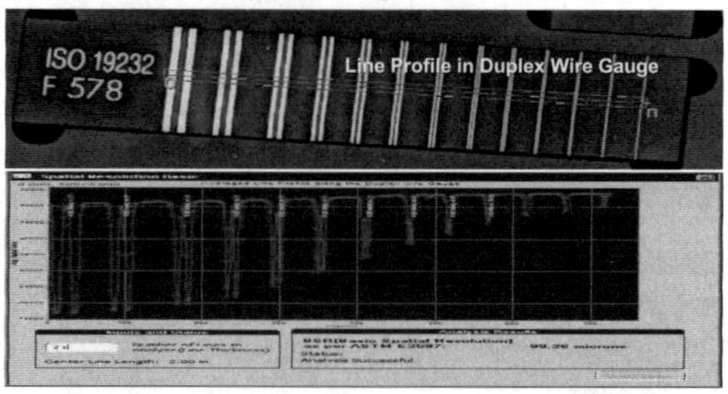

Basic Spatial Resolution (BSR) (Line Pair Gauge)

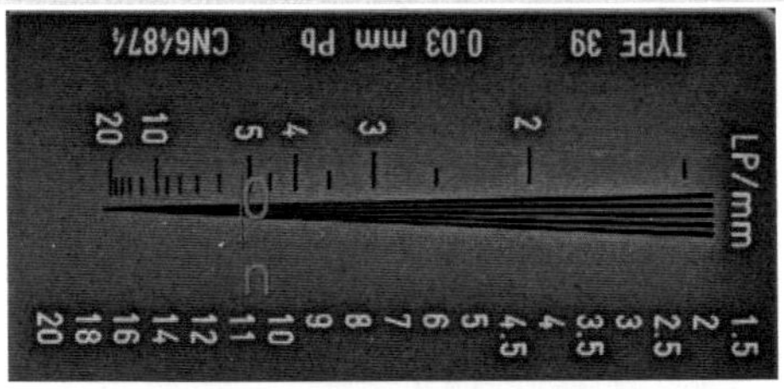

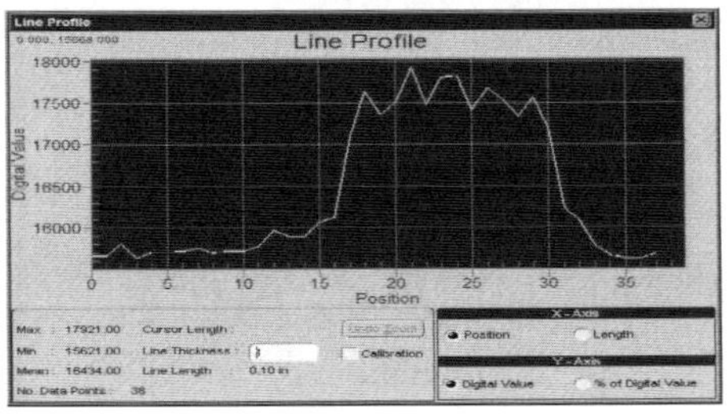

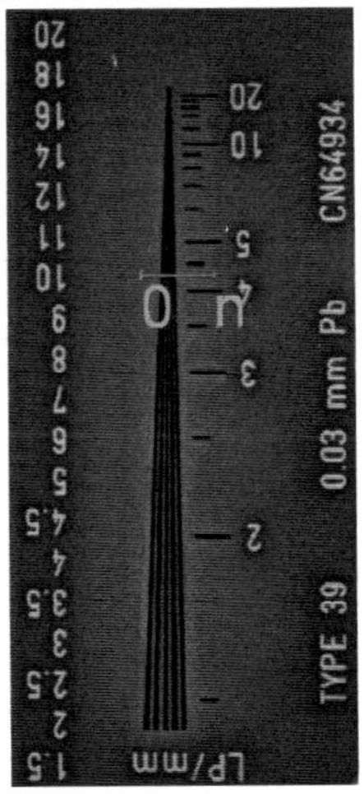

Slow Scan Direction 5.3 lp/mm (100 Micron)

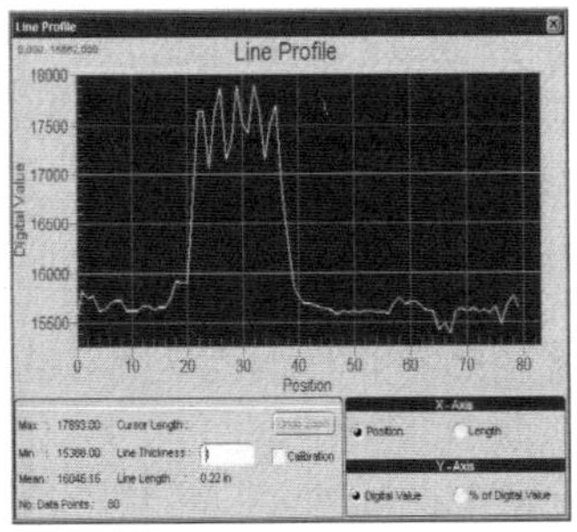

Fast Scan Direction 4.25 lp/mm 125 Micron

- ❑ Discrepancy in BSR (DW vs. LP)
- ❑ Human Factor, Aliasing
- ❑ ASTM E2445 lacks in instruction

11

Contrast Sensitivity /Contrast-to-Noise Ratio (CNR)

$$CNR = (\mu_0 - \mu_B)/\sigma_B = \Delta\mu/\sigma_B$$

- ❑ Ability to detect variations in thickness/density
- ❑ Visual evaluation of targets Al, Cu, Steel
- ❑ CNR often requires 2 - 2.5 (TBD)
- ❑ 2% sensitivity was observed even when CNR < 2
- ❑ Automated procedure showed inconsistencies in CNR
- ❑ Contrast sensitivity is key for low res. Application (e.g. composite)

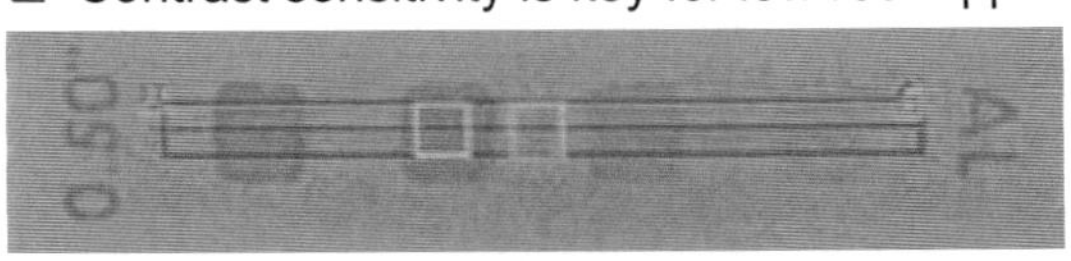

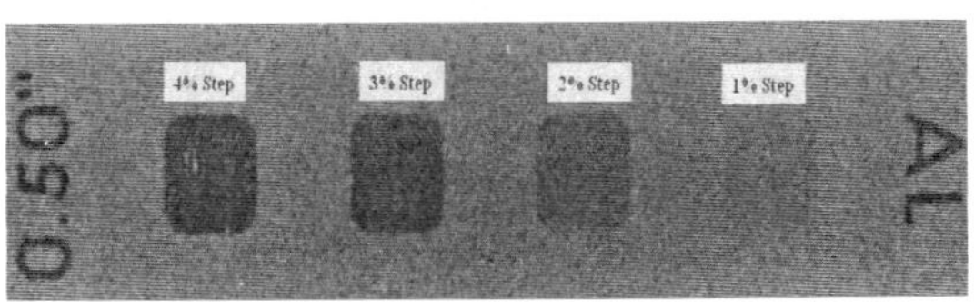

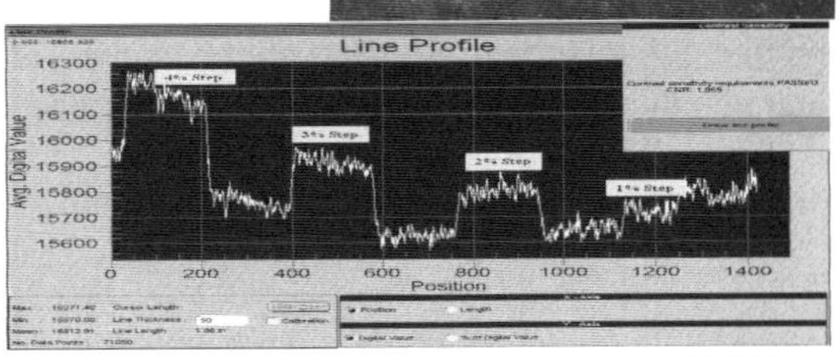

12

Signal-to-Noise Ratio (SNR) / Normalized SNR (SNR_N)

- SNR - Quantitative indication of noise level
  - (*Scatter, quantum nature of photon, IP inherent structure, PMT Gain, Laser Power, etc.*), higher the SNR, the less obtrusive the background noise
- Normalization by the effective pixel size gives an SNR_N (Independent of SR_B)
- SNR were found varied with size and location of Region-of-Interest (ROI)
- In general CR is more sensitive to scatter than film
  - Pre-filters, screens, collimator, short exposure may reduce scatter

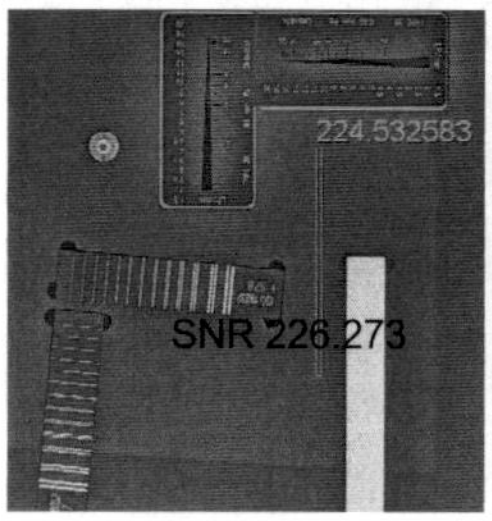

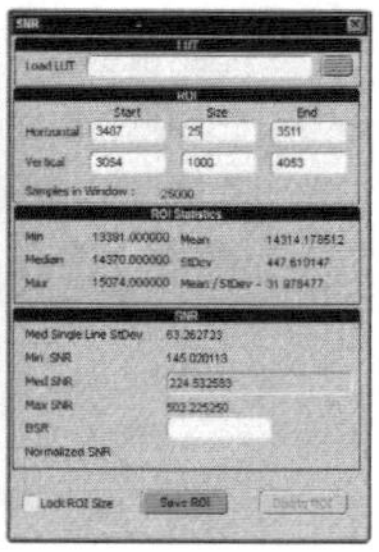

$$SNR_N = SNR_{meas}\frac{88.6\mu m}{SR_b}$$

SNR_{meas} is measured SNR

SR_b is basic spatial resolution

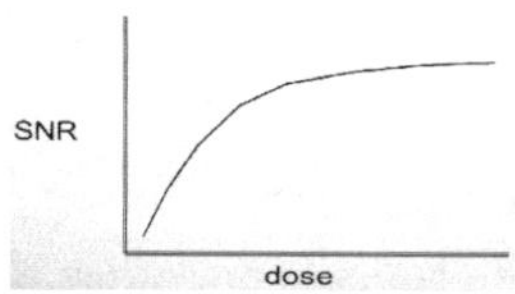

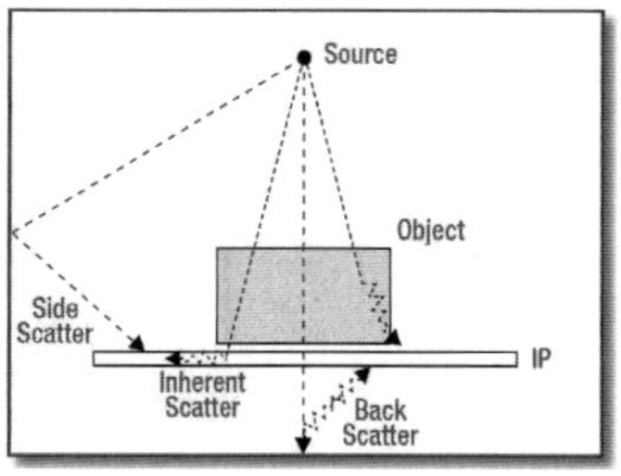

Ref: Digital and Film Radiography - Ken LaCivita

13

BAM Snail for Central Beam Alignment

- Radiation beam aligned perpendicular to the center of the (BAM-Snail)
- CR image of the image quality indicator will be a regular spiral

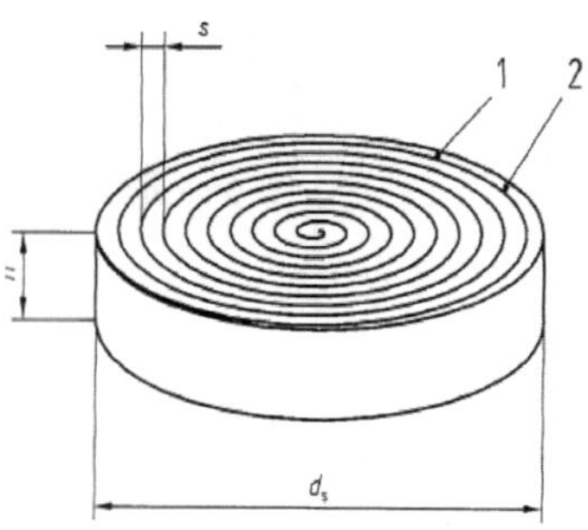

Ref: ASTM E2445

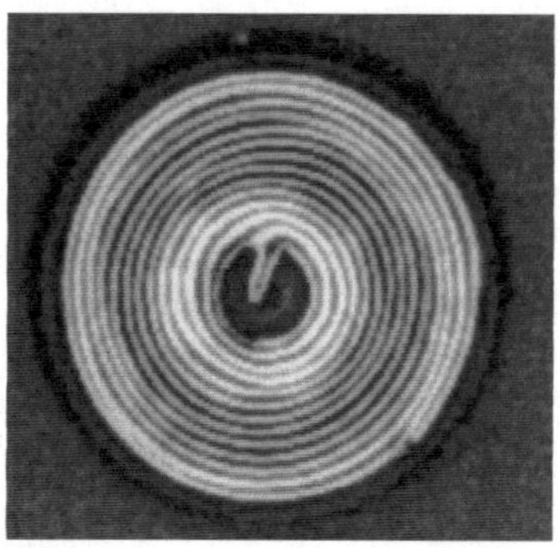

Central beam alignment @ 70 µm

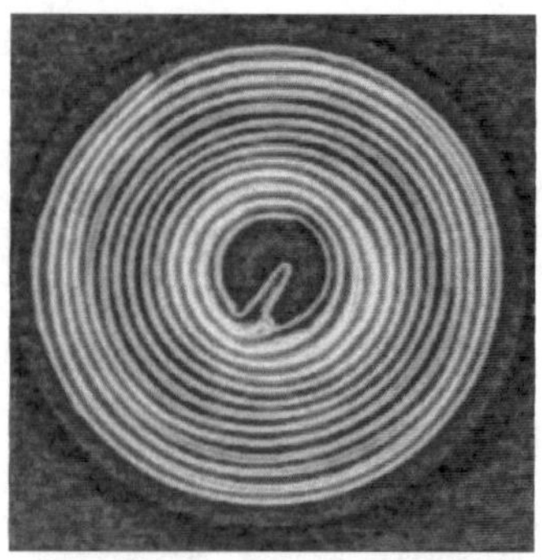

Central beam alignment @ 35 µm

14

Laser Beam Jitter

- Jitter appears if the laser beam function is not working properly *(lack of smooth movement of the imaging plate/laser scanning device)*
- Dust on the laser or mirror is the probable causes of laser malfunction
- Examining the edges of the "T-bar" on the image
- Target edges should be <u>straight and continuous</u>

Examining edge of T-beam

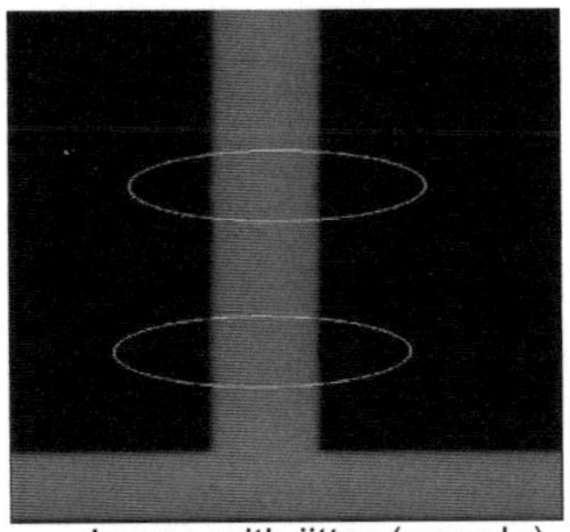

Image with jitter (sample)

Ref: LaCivita, K.J, "AFRL-RX-WP-TR-2009-4069

Image without Jitter (NRC)

15

Geometric Distortion

- **Checks the <u>spatial linearity</u> of the CR system to find image distortion**
- **Ruler of high absorbing materials located in the perimeter (x- and y- direction)**
- **Tilting / twisting of IP during the scan**
- **Maximum geometric distortion was found <0.045% [Require < 5%]**

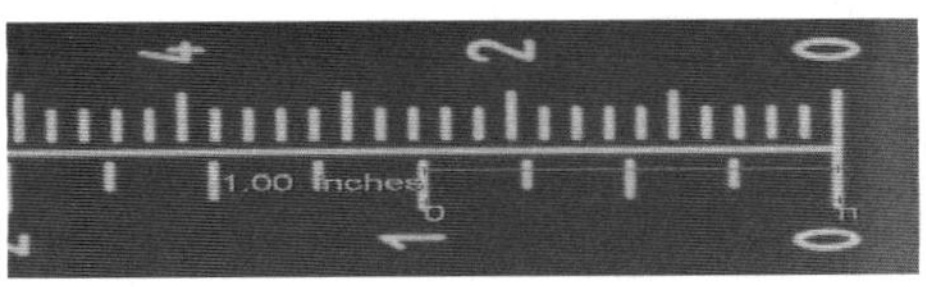

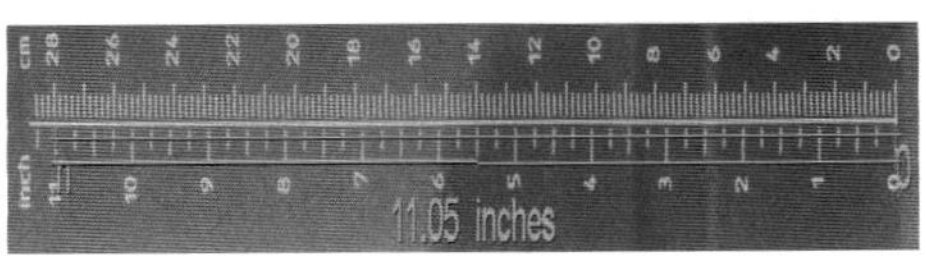

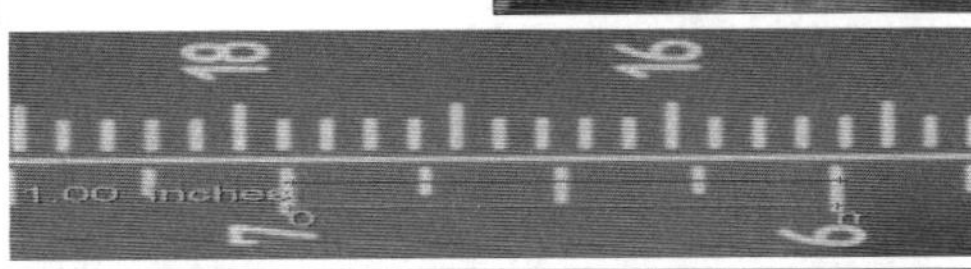

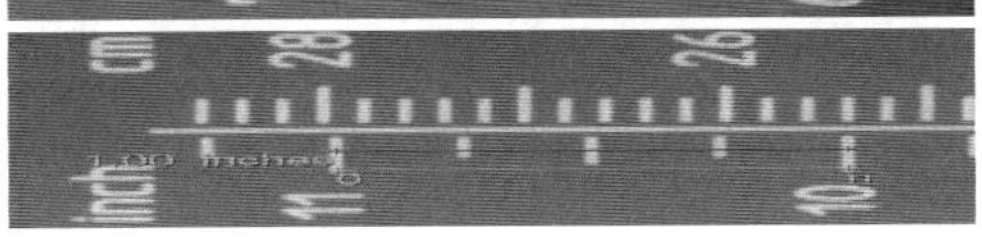

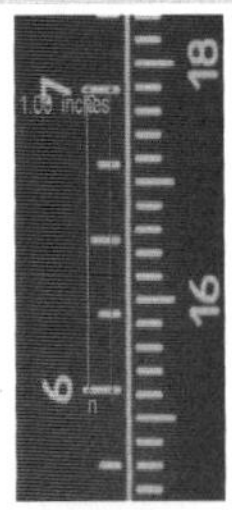

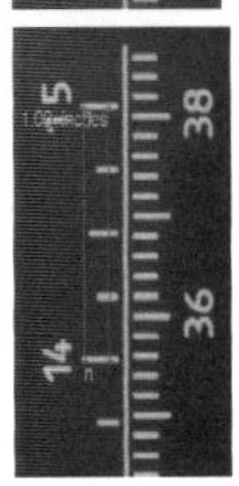

16

Shading Effect

- ❑ **Ensure scanning laser intensity is uniform across scanning width of IP**
- ❑ **Check for proper alignment of light-guide/photo-multiplier tube assembly**
- ❑ **ASTM E2445 ± 15%, Centre hole vs. Corner hole)**
- ❑ **NRC ± 1%, 8% and ± 3%, 1%**

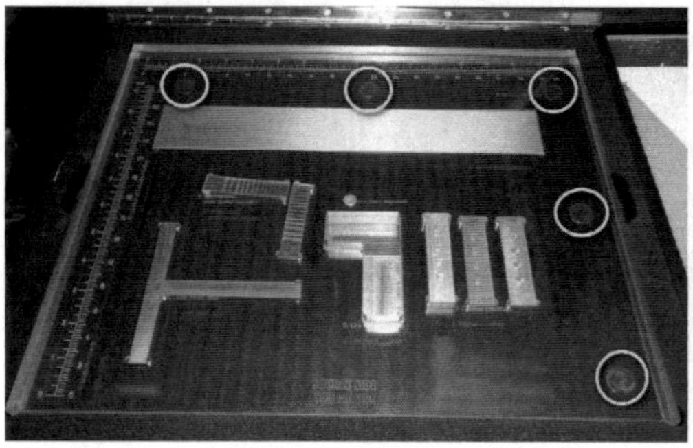

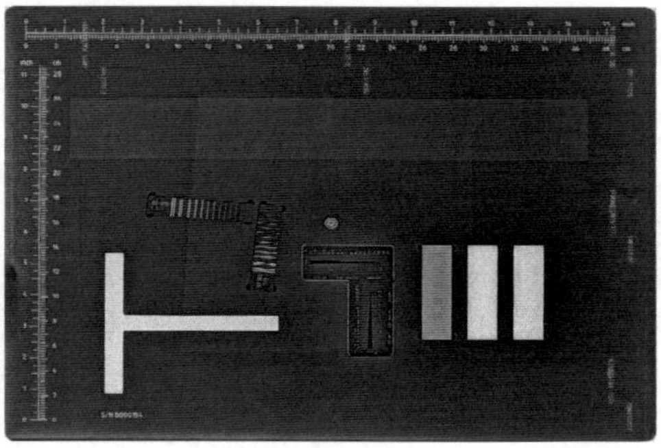

17

Scan Column Dropout

- ❑ **No test object required**
- ❑ **Obserable as zero pixel value linear image artifact**
- ❑ **Straight bright white line parallel to the transport direction**
- ❑ **Probable cause is internal obstruction (e.g. contaminates]**

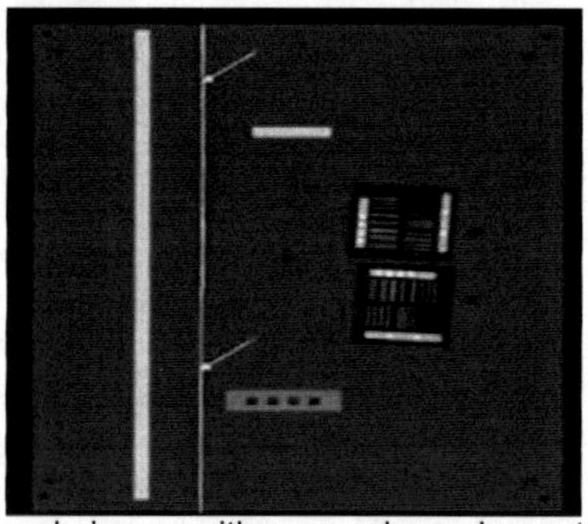

Sample image with scan column dropout
Ref: ASTM E2445M-14

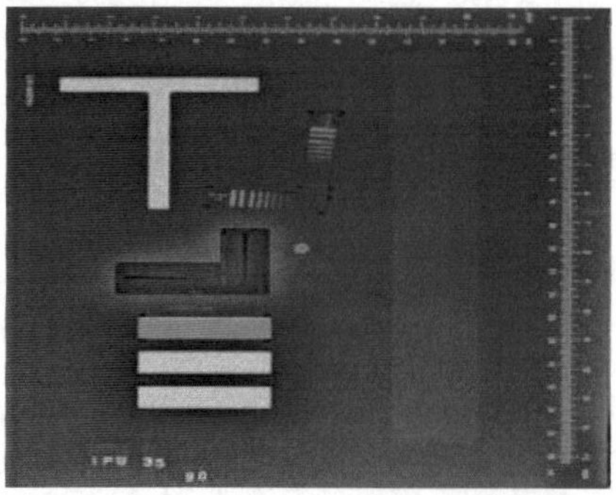

No scan column dropout white line (NRC study)

18

Scanner slippage

- Slipping of an IP in a scanner transport system
- Slipping would result in deviations in the intensity of scanned lines
- Distortion in homogeneity of scanning/reading system
- Deviation showed as intensity is greater than the system noise

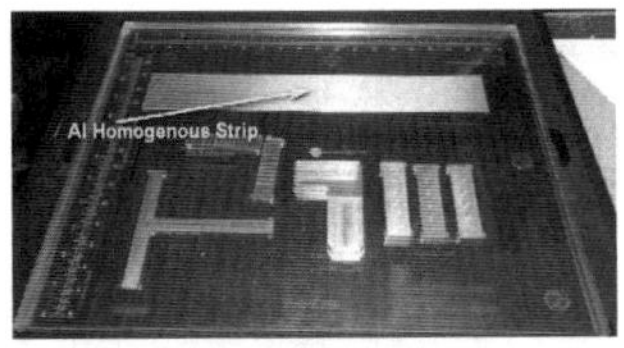

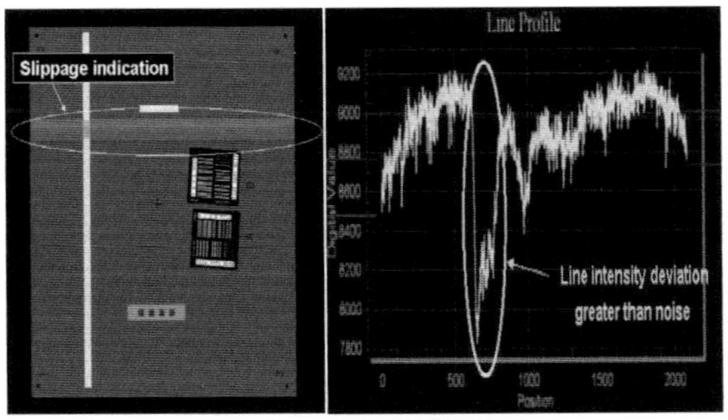

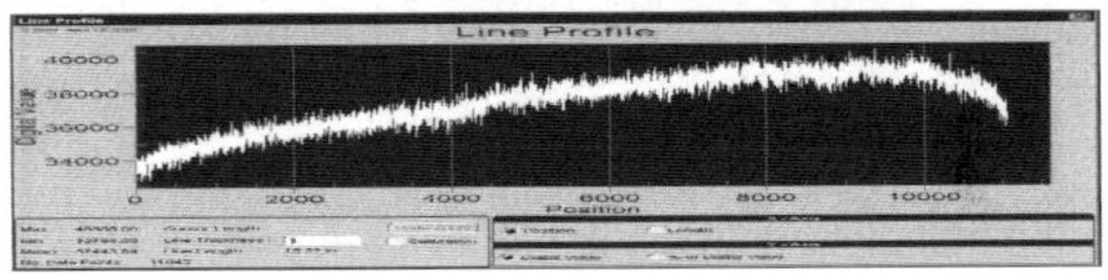

Sample Image with Slipping
Ref: LaCivita, K.J, "AFRL-RX-WP-TR-2009-4069"

19

Blooming or Flare or Afterglow

- Overshoot or streaking (dark shadow immediately adjacent of a high density target)
- IP oversaturated around target may leave a latest image or ghost
- Should not be visible, BAM okay 1.5 %, NRC no blooming
- Need to perform again with changed settings (High / Low PMT Gain)
- May require use of shielding or rotate IP to prevent local "blooming" artifacts and/or latent images

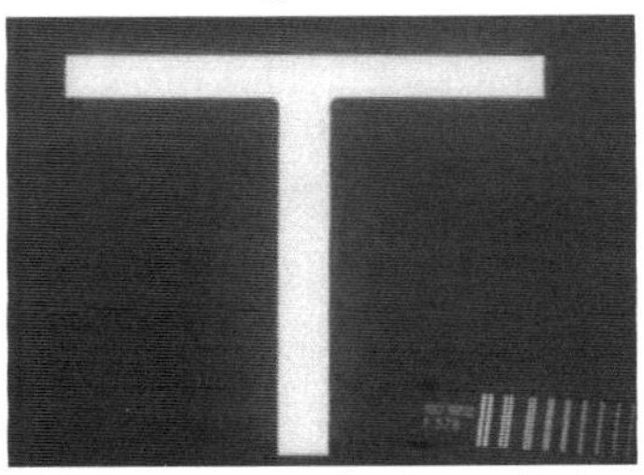

NRC Experiment (No blooming)

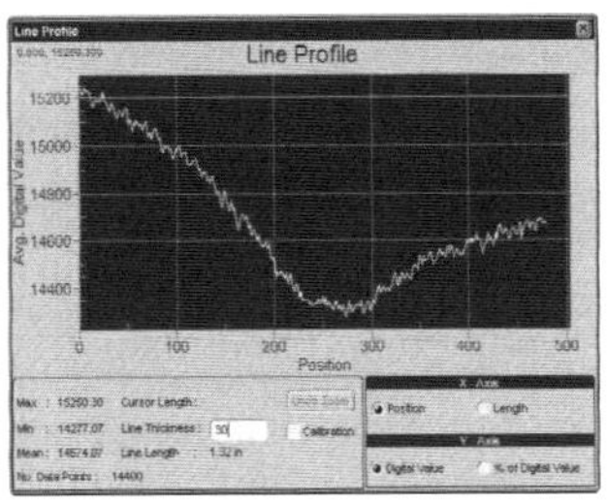

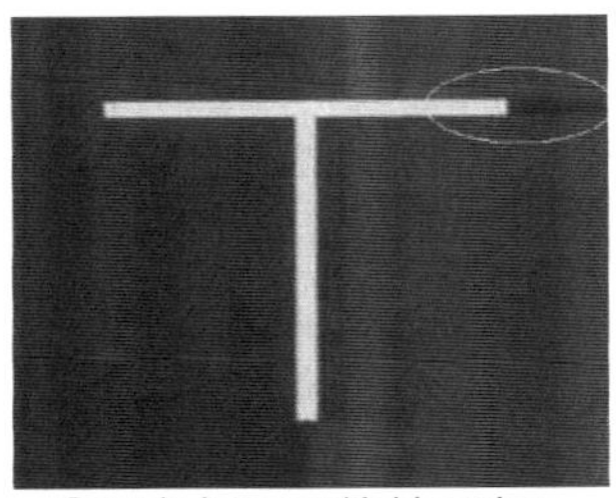

Sample Image with blooming
Ref: LaCivita, K.J, "AFRL-RX-WP-TR-2009-4069"

20

Image Artifacts (Ghost)

- ❑ **Unwanted image features (unanticipated and unexplainable)**
- ❑ **Hardware, cracked plates, laser beam head, wear/scratch or moisture**
- ❑ **Harm the quality and may obscure the area of interest**

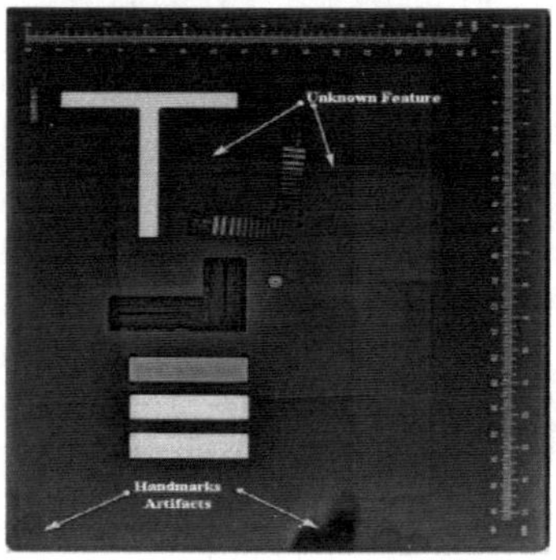

- ❑ Few artifacts were attributable to residual permanent image of reference specimen.
- ❑ High exposure in unshielded part of IP plate, scatter radiation, etc. which eventually contributed to form a hard to erase latent image
- ❑ Erasure test can help to determine if erasure is adequate
- ❑ Masking unexposed part with lead and additional erasure cycles are few options to remedy latent image

Equivalent Penetrameter Sensitivity (EPS)

- ❑ **EPS plaques discern subtle differences in image quality as radiographic parameters are changed**
- ❑ **Determine the exposure levels necessary to ensure an adequate signal-to-noise ratio (SNR)**
- ❑ **Generate the acceptable gray value working range of the CR system and IP plate combination**
- ❑ **EPS is a subjective test (Related to inspector eye capability to view & count no. of visible holes)**
- ❑ **EPS curve is used for establishing & monitoring critical process control metric of CR/IP/procedure**
- ❑ **EPS is inversely proportional to SNR (See ASTM E2445 or E746 for details)**

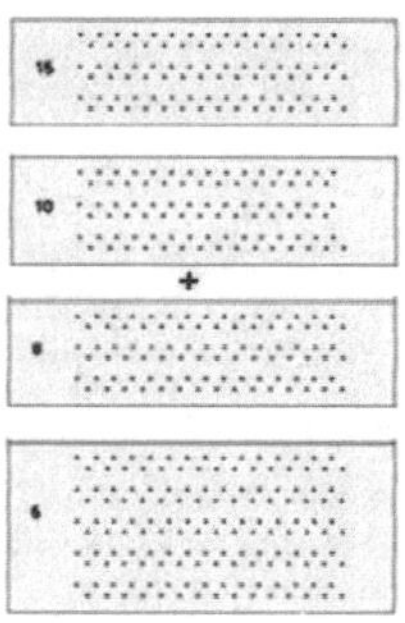

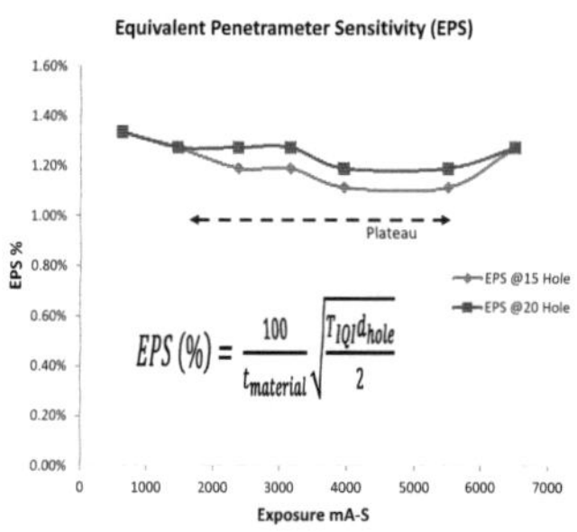

$$EPS(\%) = \frac{100}{t_{material}}\sqrt{\frac{T_{IQI}d_{hole}}{2}}$$

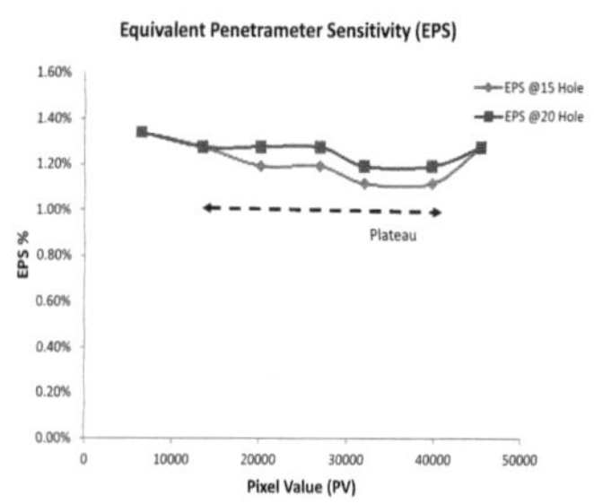

Pixel Value Linearity

- Signal//Pixel value required to be proportional to the radiation dose or exposure
- Data point is within ± 10% of a straight line best fit
- ASTM E2445 lacks in guideline

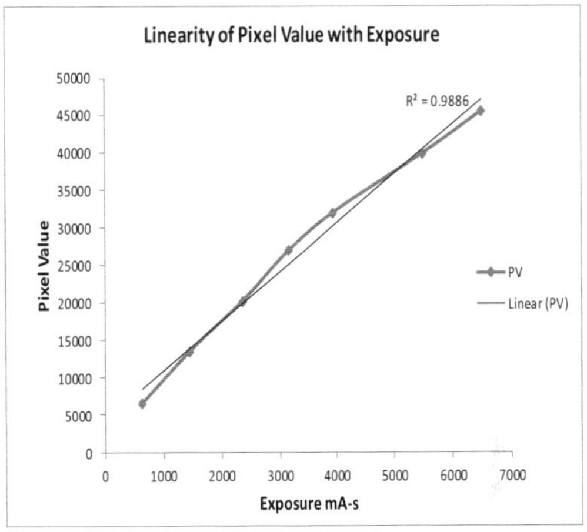

23

CR Comparison (Electronic Board)

Specimen

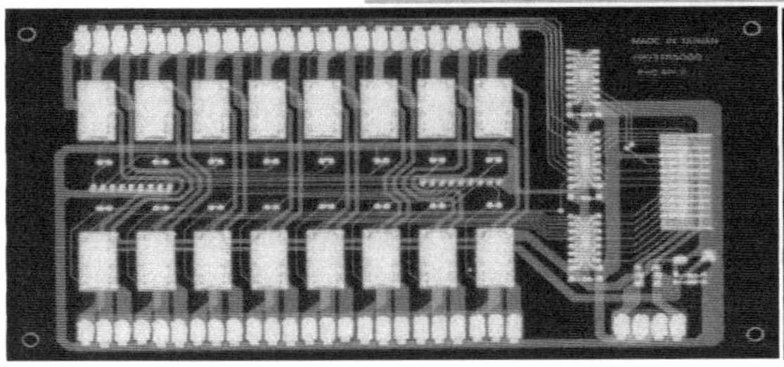

Film

CR

24

Film vs CR (Sandwich honeycomb core panel cupper mesh

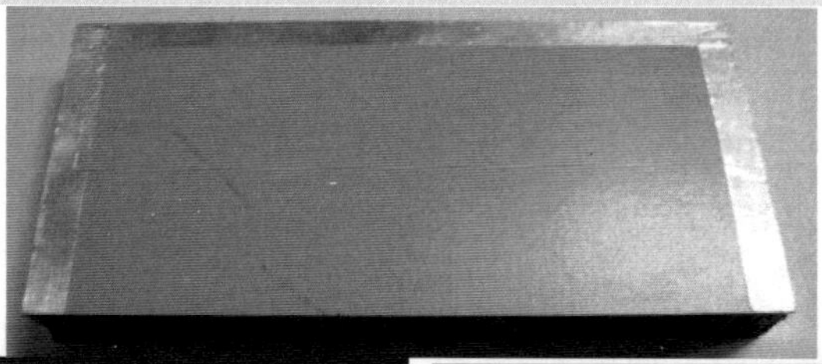

Specimen

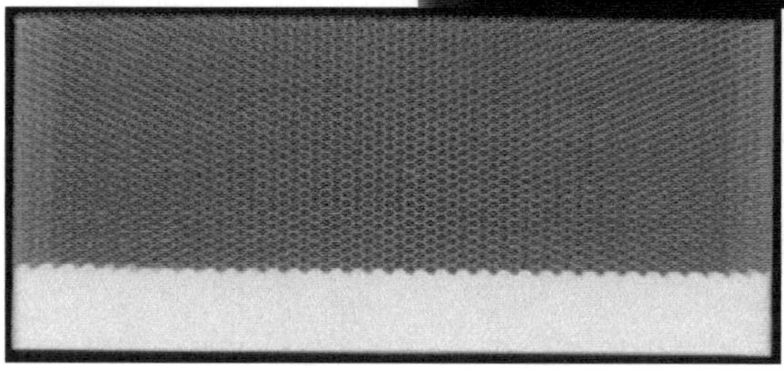

Film

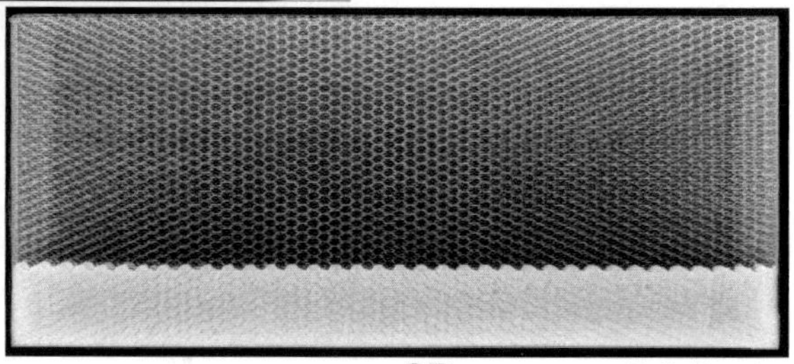

CR

25

CR for High Resolution Application (Cracks, Weld Defects)

- **Much more challenging to match detection limits of film!**
  - Should have good contrast sensitivity
  - Spatial resolution and image noise are also critical
  - Spatial resolution (primarily affected by sampling rate or scanning resolution (100μm, 50μm, 25μm), imaging plate grain size
  - Image noise primarily affected by imaging plate characteristics, exposure (mA x time)

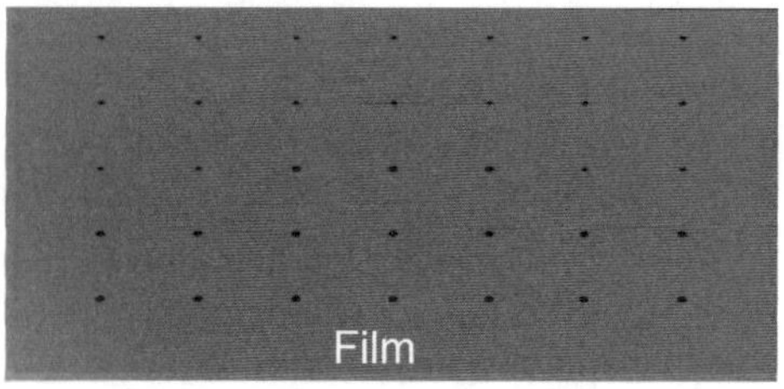

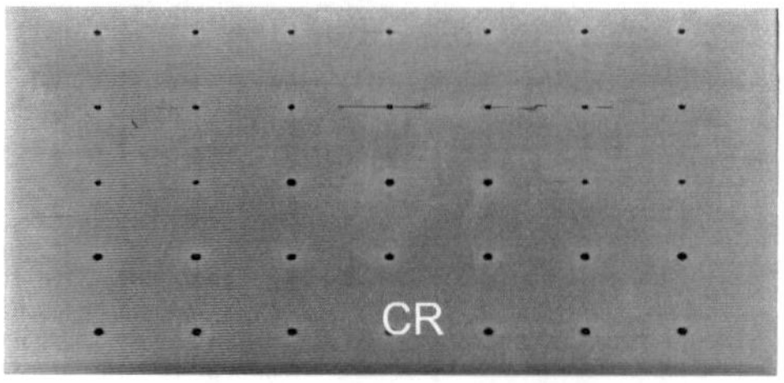

26

Film vs CR (Nozzle Guided Vane)

Specimen

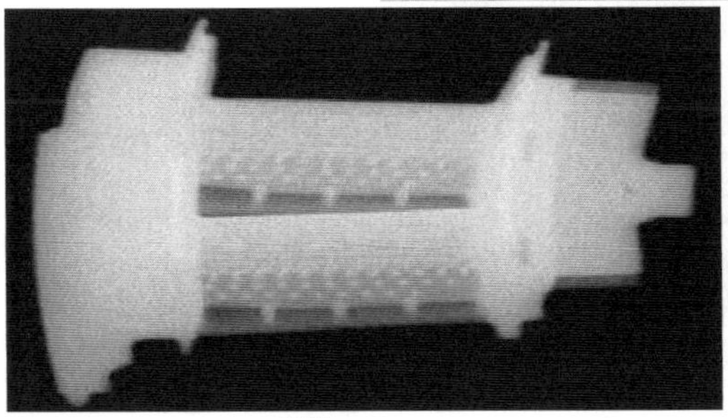

Film

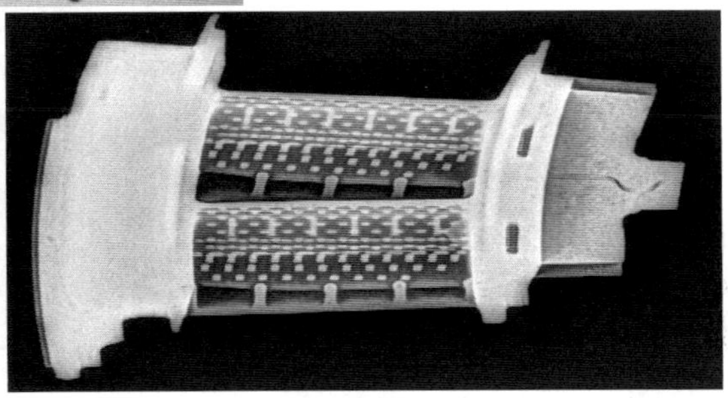

CR

Summary

- ❑ SNR_N exceeded the ASTM highest system class (130 IP special)
- ❑ Resolution, contrast, SNR, shading, beam alignment, jitter, slippage, geometric distortion etc.) met the current ASTM requirements
- ❑ SNR or EPS improves with exposure up to certain point.
- ❑ Best image resolution – blue plates & lower scanning resolution
- ❑ Ghosting-like artifacts were observed which required engineering judgement to overcome
- ❑ EPS curve is not typical but identify min. exposure (key requirement)
- ❑ CR image is <u>not</u> the same as film, many variables affect image quality and many unique phenomena need to consider

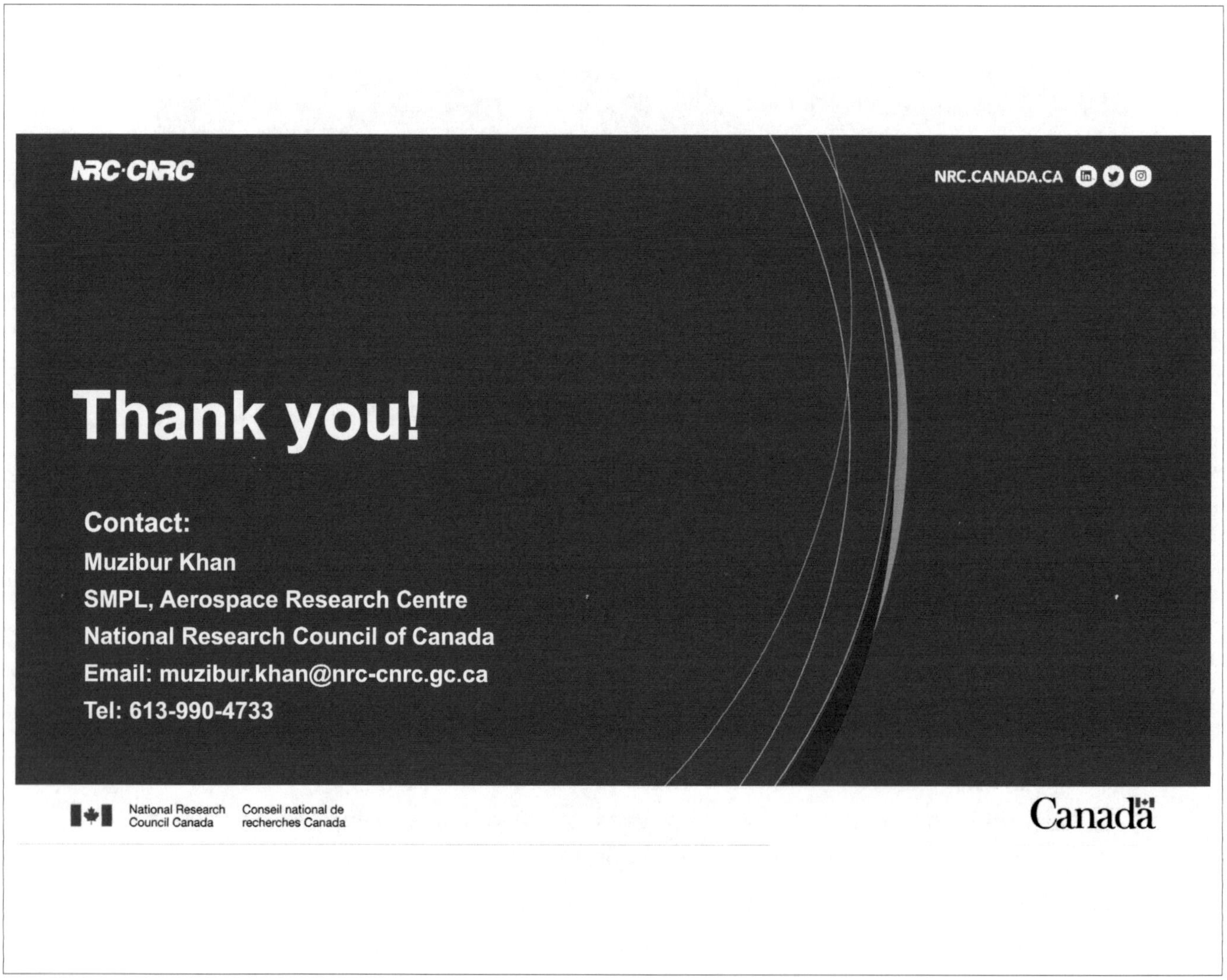
NRC·CNRC
NRC.CANADA.CA
Thank you!
Contact:
Muzibur Khan
SMPL, Aerospace Research Centre
National Research Council of Canada
Email: muzibur.khan@nrc-cnrc.gc.ca
Tel: 613-990-4733
National Research Council Canada
Conseil national de recherches Canada
Canada

Mechanical & Industrial Engineering
UNIVERSITY OF TORONTO

NDT in Canada NDTC 2019 Inform - Engage - Advance

Non-destructive and Non-contact Stress-Strain Characterization of Aerospace Alloys and Coatings using Laser Infrared Photo-Thermo-Mechanical Radiometry (PTMR)

Andreas Mandelis[1,2] , Huiting Huan[1,2] and Alexander Melnikov[1]

[1] *Center for Advanced Diffusion-Wave and Photoacoustic Technologies (CADIPT), Department of Mechanical and Industrial Engineering, University of Toronto, Toronto, Ontario M5S 3G8, Canada*

[2] *School of Optoelectronic Information, University of Electronic Science and Technology of China, Chengdu, Sichuan 610054, China*

1

CADIPT

Outline

NDT in Canada NDTC 2019 Inform - Engage - Advance

1. Background: Mechanical strength evaluation for aerospace materials
2. Methodology: Laser Infrared Photo-Thermo-Mechanical Radiometry (PTMR)
3. Experiment: Non-contacting stress-strain relation characterized by the PTR signal
4. Theory and analysis: Quantification of experimental results through a 1-D thermo-mechanical-wave model
5. Results and Outlook: The present work gives rise to Photo-Thermo-Mechanical Radiometry (PTMR) as a non-destructive, non-contact strain gauge for the evaluation of mechanical strength of materials

Background

- Hidden fatigue underlies threat to safety in aerospace components

Intact material ⇒ (Overloading or cyclic loading) ⇒ Stress residue/ fatigue ⇒ (Continued loading) ⇒ Cracks and failure

It will be of great value if the strength condition of the material can be evaluated before fatigue actually occurs!

NDT in Canada NDTtC 2019 Inform - Engage - Advance

Material Property

- Strength evaluation by FEM

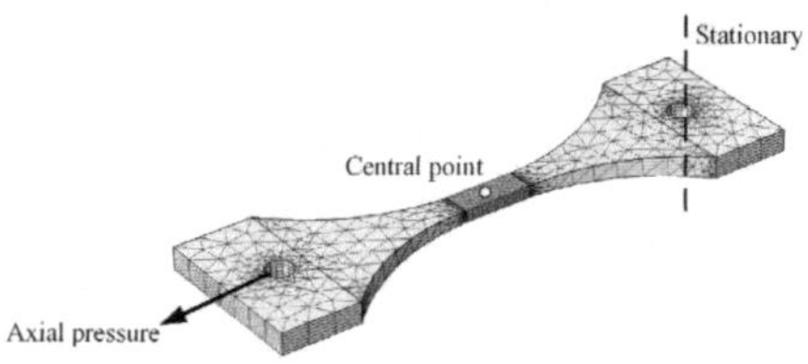

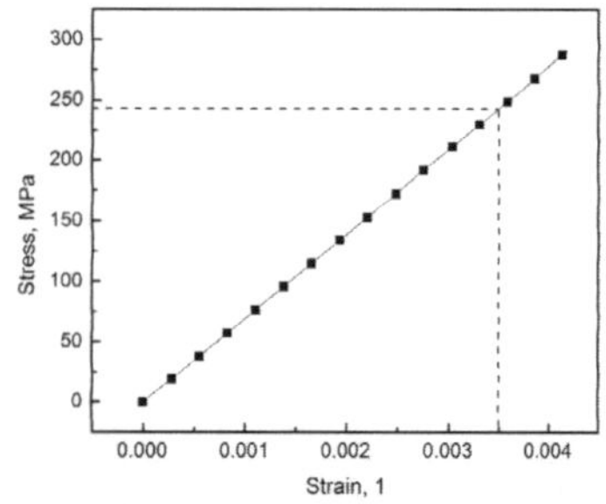

- The sample is made of aluminum 6061-T6, a general material in aerospace industry
- Use linear elastic stress-strain constitutive equation and balance of force:

$$\boldsymbol{\tau} = \mathbf{C} : \boldsymbol{\varepsilon}$$

$$\nabla \cdot \boldsymbol{\tau} = -\mathbf{F}$$

$\boldsymbol{\tau}$ – stress tensor, $\boldsymbol{\varepsilon}$ – strain tensor, $\mathbf{C} = C_{ijkl}$ – modulus tensor of rank four (i,j,k,l = 1,2,3,4), **F** – external force.

- According to ASTM 308, the elastic limit of this material is at least 240 MPa, which yields:

$$\varepsilon \leq 0.0035 (3500\ \mu\text{m/m})$$

in terms of strain representation.

H. Huan, *et al.* NDT & E Int. **84,** 47-53 2016

NDT in Canada NDTtC 2019 Inform - Engage - Advance

Methodology

CADIPT

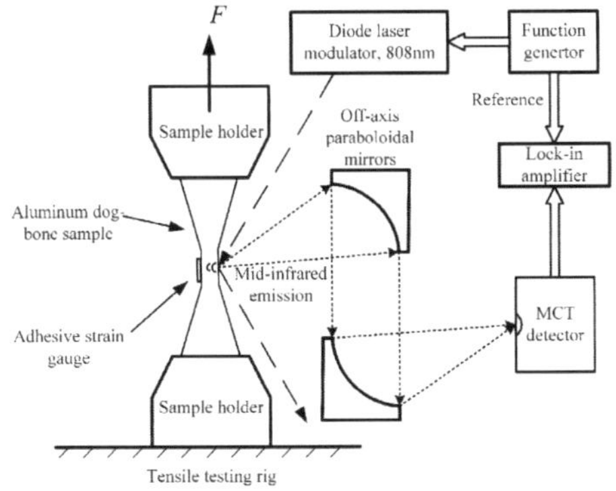

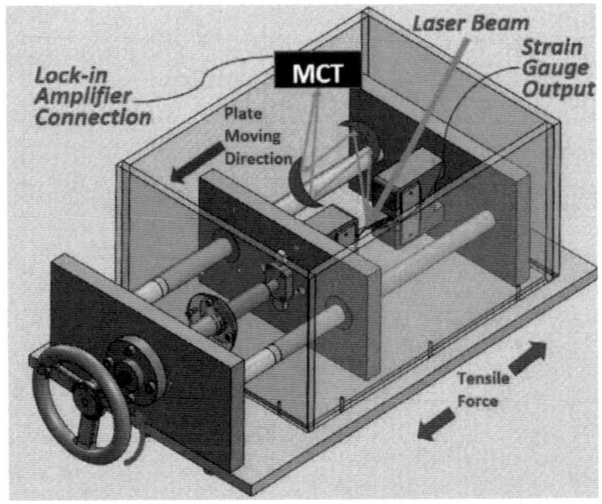

Main advantages:

- Non-destructive when operated below the elastic ceiling.
- Totally non-contact and localized detection.

PTR signal: $S = A\exp[i(\omega t + \Phi)]$

Signal amplitude: $A = \|S\|$

Signal phase: $\Phi = \arg(S)$

NDT in Canada NDTC 2019 Inform - Engage - Advance

Experimental

CADIPT

Frequency-scan signal: 1-30 Hz (low frequency) $\mu(f) = \sqrt{\alpha / \pi f} >> \Delta L$

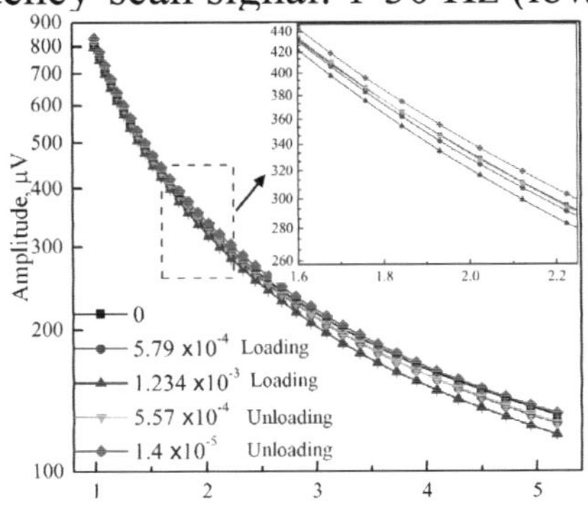

(a)

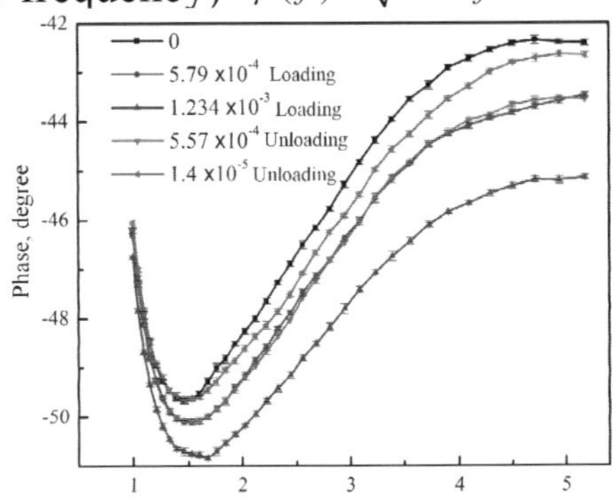

Strain Unit: [1]

(b)

Stress condition: Within elastic regime.

Stress procedure: Loading and relaxing.

Good reproducibility and reversibility of the PTR signal within the elastic regime!

NDT in Canada NDTC 2019 Inform - Engage - Advance

Experimental Results

CADIPT

- Strain-scan result: stress free to ~1400 μm/m, within the elastic regime
  - Set frequency at 2.5 Hz. Involving multiple loading and unloading process

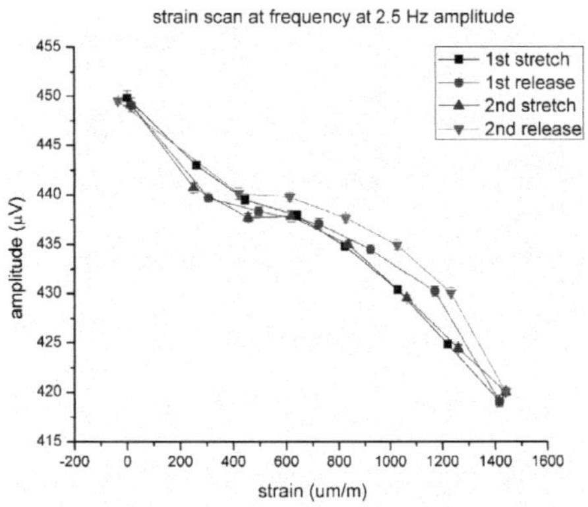

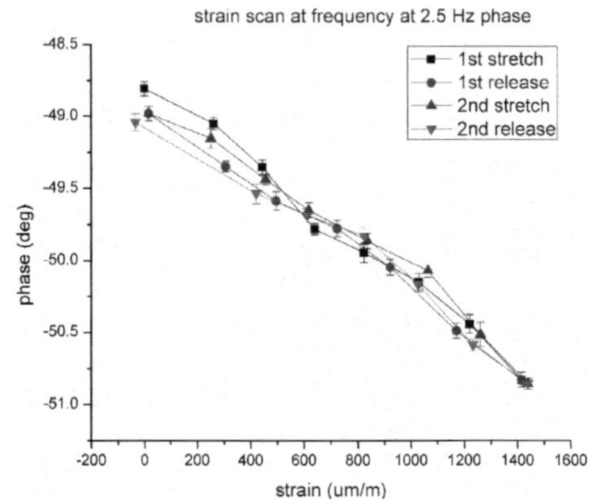

The elastic loading and unloading test shows a good repeatable and reversible pattern. This indicates that the sample recovers its thermal properties after removal of tension within the elastic regime.

NDT in Canada NDT&C 2019 Inform - Engage - Advance

Experimental Results (cont'd)

CADIPT

- Signal strain dependence: beyond the elastic regime

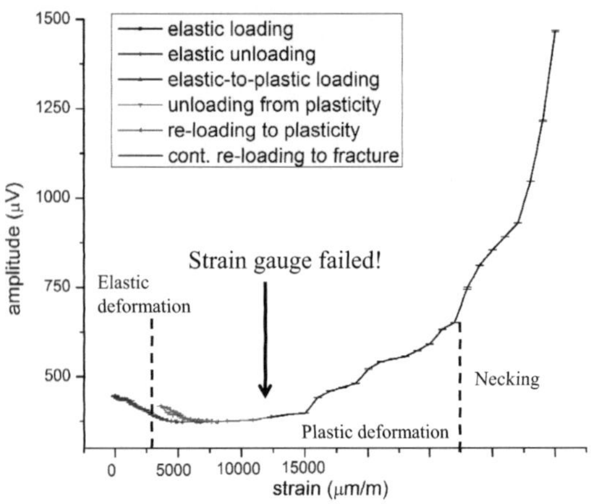

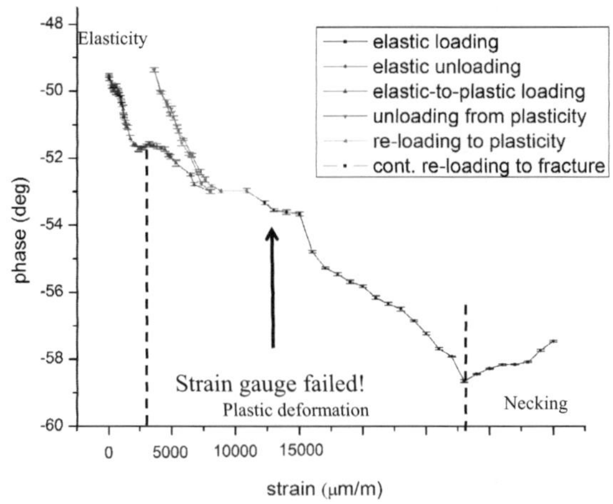

- ➢ In the necking regime, the sample undergoes large deformation and thus the surface is deformed significantly, which drastically changes the PTR amplitude. Phase is less sensitive to this change and is more reliable as it is less affected by extraneous factors like surface curvature and shape change.

NDT in Canada NDT&C 2019

Theory and Analysis

CADIPT

- Frequency dependence of PTR signal

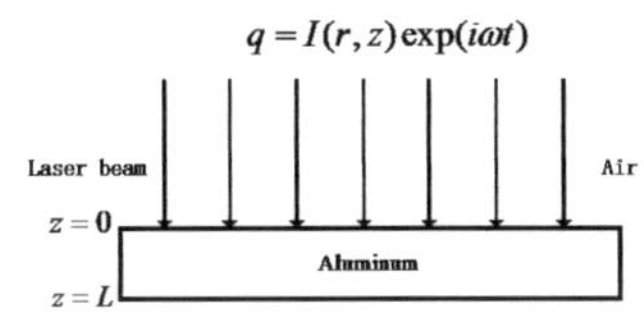

$$T(0,\omega) = \frac{\beta I(\omega)(1+e^{-2\sigma_1 L})}{k_1\sigma_1(1-e^{-2\sigma_1 L})} \quad (1)$$

$$\text{Amplitude:} \quad A(\omega) \sim \|Y(\omega)\| I(\omega) \frac{1+e^{-2\sigma_1 L}}{k_1\sigma_1(1-e^{-2\sigma_1 L})} \quad (2)$$

$$\text{Phase:} \quad \Phi(\omega) = \arg\left(\frac{1+e^{-2\sigma_1 L}}{1-e^{-2\sigma_1 L}}\right) - \frac{\pi}{4} + \Phi_0(\omega) \quad (3)$$

$$\Phi_0(\omega) = \arg[Y(\omega)]$$

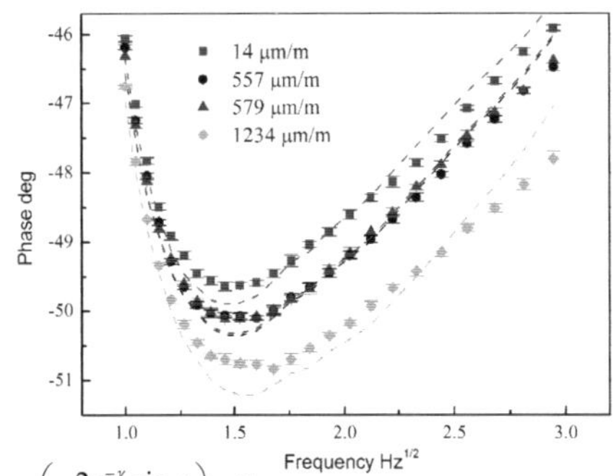

Normalized to:

$$A(f) \sim \|Y(f)\| I_0(f)\left(\frac{1}{e_1\sqrt{f}}\right)\frac{\left[(1-e^{-2\gamma})^2+4e^{-2\gamma}\sin^2\gamma\right]^{1/2}}{(1-e^{-\gamma}\cos\gamma)^2+e^{-2\gamma}\sin^2\gamma},\ \Phi(f) = \arctan\left(\frac{-2e^{-\gamma}\sin\gamma}{1-e^{-2\gamma}}\right)-\frac{\pi}{4}, \gamma \equiv 2\sqrt{\pi f}\kappa(\tau), \kappa(\tau) \equiv L/\sqrt{\alpha(\tau)} \quad (4)$$

κ is the only parameter that affects the phase signal Eq. (4) while amplitude relies on both κ and e_1. Fitting the phase curve can extract κ; subsequent fitting the amplitude can yield e_1.

NDT in Canada NDT C 2019 Inform - Engage - Advance

Quantification Results

CADIPT

- Applying 1-D single layer thermal-wave model to the tensile test results, we can obtain diffusivity and effusivity as functions of strain within the elastic regime:

FS : Frequency scan
SS : Strain scan

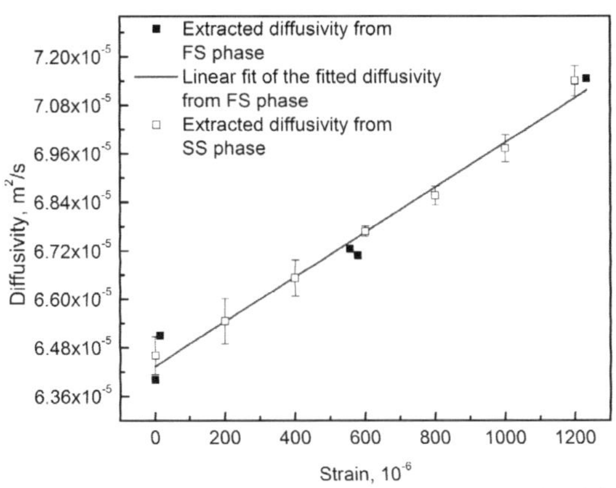

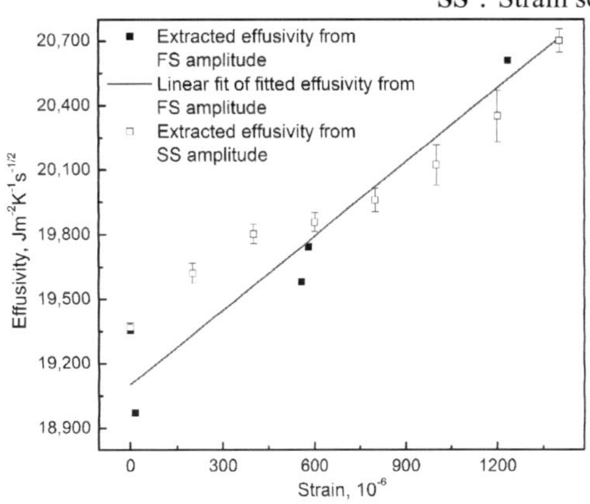

- Diffusivity increases with stress/strain: $\alpha(\varepsilon) = k/\rho C = 5.536\times10^{-3}\times\text{strain} + 6.434\times10^{-5}\ \text{m}^2/\text{s}$
- Effusivity increases with sress/strain: $e_1(\varepsilon) = \sqrt{k\rho C} = 8.17\times10^5\times\text{strain} + 19389.25\ \text{J/(m}^2\text{Ks}^{1/2})$

NDT in Canada NDT C 2019 Inform - Engage - Advance

Results (cont'd)

- Comparison between effusivity- and diffusivity-derived thermal conductivity:

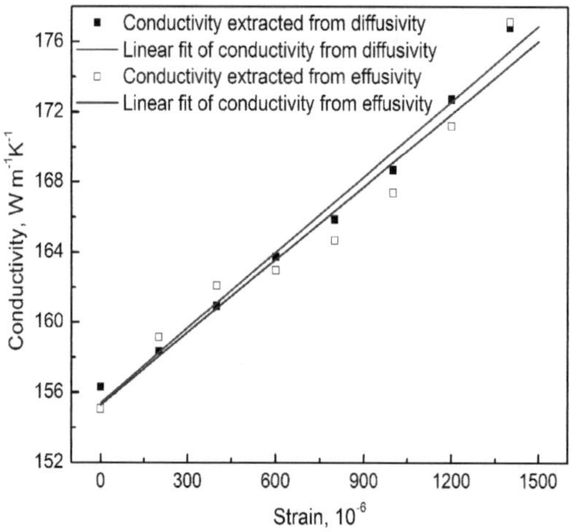

H. Huan, *et al*. NDT & E Int. **84,** 47-53 2016

- ➢ Conductivity values obtained from two approaches show very good agreement.
- ➢ Conductivity shows linear dependence on strain within the elastic regime
- ➢ Thermal conductivity dependence on stress is the primary effect within the elastic regime

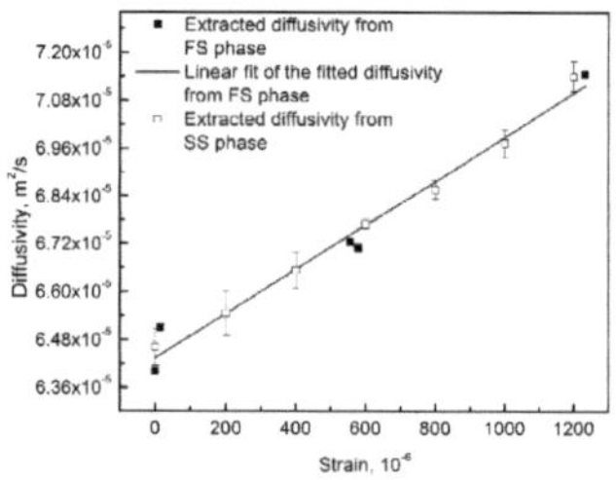

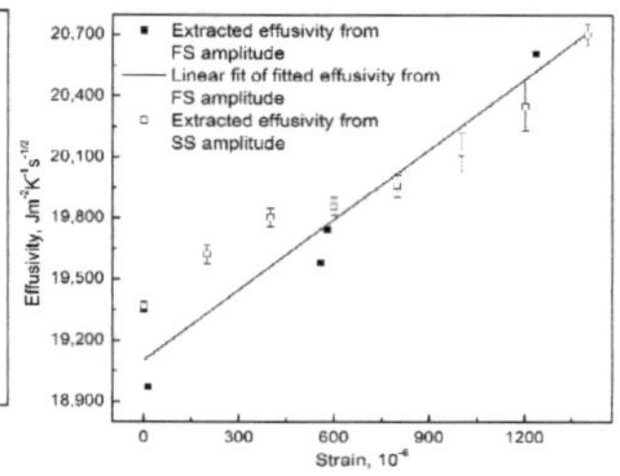

Analysis of Results

- Analogy: PTR phase-measured diffusivity-strain vs. stress-strain relation[2]:

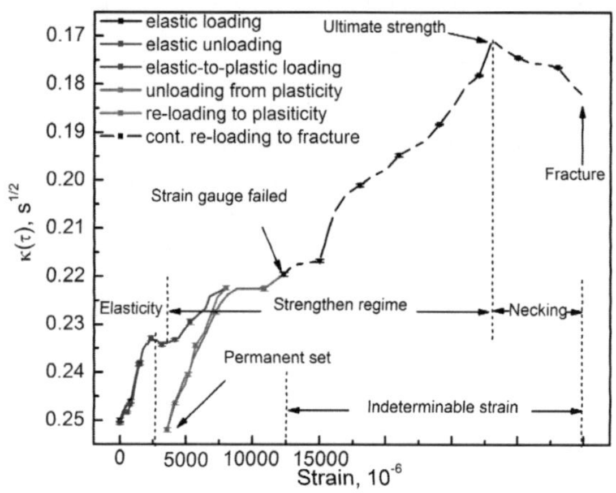

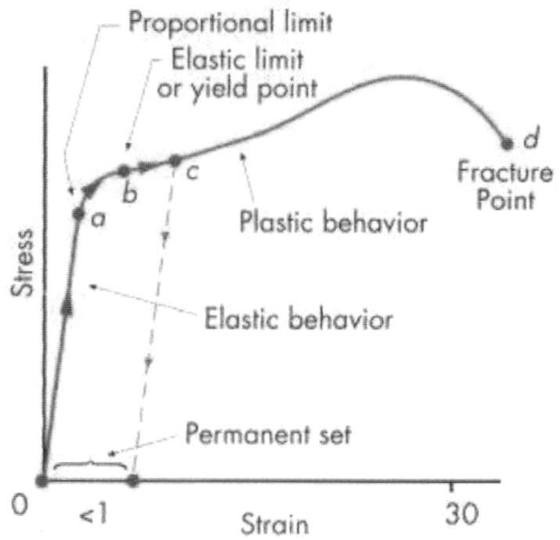

[2] Figure on the right from: http://www.leonghuat.com/articles/civil%20engineering.htm

Tests on Coated Samples

- The coated samples are:

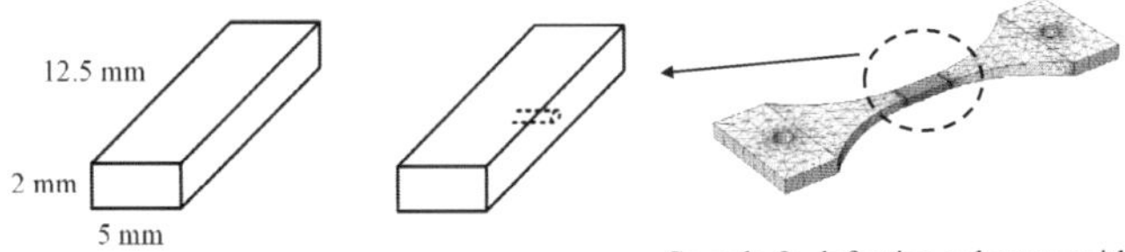

Sample 1: Substrate: intact dog-bone
Coating thickness: 0.005"

Sample 2: defective substrate with coating (one hole at center, diameter: 1 mm, depth: ~1 mm)
Coating thickness: 0.005"

Intact coated sample

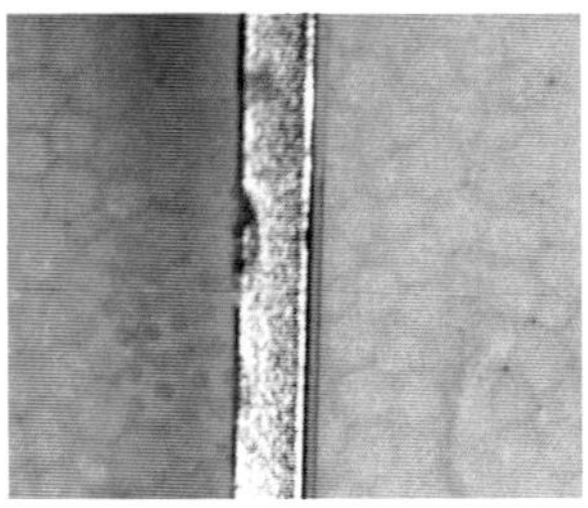

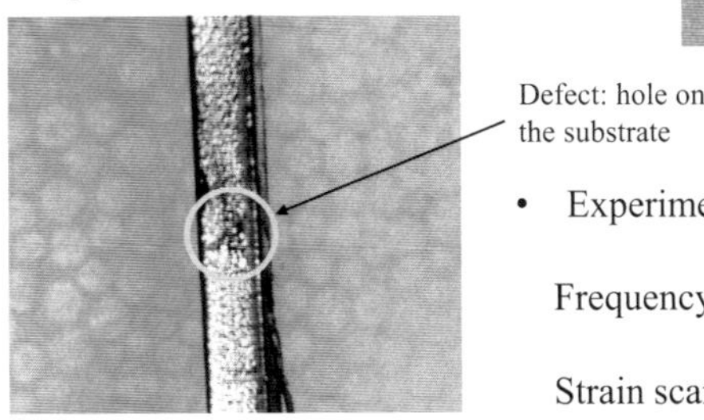

- Experiments:
  - Frequency scan
  - Strain scan
- Analysis:
  - Single layer model
  - Three-layer model

Coated Samples (cont'd)

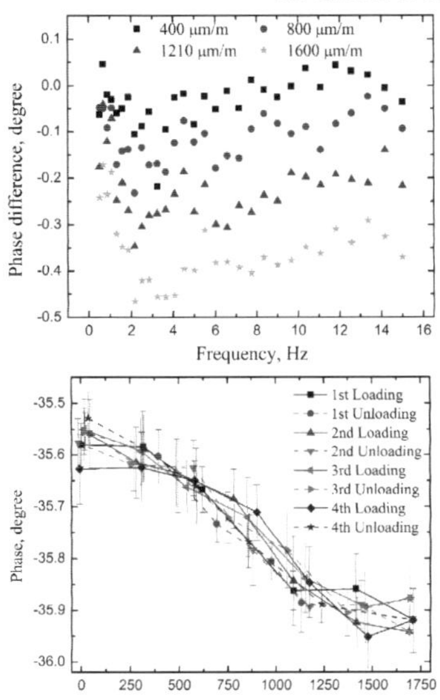

Sample 1 (intact substrate)

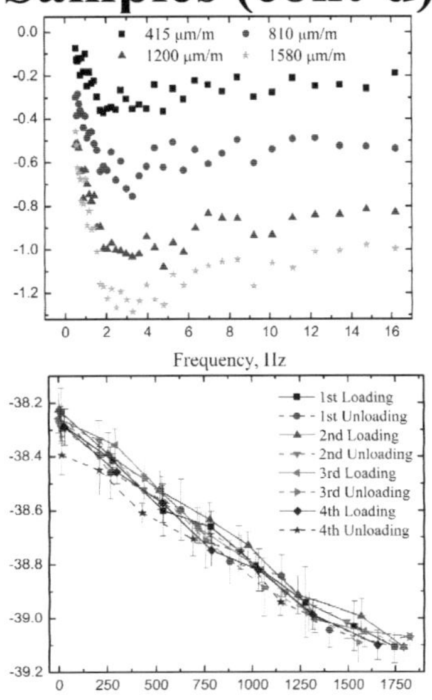

Sample 2 (defective substrate)

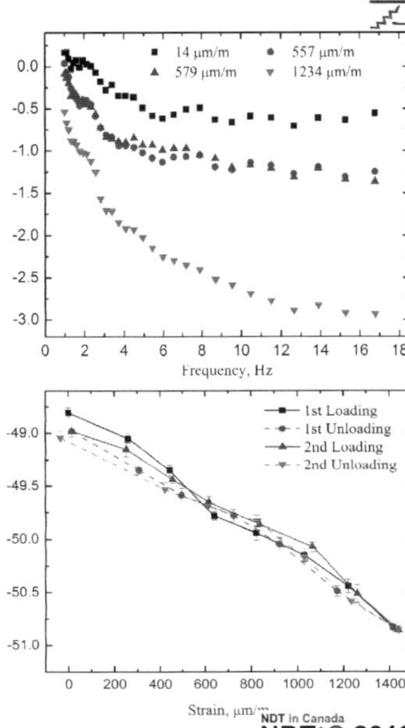

Bare aluminum

Coated Samples (cont'd)

CADIPT

- Quantification: single layer model (results)

Because the coated sample is not a single layer, the frequency range used for quantification should be low enough (0.5 Hz-5Hz). The results yield averaged overall thermal parameters for both coating and substrate.

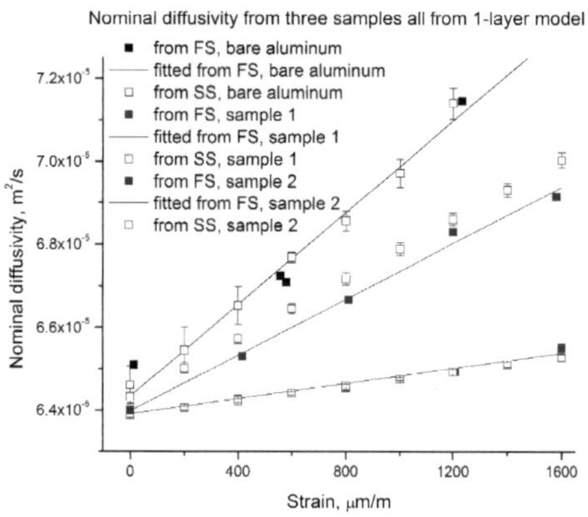

- As the coated samples have different materials for substrate and coating, this single-layer model can only derive a nominal diffusivity which represents an approximate average diffusivity of the samples. The diffusivity of the aluminum substrate is chosen to be the nominal diffusivity for all three samples.
- Due to the existence of defects on the substrate of sample 2, best fits from FS and SS show larger differences than the other samples.

NDT in Canada NDTC 2019 Inform - Engage - Advance

Coated Samples (cont'd)

CADIPT

- Quantification: three-solid-layer model (results)

Frequency was scanned from 0.5 Hz to 15 Hz, and was fixed at 2.3 Hz for the intact sample and at 1.07 Hz for the defective substrate sample. For the coating layer and substrate:

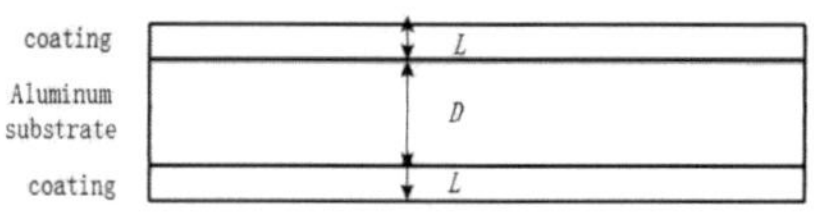

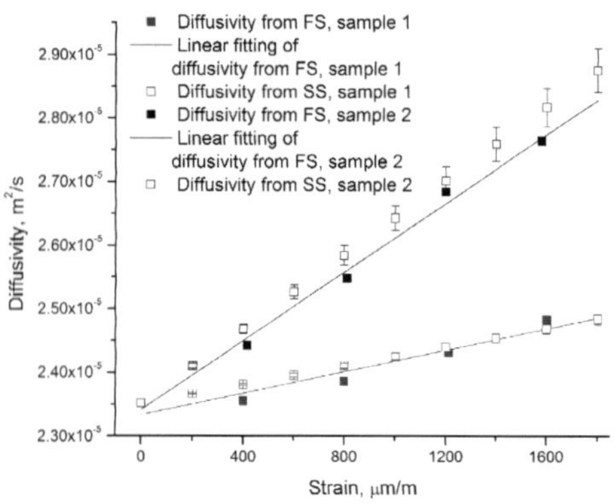

Coating diffusivity of the two coated samples

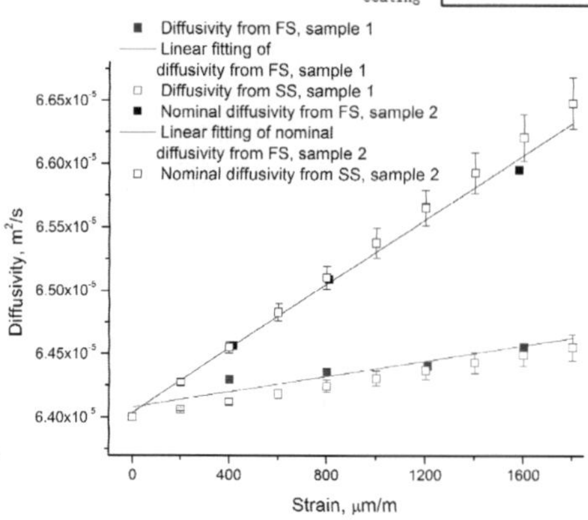

Substrate diffusivity of the two coated samples

- Compared with the single-layer results, the three layer model indicates diffusivity changes of both coating and substrate. For both, the changes are larger for the defective substrate.

NDT in Canada NDTC 2019 Inform - Engage - Advance

Coated Samples (cont'd)

CADIPT

Discussion:

- Compared with the single-layer model, the three-layer model reveals more detailed information on the thermal conductivity strain dependencies of *both* coating and substrate materials
- The coated samples perform much better at the same strain than the bare aluminum sample: A comparison between coated and uncoated sample at fixed strain is shown below:

Strain at 1200 μm/m

Sample	Effective Diffusivity Change	NiCo coating Diffusivity Change	Aluminum Substrate Diffusivity Change
Bare Aluminum Alloy	11%	-	-
Sample1 (intact and coated)	1.6%	3.4%	0.55%
Sample 2 (Defective substrate coated)	6.1%	17%	2.3% (Effective)

It is hypothesized that for the defective aluminum substrate, the stretch is larger for both substrate and coating at the same strain, so the coating undergoes more deformation and thus larger thermal conductivity / diffusivity change. At the same level of strain, the defective substrate sample undergoes larger tensile loading because its "waist" is more "yielding" than the intact substrate. Thus, it elongates more and so does the coating.

H. Huan, *et al.* NDT & E Int. 87, 44-49 2017

NDT in Canada NDTtC 2019 Inform - Engage - Advance

Conclusions: Significance and Outlook

CADIPT

- A) PTMR analysis proved to be able to quantify mechanical property relations of aerospace-relevant metallic components under stress
- B) PTMR emerges as a non-contacting, non-destructive, quantitative "strain gauge" with a much expanded strain range compared to conventional contacting mechanical strain gauges. It works instantaneously and does not require long adhesive curing times (usually overnight).
- C) PTMR can quantify the mechanical performance of multilayer (coated) samples:
  - 1) It can assess the mechanical strength of NiCo coatings toward the protection of coated substrates through measurements of PTMR signals at fixed strain.
  - 2) It can assess the mechanical strength or improvement of defective substrates through coating and can quantify thermophysical changes of both coating and substrate upon mechanical stress application using stress scans and frequency scans.
  - 3) The elastic limits of solids can be identified and studied as functions of geometric shape, material and coating.
- D) PTMR can map the entire stress-strain cycle for uncoated and coated samples from the unstressed state through the elastic, plastic and fracture stages. *This is not possible for attached mechanical strain*

NDT in Canada NDTtC 2019 Inform - Engage - Advance

Future work

CADIPT

Having proven the feasibility of non-contact evaluation of mechanical property relations by the PTMR approach, further work for this methodology will include:

- Apply PTMR test with the application of mid-infrared camera to quantify the three conductivity components under multi-directional loading.
- Develop more complex, applicable 3-D quantitative thermal-wave theory and inverse algorithms to reconstruct the internal thermal conductivity tensor distribution from fitting the contours at various surfaces of the sample.
- Testing samples with known (or unknown) residual stresses

NDT in Canada NDTiC 2019 Inform · Engage · Advance

Acknowledgments

NSERC CRSNG / People. Discovery. Innovation.

The Natural Sciences and Engineering Research Council of Canada

Ontario MINISTRY OF RESEARCH AND INNOVATION

Canada Research Chairs Chaires de recherche du Canada

The Canada Research Chairs Program

20

TOFD Inspection of ERW in thin walled piping

A specialised inspection technique

presented by Tim Poudrier

CINDE
CANADIAN INSTITUTE FOR NON-DESTRUCTIVE EVALUATION

NDT in Canada
NDTiC 2019
Inform - Engage - Advance
June 18 - 20
River Cree Resort & Casino
Edmonton, Alberta

What we cover

1. Identify the concerns regarding Lack of Fusion (LoF) and cracks in ERW longseams.
2. Define the requirements of inspecting ERW longseams in thin walled pipe.
3. Define the limitations of various inspection methods.
4. Show how a specialized TOFD inspection answers the requirements to properly inspect ERW longseams in thin walled pipe.
5. Answer questions.

NDT in Canada 2019 | June 18-20 | Edmonton, AB

Identify the concerns

- OD surface-breaking LoF or cracklike anomalies in Electronic Resistance Welds(ERW) cause great concern for catastrophic failure.

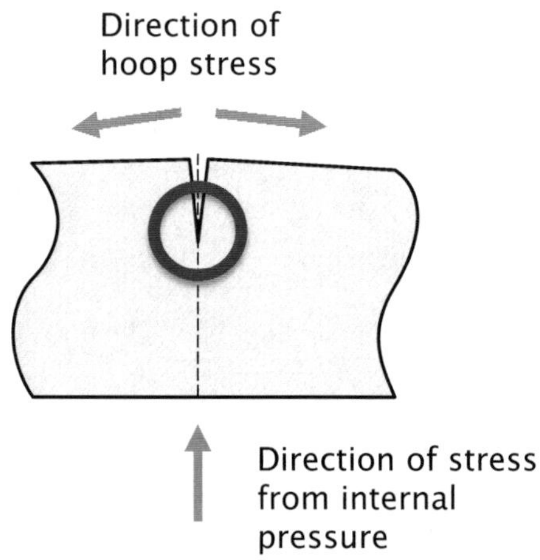

Identify the concerns

- Buried (midwall) anomalies are less likely to cause failure.

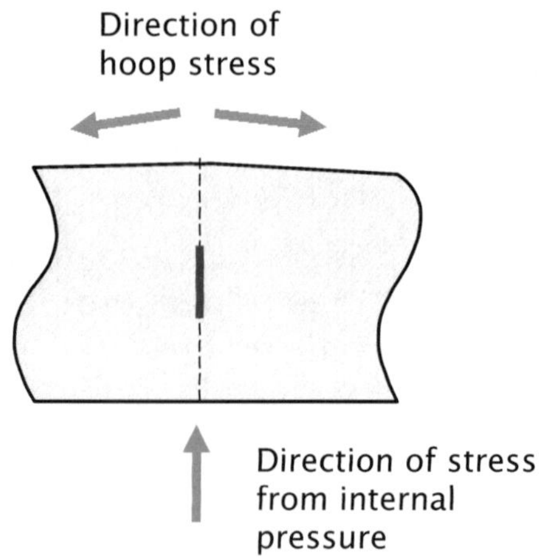

Identify the concerns

- Many LoF anomalies in the ERW develop into hook cracks after cyclic loading.

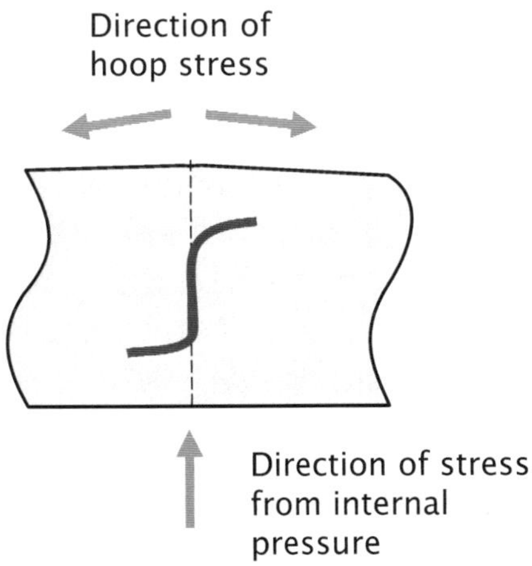

NDT in Canada 2019 | June 18-20 | Edmonton, AB

Identify the concerns

- Buried anomalies of sufficient thru-wall height can pose a dangerous risk of failure.

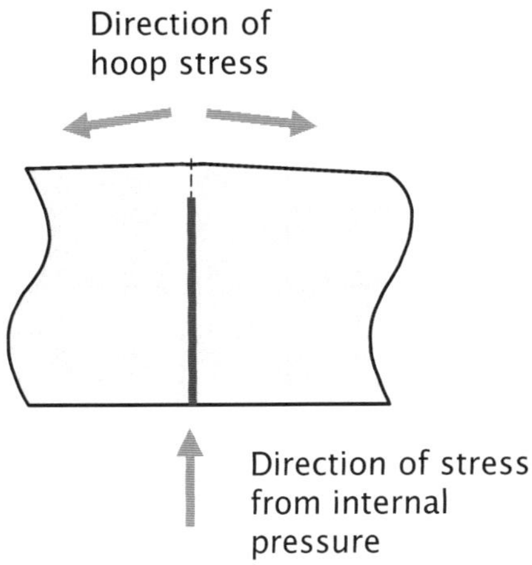

NDT in Canada 2019 | June 18-20 | Edmonton, AB

Identify the concerns

- ID connected anomalies located beneath OD connected anomalies pose a serious danger of failure during OD crack removal.

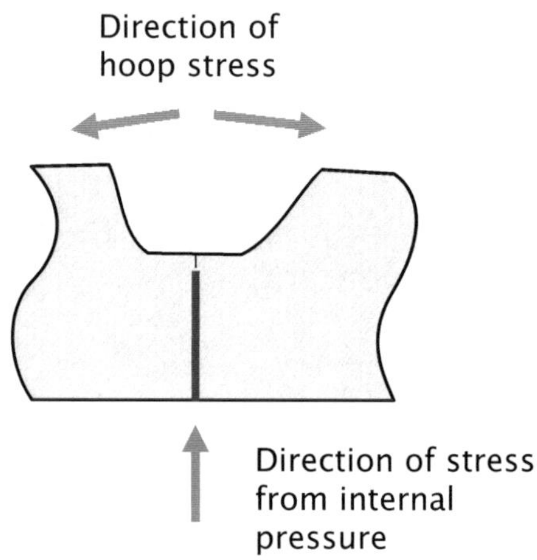

Identify the concerns

- Complete removal of crack-like anomalies by grinding may still require a compression sleeve if the material removed is too deep and/or too long.

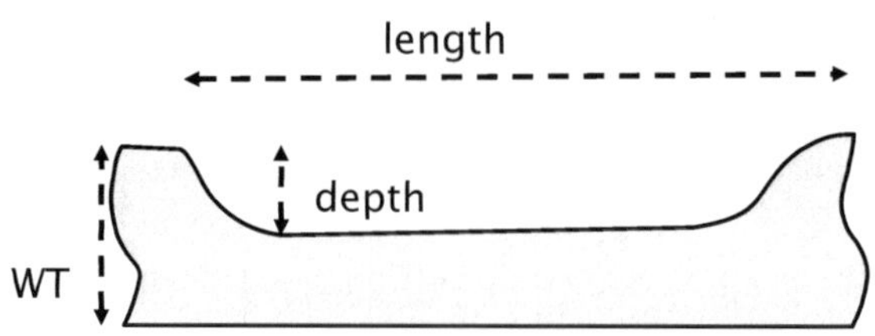

If calculated BP<MAOP x SF, still requires compression sleeve

NDT in Canada 2019 | June 18-20 | Edmonton, AB

Recap the concerns

1. OD connected LoF or cracklike defects pose the highest risk of failure.
2. Midwall anomalies are less likely to cause failure, unless they are of a dangerous height
3. Repairing the pipe by grinding out the LoF or crack may pose a danger to connecting with a buried defect.
4. After repair by grinding, it's possible a steel compression sleeve will still need to be installed, due to low BP.
5. Cyclic loading can cause relatively small buried defects to develop into potentially hazardous hook cracks.

NDT in Canada 2019 | June 18-20 | Edmonton, AB

What we need

1. A method with a high probability of detection (POD) of anomalies which pose a threat and interact with the OD.
2. A ***VERY HIGH POD*** of anomalies that have a threatening through-wall height, even if they don't break the OD surface.
3. To know that removing a surface anomaly will not endanger the people or environment due to an ID connected flaw.
4. To make this inspection in a timely manner, in order to ***reduce the total overall cost of defect mitigation***.

NDT in Canada 2019 | June 18-20 | Edmonton, AB

What we need

The proposed specialized TOFD inspection technique will do what we need.

NDT in Canada 2019 | June 18-20 | Edmonton, AB

TOFD has advantages

- Intuitive
- NOT angle dependent
- NOT amplitude dependent
- Relatively small file sizes from scanning
- BEST at avoiding false positives
- In ERW can clearly show misalignment

NDT in Canada 2019 | June 18-20 | Edmonton, AB

TOFD has advantages

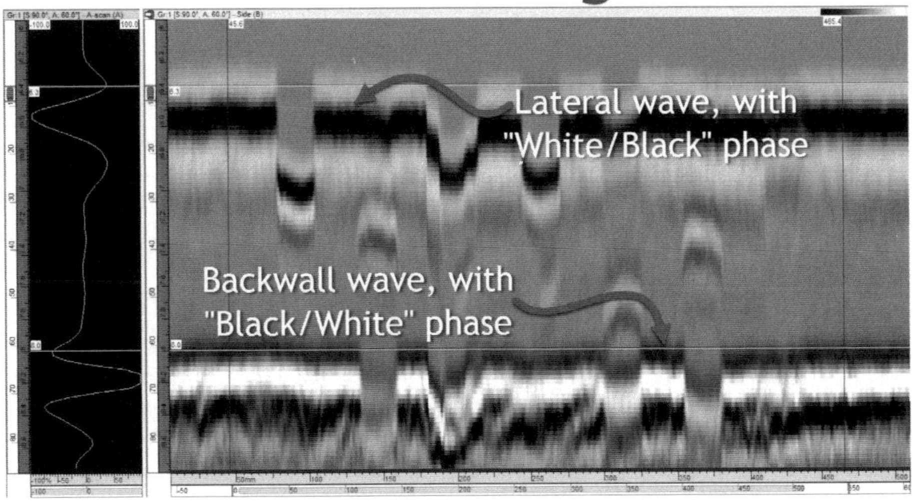

NDT in Canada 2019 | June 18-20 | Edmonton, AB

TOFD has advantages

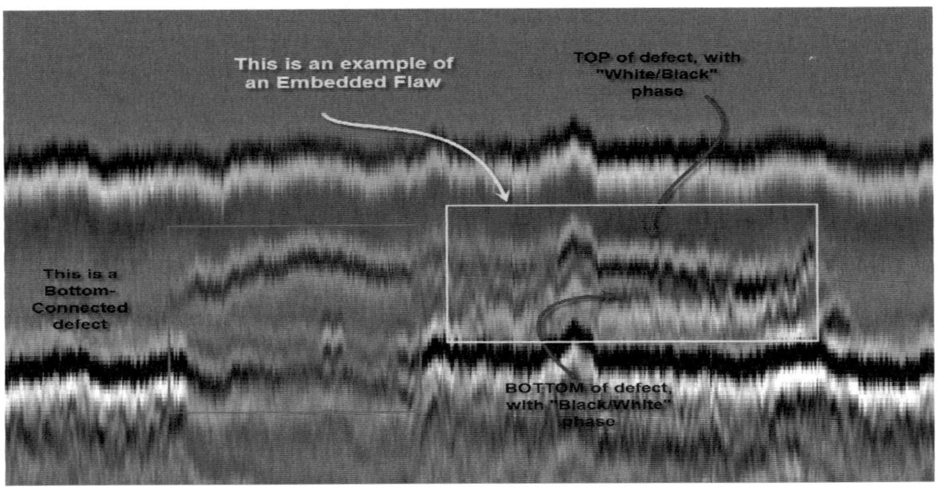

NDT in Canada 2019 | June 18-20 | Edmonton, AB

TOFD has limits

- Cannot tell which side of the weld the defect is on. ①
  - In thick material, this matters for a repair location.
- Some indications are difficult to characterize. ②
  - Often cannot be done with TOFD data alone.
- TOFD display magnifies small indications. ③
  - Distracting, may cause unnecessary repairs.
- Lateral Wave and Backwall Reflector have "dead zones". ④
  - If they're truly dead, there's no way to inspect these areas.

These limitations are usually addressed in a regular butt weld by combining TOFD with Phased Array UT (PAUT)

NDT in Canada 2019 | June 18-20 | Edmonton, AB

TOFD has limits

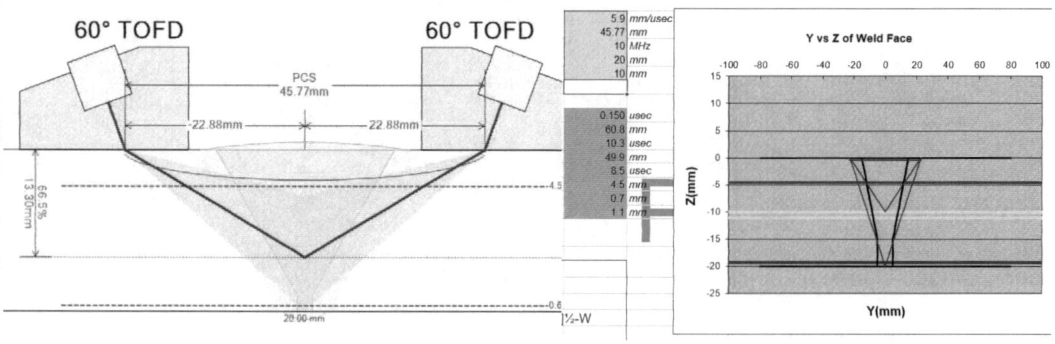

Lateral Wave dead zone = 4.5mm

Backwall dead zone = 0.7mm

TOFD inspection covers 79% of volume outside dead zones

NDT in Canada 2019 | June 18-20 | Edmonton, AB

Butt welds use TOFD + PAUT

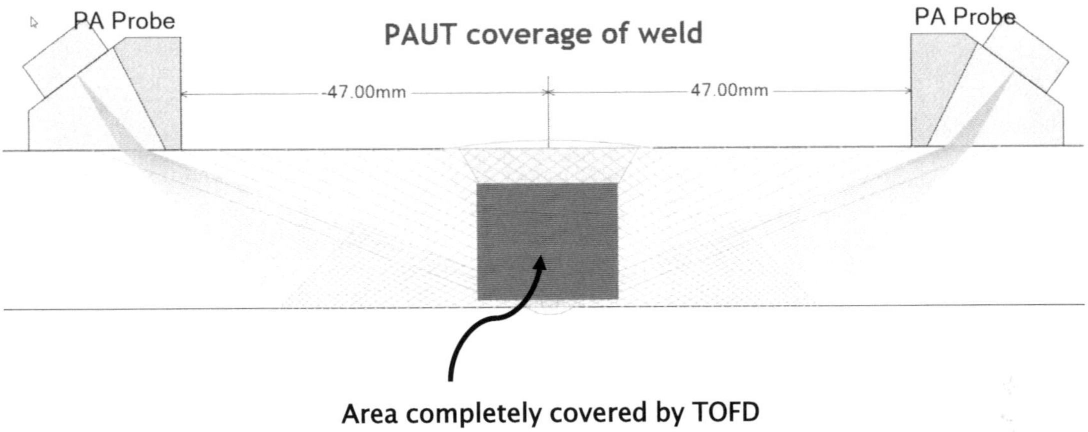

Butt welds use TOFD + PAUT

- ① By using both techniques, it's easy to measure a reflector's location in relation to the centerline of the weld
- ② The "A" Scan data of a PAUT scan aids in characterization of a reflector
- ③ Data from the angles of a Sectorial scan or the beams of a Linear scan help in sizing reflectors
- ④ A Scan Plan will make sure any dead zones in TOFD are completely covered by the PAUT setup(s).

Butt welds use TOFD + PAUT

ERW longseams present a different set of variables, and are not subject to the same limitations.

NDT in Canada 2019 | June 18-20 | Edmonton, AB

ERW welds are different

- ① ERW has no width. The two faces are pressed together, current is applied, heat is generated, and they melt.
- ② The indications that will be found in an ERW are limited to LoF, cracking, and inclusions. There may be a lamination or "stringer" so close to the ERW that TOFD will detect it.
- ③ Qualified TOFD technicians have been taught to recognize indications that are superficial, and those that are relevant.
- ④ Because the ERW has a specific shape, we are able to change our setup and inspect the weld with TOFD.

NDT in Canada 2019 | June 18-20 | Edmonton, AB

What TOFD sees

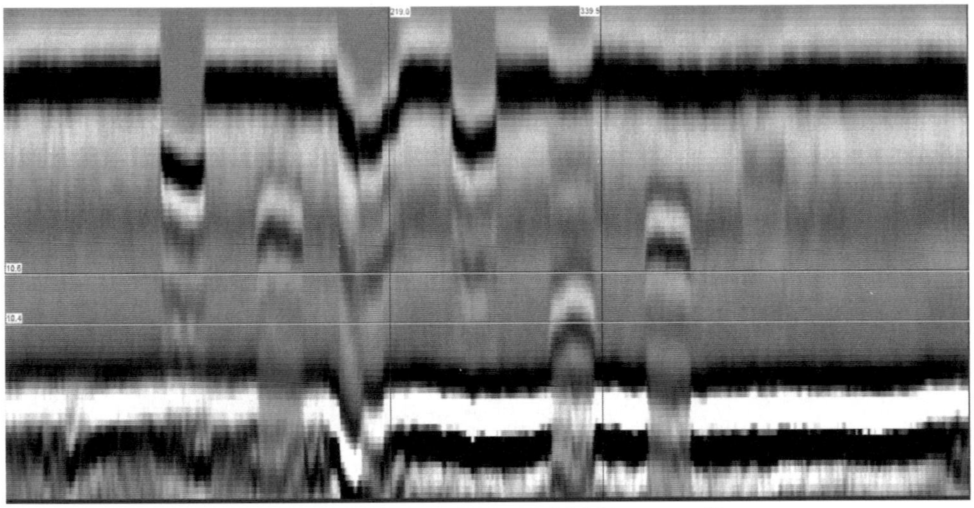

NDT in Canada 2019 | June 18-20 | Edmonton, AB

What TOFD sees

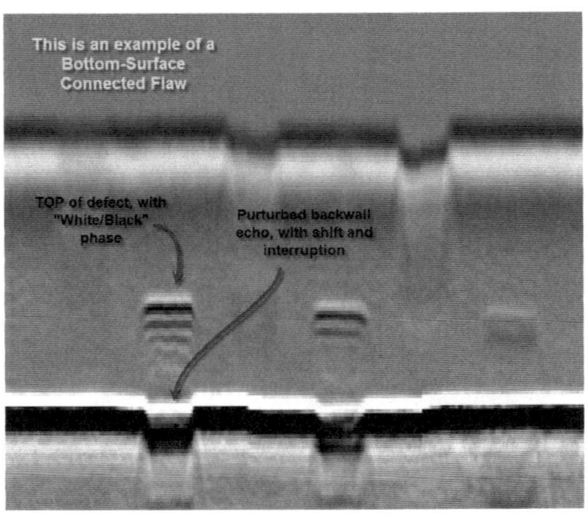

ID Connected

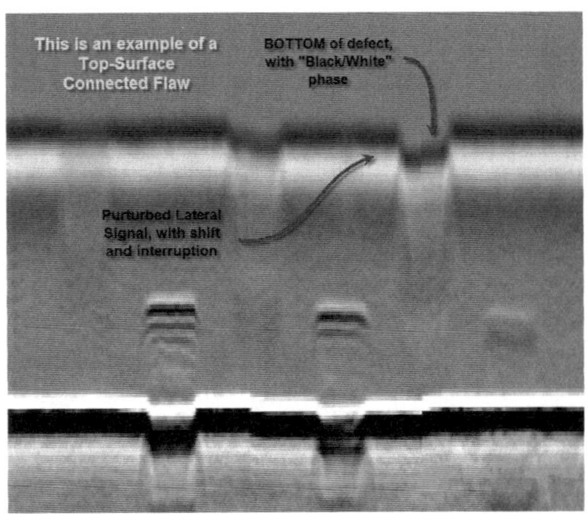

OD Connected

NDT in Canada 2019 | June 18-20 | Edmonton, AB

What TOFD sees

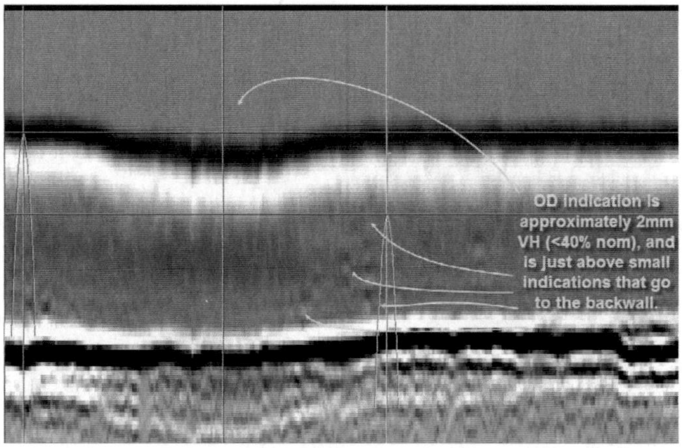

OD with indications below

What TOFD sees

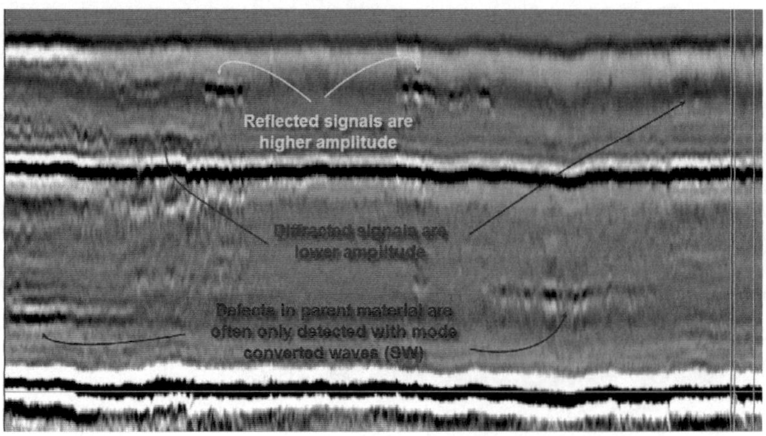

Laminations/stringers, hook cracks, and LoF

What TOFD sees

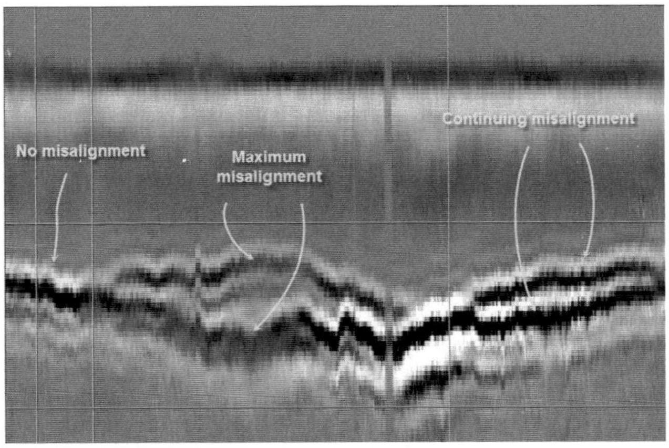

Alignment and misalignment

Results

- Because we normally inspect 100% of bare pipe with MT and/or ET Arrays, we will detect surface anomalies.
  - Shallow anomalies <10 - 15% WT easily removed by grinding.
- TOFD setup will detect and measure the length, depth, and VH of surface-breaking (ID/OD) *AND* midwall anomalies.
  - Prevent leaks during grinding by being able to see other indications.
  - Detects defects with a threatening height that don't break the "skin".
- Encoded and recorded data allows for auditing, training, review, etc.

Results

- Small file size allows for sharing and remote interpretation of scan data.
  - Files are small enough to send via email.
  - No delay for decisions to be made.
- Easy to avoid misinterpretation of Hi-Low as a LoF defect.
- Full length of ERW can be scanned and analyzed in a short time.
- Shows if a compression sleeve would be required after defect removal, saving time – and money – for the owner.
- Confirmation of hook crack shape/depth is easily done with 0° UT, as necessary.

TOFD for ERW – Yes or No?

- When combined with MT, or ET arrays, the probability of detection of surface anomalies is very high.
- Subsurface anomalies which could pose an immediate risk of failure are detected and measured.
- False positives from misalignment are avoided.
- Files are shareable, auditable, and can be used for review and training.
- TOFD limitations are managed by the specialized setup.
- Training is straightforward, and regular technicians get excellent repeatable results.

TOFD for ERW – Yes or No?

What say you?

Questions?

NDT in Canada 2019 | June 18-20 | Edmonton, AB

Remote Magnetic Sensing of Plastically Deformed Steel

Aroba Saleem, P. Ross Underhill and Thomas W. Krause

Department of Physics and Space Science
Royal Military College of Canada

NDT in Canada
NDTiC 2019
Inform - Engage - Advance
June 18 - 20
River Cree Resort & Casino
Edmonton, Alberta

CINDE
CANADIAN INSTITUTE FOR NON-DESTRUCTIVE EVALUATION

Outline

- Motivation
- Introduction
- Materials and Methodology
- Results and Discussion
- Conclusion
- Acknowledgements

Motivation

https://www.google.ca/search?q=pipe+digging+for+testing&rlz

Collection of above-ground survey data.

Pipes- Residual Magnetization

- Underground steel pipes - carrying high pressure fluid or gas - Stress from internal pressure, placement and ground settling produces residual magnetization
- Orientation of pipes with respect to Earth's magnetic field
- Pipe manufacturing process may also cause residual magnetization
- Stress concentration zones and anomalies cause a local change in magnetic permeability - Self Magnetic Flux Leakage (SMFL)
- Metal Magnetic Memory (MMM) - Contact and Non-Contact methods

Advantages of MMM

- No external applied magnetic field is required
- Larger detection distance compared to other currently used methods
- Easy to use and time saving process
- Effective in detection of stress induced magnetization in steel pipeline material

NDT in Canada 2019 | June 18-20 | Edmonton, AB

5

Magnetization due to Stress

Metal Magnetic Memory (MMM): Based on the principle of Inverse Magnetostriction (Piezomagnetic effect or Villari effect)

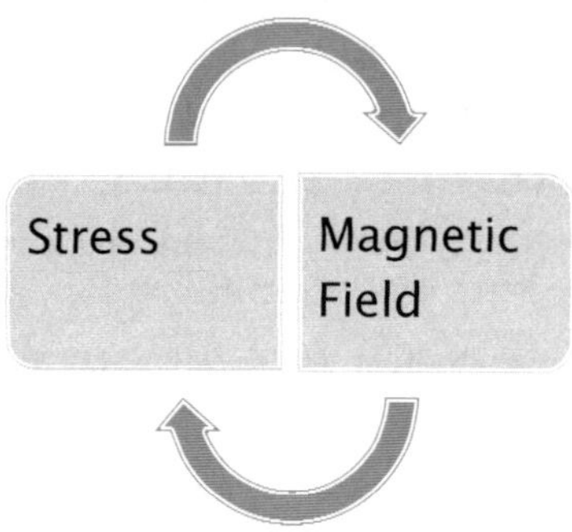

NDT in Canada 2019 | June 18-20 | Edmonton, AB

6

Magnetic Domains

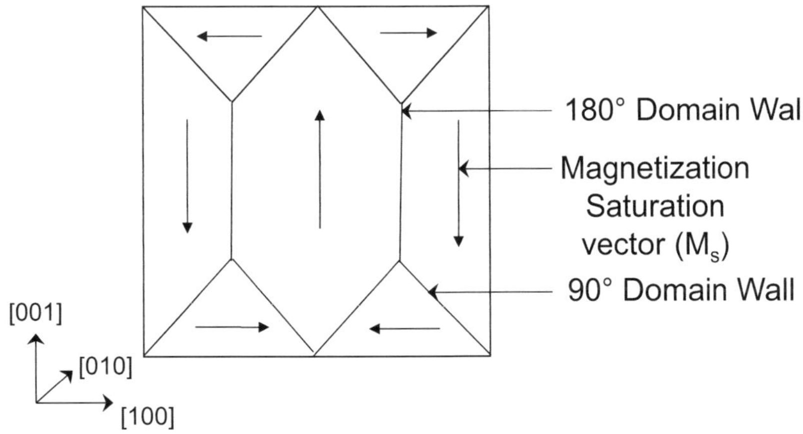

T.W. Krause and A. Samimi, "Micromagnetic Techniques", in ASM HandbookVolume17: Nondestructive Evaluation and Quality Control, 515-530, 2018.

Magnetic Domains under Stress

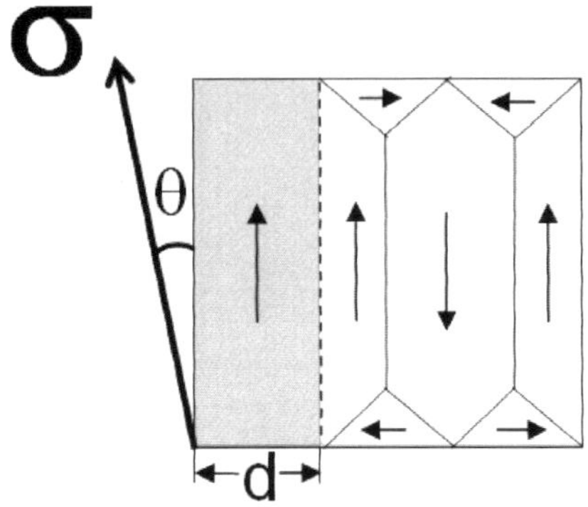

$$E_\sigma = \frac{3}{2}\lambda_s \sigma cos^2\theta$$

E_σ - magnetoelastic energy
θ - angle between saturation magnetization of the domain and applied stress (σ)
λ_s - saturation magnetostrictive constant

M. Kashefi, T. W. Krause, L. Clapham, P. R. Underhill and A. K. Krause, "Stress Induced Self Magnetic Flux Leakage at Stress Concentration Zone".

Materials and Methodology

Sample	Shape	Length(mm)	Width/Outer diameter (mm)	Thickness (mm)
Sample 1	Bar	275	19	3
Sample 2	Bar	560	19	3
Sample 3	Tube	1170	32	1.8

Sensor used

- Honeywell 3-Axis (HMC 5883L) anisotropic magnetoresistive (AMR) sensor
- 4.5 milliGauss (mG) resolution
- ±8 Gauss maximum field range

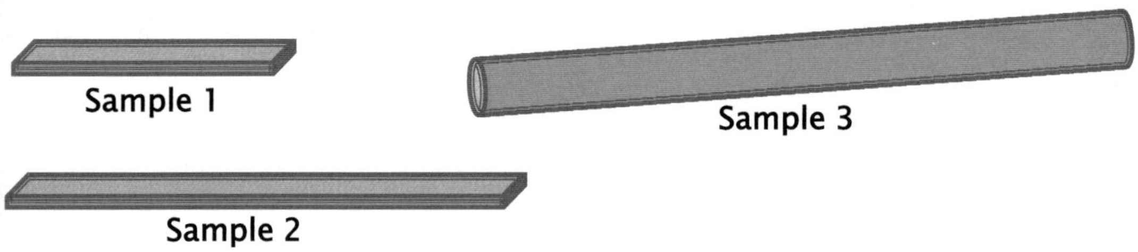

Sample 1 - Short Bar

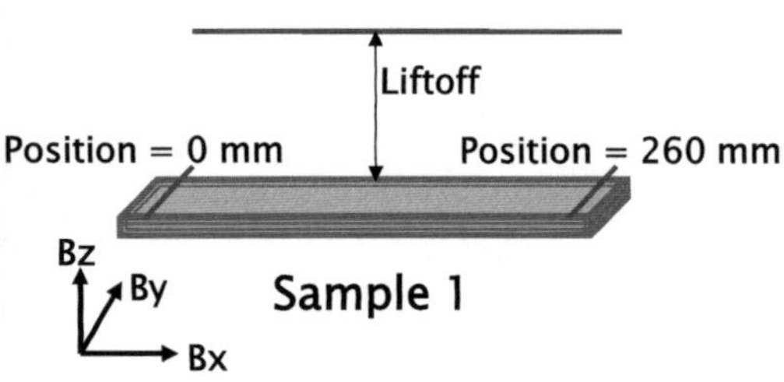

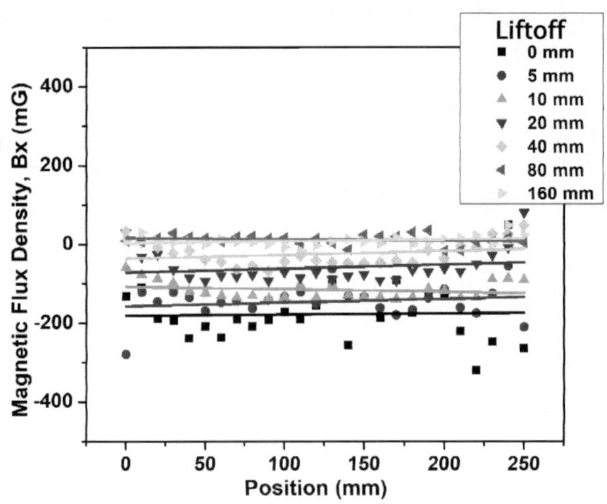

Sensor

X - along the long edge of the sample

Y - perpendicular to the paper

Z - normal to the sample surface

Sample 1 – Short Bar

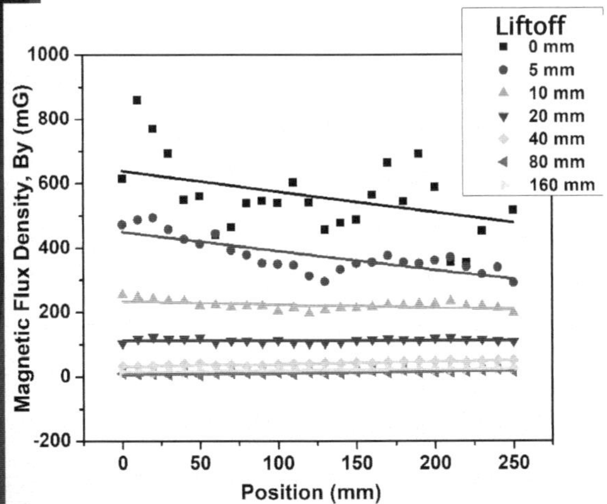

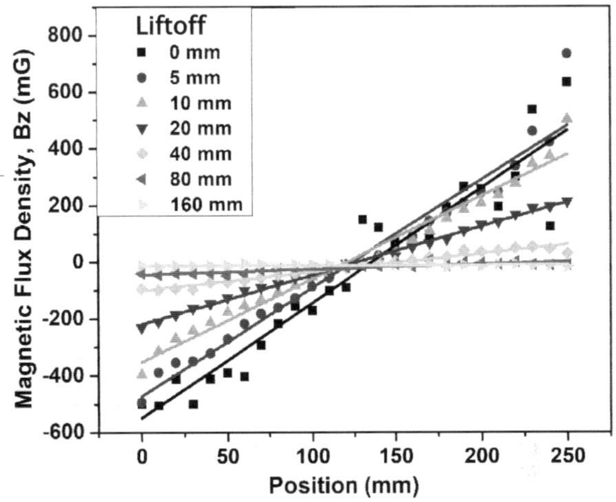

Sample 2 – Long Bar

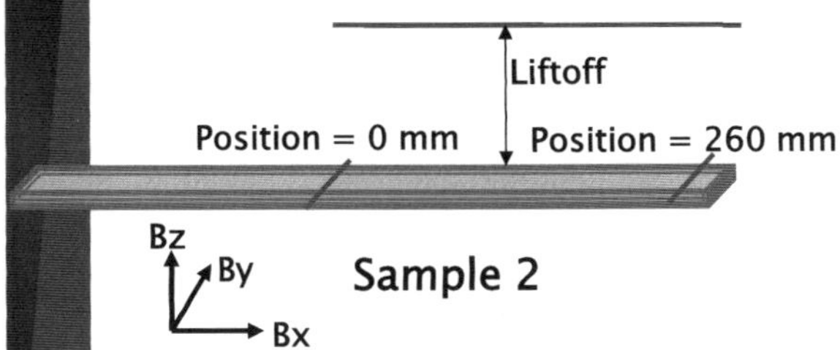

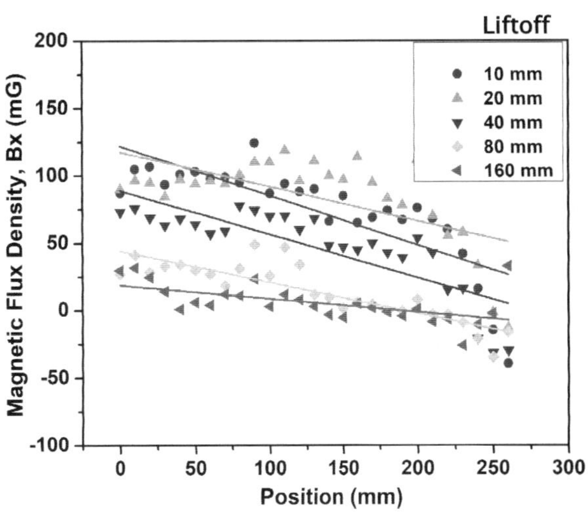

Sample 2 - Long Bar

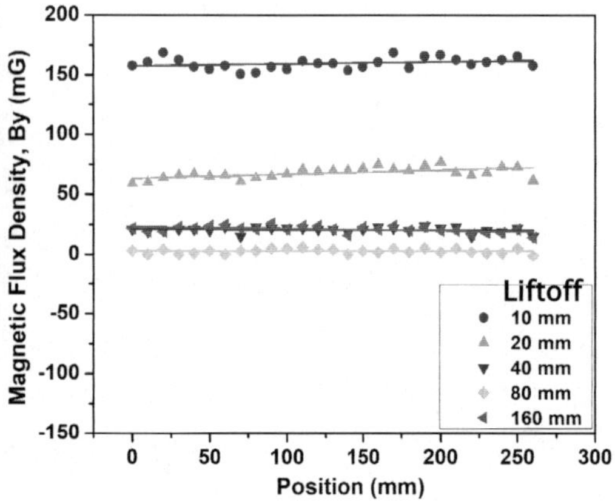

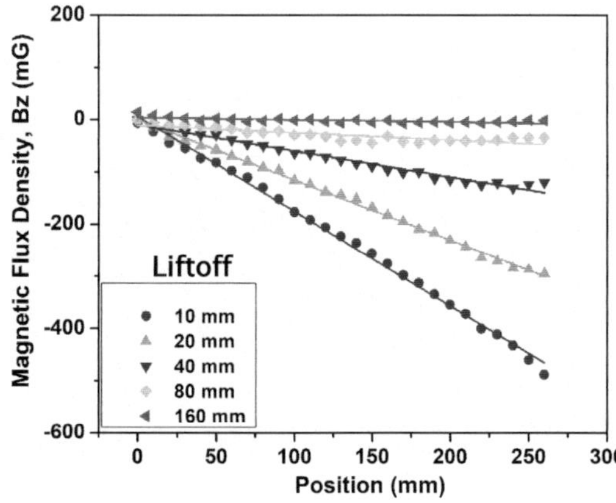

NDT in Canada 2019 | June 18-20 | Edmonton, AB

13

Sample 3 - Tube

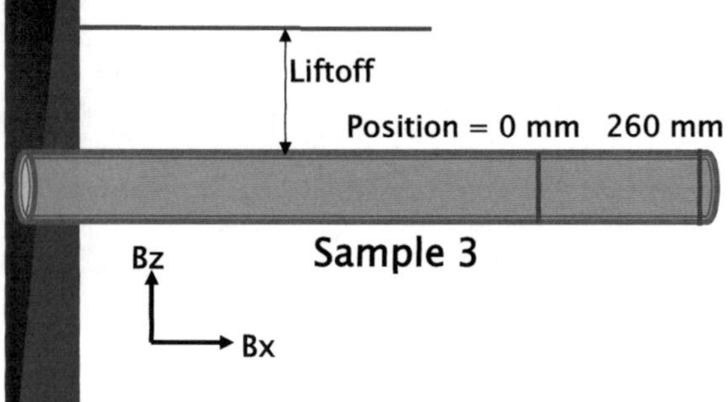

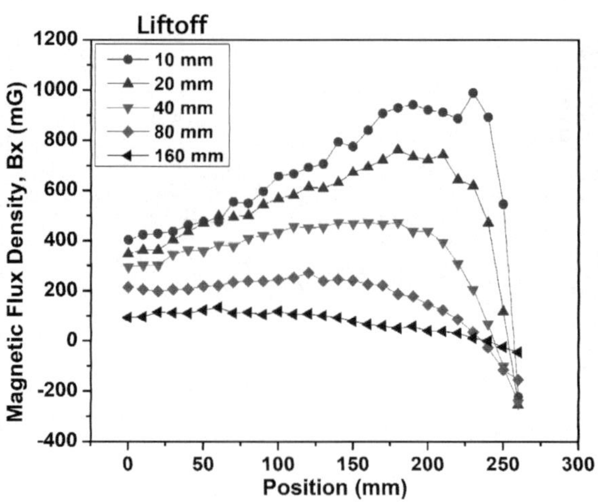

NDT in Canada 2019 | June 18-20 | Edmonton, AB

14

Sample 3 - Tube

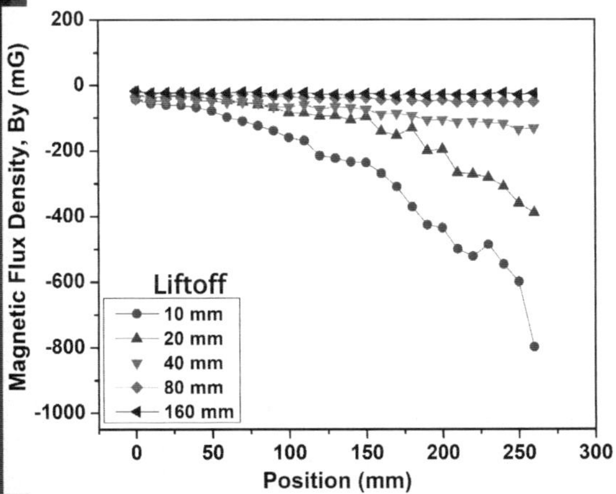

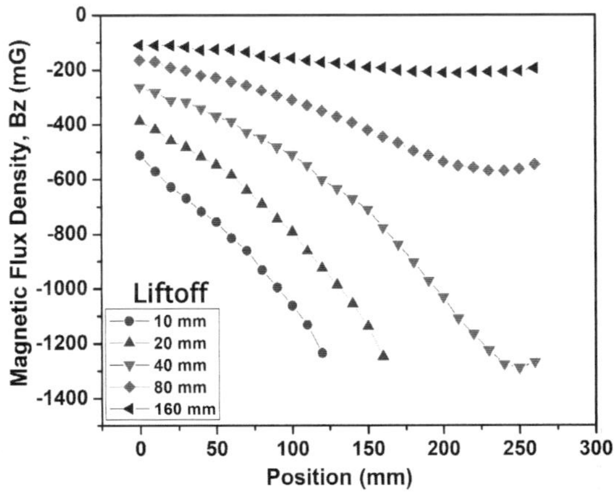

Bending Stress - Short Bar Sample

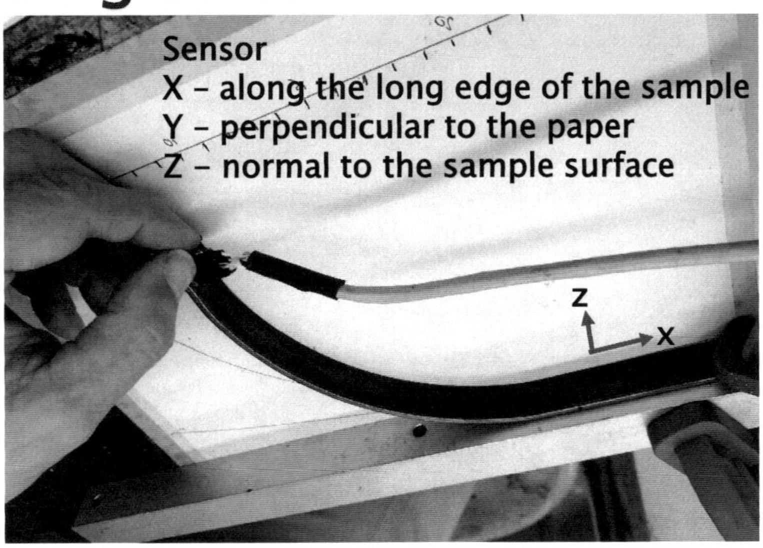

Bent Short Bar Sample 1

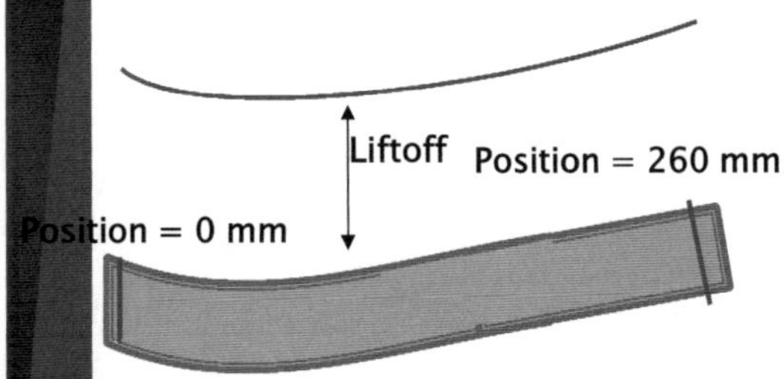

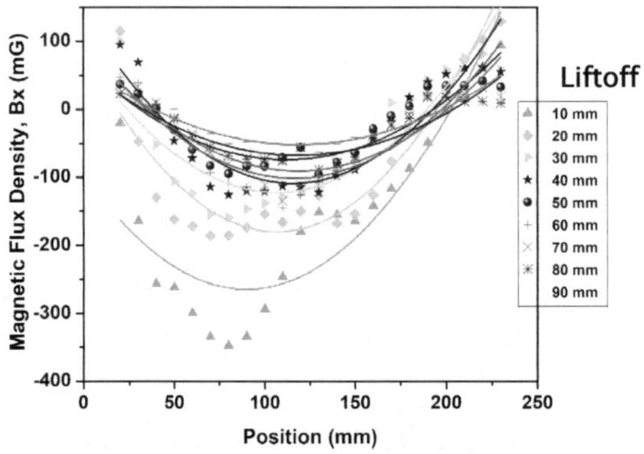

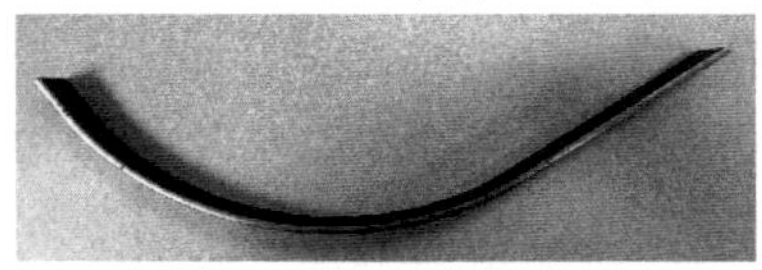

Bent Short Bar Sample 1

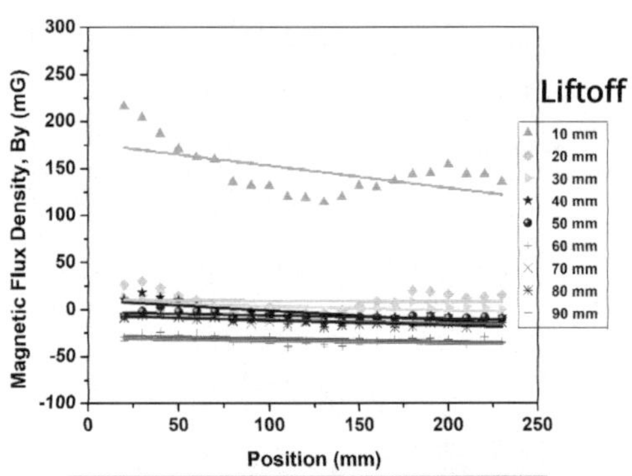

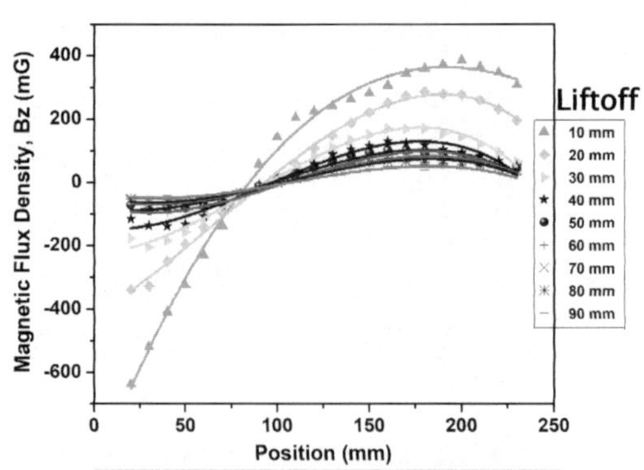

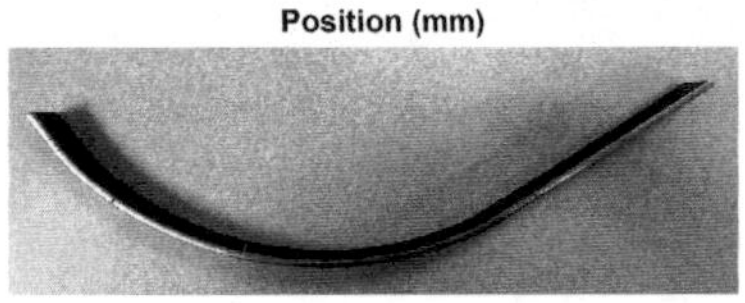

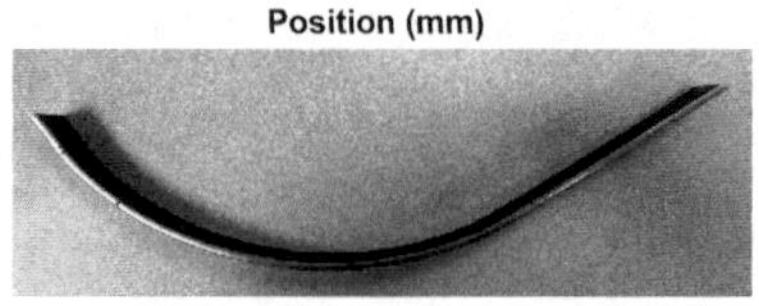

Bent Tube Sample 3

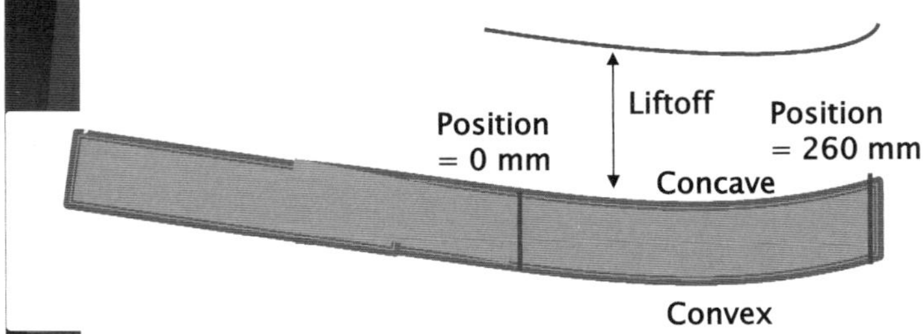

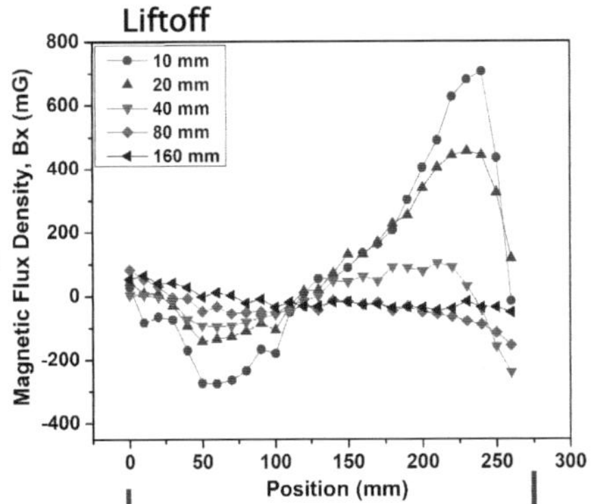

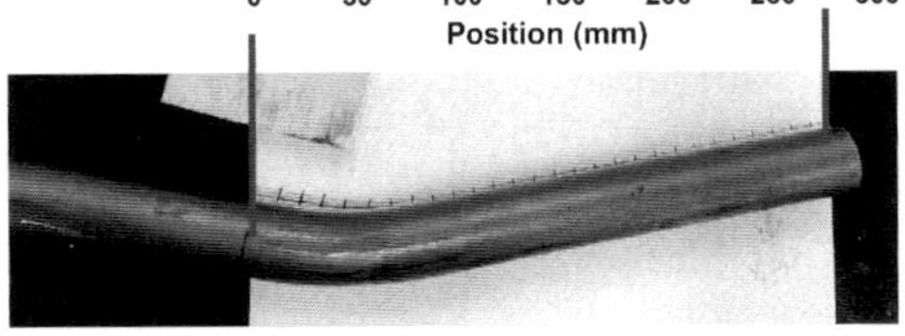

Bent Tube Sample 3

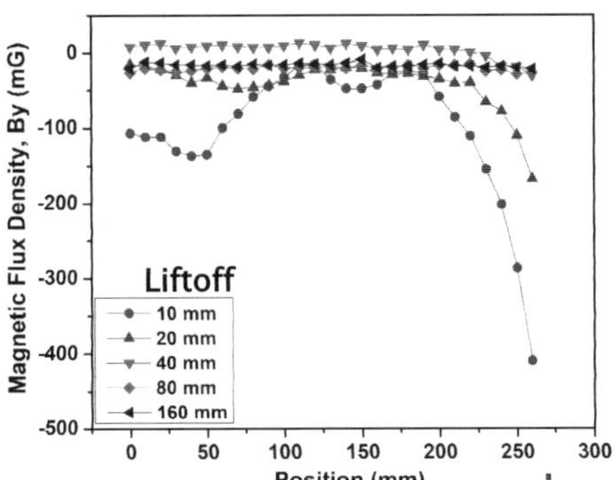

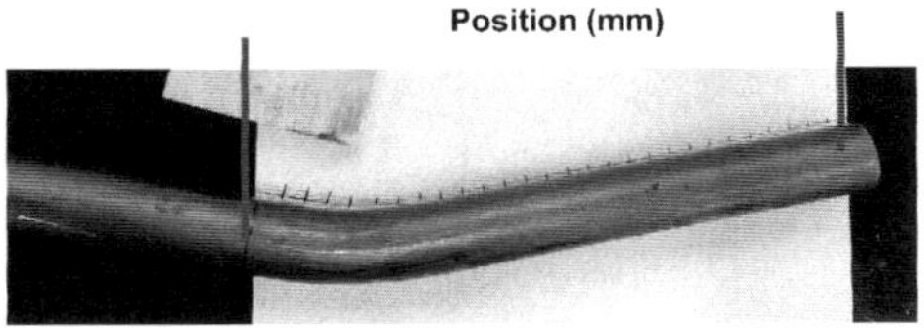

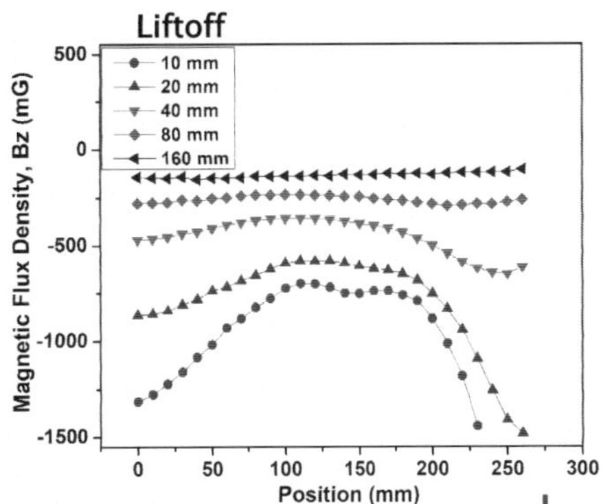

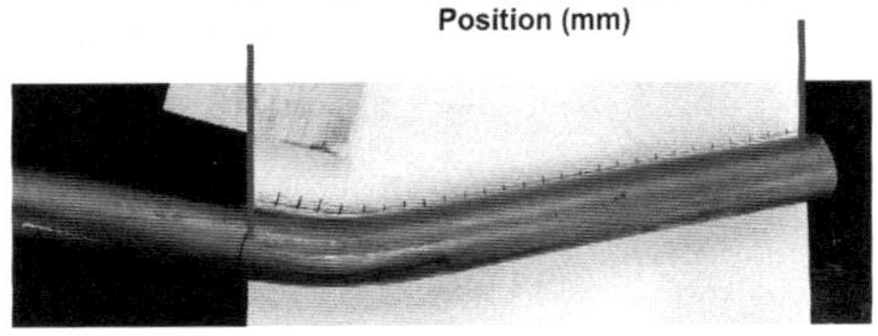

Effect of Bending

Before bending

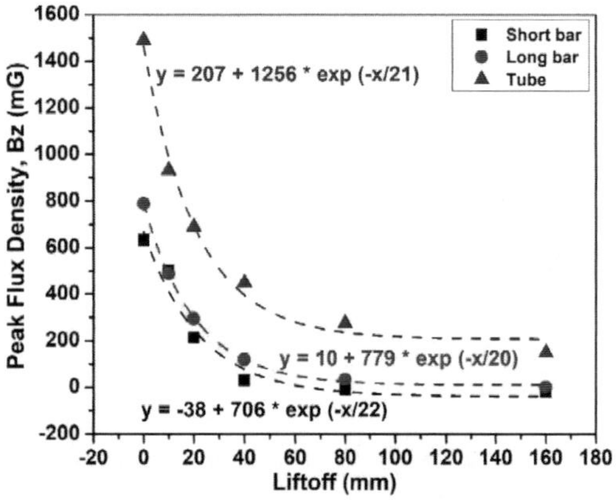

After bending

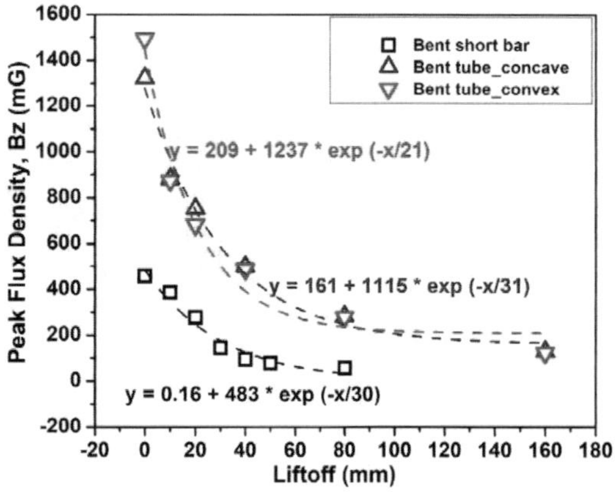

Effect of Bending_Short Bar Sample 1

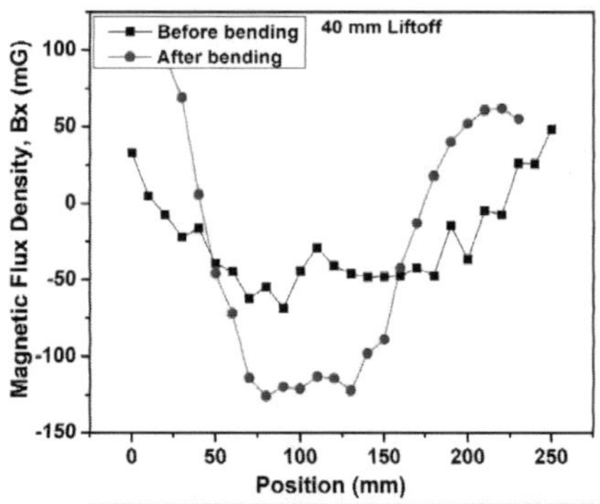

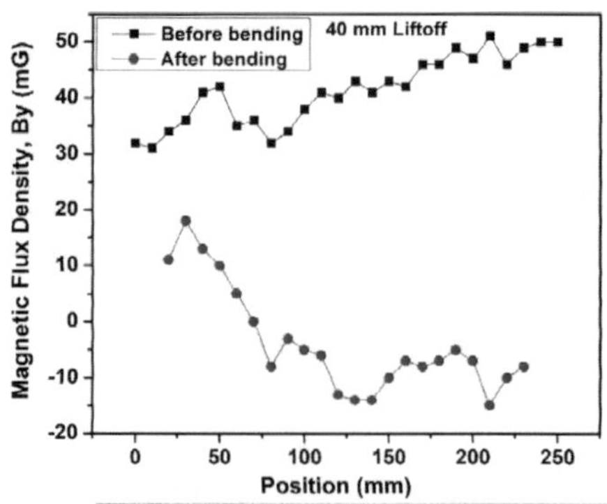

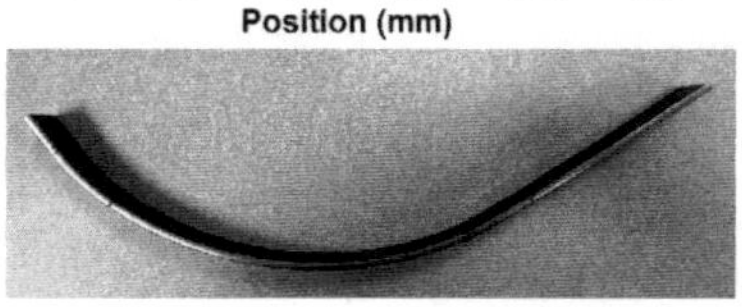

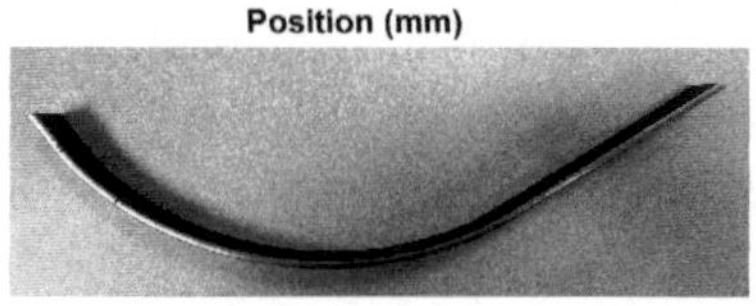

Effect of Bending_Short Bar Sample 1

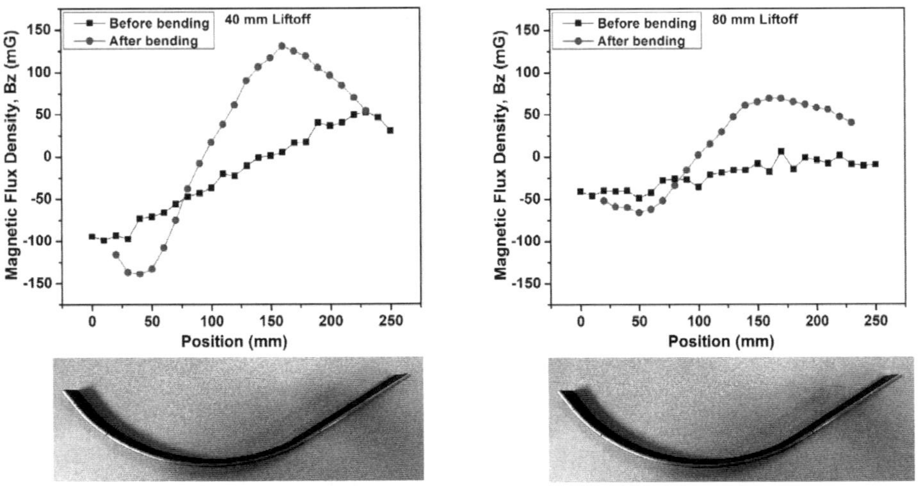

Effect of Bending_Tube Sample 3

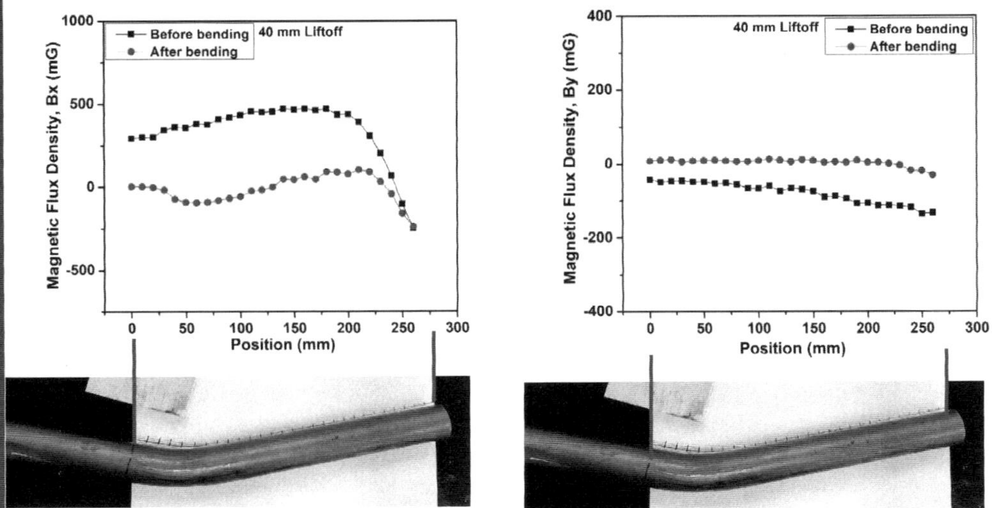

Effect of Bending_Tube Sample 3

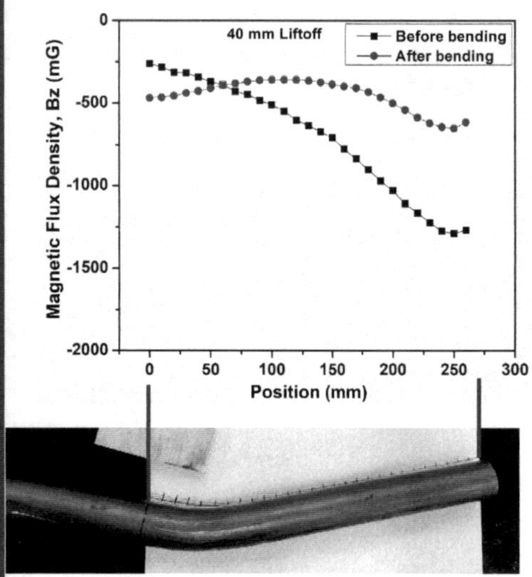

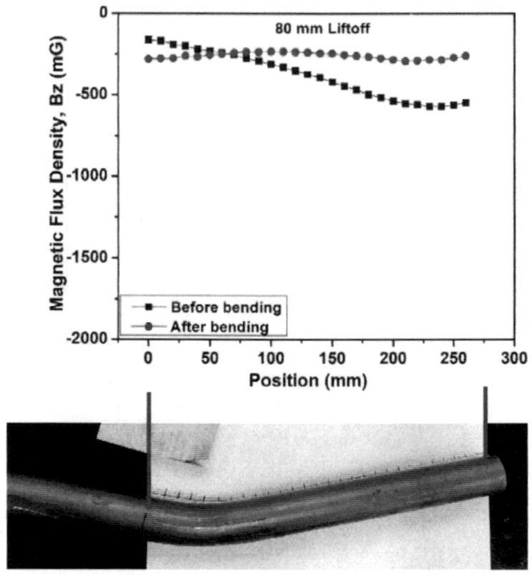

Discussion

- Larger signal from long bar associated with larger dipole
- More material appears to generate larger residual magnetization before and after bending
- Residual stress components in tube are more complex and generate a larger remnant signal
- Results have implications for underground oil and gas pipeline, which is continuous (no end effects), and larger and deeper (scaling effects may be assumed).

Conclusions

- MMM technology has the potential to be used to characterize pipeline magnetic signature
- Magnetic flux density is affected by residual stress in all three directions
- Bending changes the residual magnetization of the sample, which is most prominent in z-component (normal to the surface) of the field
- The decay constant for unbent bar and tube samples was about 22 mm, whereas it was 30 mm for bent samples.

Acknowledgements

- Peter Snell for assistance with sample preparation
- Ontario Centres of Excellence
- Natural Sciences and Engineering Research Council of Canada

Thank you

NDT in Canada 2019 | June 18-20 | Edmonton, AB

29

Advanced Ultrasonic Alternatives for Inspecting Coarse Grained Stainless Steel Components

Mike Wechsler
Technical Director, MISTRAS Group, Inc.

CINDE Canadian Institute for Non-Destructive Evaluation

NDT in Canada
NDTiC 2019
Inform - Engage - Advance
June 18 - 20
River Cree Resort & Casino
Edmonton, Alberta

ULTRASONIC INSPECTION

UT has become a very effective inspection tool for a variety of things including:

- ✓ Thickness Points (TMLs) for establishing corrosion rates.
- ✓ Code Inspection for New Weld Integrity (In-Lieu of Radiography)
- ✓ In-Service Crack Sizing and Detection
- ✓ Non-Intrusive Alternatives
- ✓ Supporting RBI Needs or Fitness-for-Service (FFS) evaluations

Bottom Line:

Ultrasonics are an integral part of our programs. These inspections are being used to allow for longer run times, extended life, and fewer internal inspections.

ULTRASONIC TECHNIQUES

Available Tools in the box vary by application:

- ✓ Standard 0 degree
- ✓ Shear Wave
- ✓ High Angle Longitudinal Waves
- ✓ Time of Flight Diffraction
- ✓ Phased Array
  - ✓ Sectorial
  - ✓ Linear
  - ✓ Dual Matrix
  - ✓ TFM
  - ✓ Full Matrix Capture
  - ✓ Adaptive Focusing

All great advancements and tools to help us verify integrity, reduce risk, and increase profitability.

ULTRASONIC VARIABLES

Variables that can affect the effectiveness of an inspection include:

- ✓ Velocity
- ✓ Material Type
- ✓ Age
- ✓ Thickness
- ✓ Heat Treatment
- ✓ Temperature
- ✓ Acoustic Impedance
- ✓ Wave Modes
- ✓ Plate Differences
- ✓ Grain Structure

Some of these are controllable and verifiable, but which ones?
And is this being done on all inspections?

CODE

Some variables that the code addresses are:

- ✓ Calibration Block Material
  - ✓ Acoustically similar
  - ✓ Same product form, and/or equivalent P Number
- ✓ Heat Treatment
  - ✓ Minimum tempering by material spec for type, grade, and PWHT if there are welds involved
- ✓ Surface Finish
  - ✓ Shall be representative
- ✓ Austenitic Materials
  - ✓ Those that exhibit coarse grained or directionally-orientated grain structure that can cause reflection and refraction at grain boundaries as well as velocity changes within the grains……

NDT in Canada 2019 | June 18-20 | Edmonton, AB

STAINLESS STEEL APPLICATION

Customer request inspection in-service cracking of typical 304SS Material.
Typical approach would be:

- ✓ Calibration Block Material
  - ✓ Acoustically similar
  - ✓ Range of Diameters
- ✓ Calibration Flaw Sizes
  - ✓ EDM Notches, and SDH's
- ✓ Reference Blocks
  - ✓ 304 SS Naviships/IIW
- ✓ Probe/Frequency Selections
- ✓ Procedural Development/Scan Plans
- ✓ IF coarse grained – weld mockups according to T150

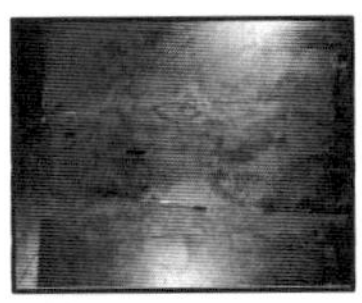

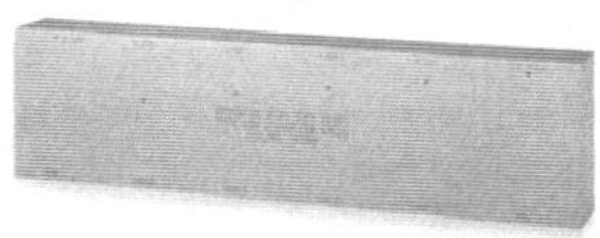

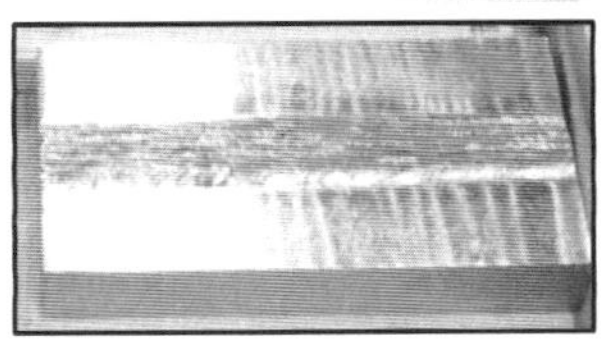

NDT in Canada 2019 | June 18-20 | Edmonton, AB

STAINLESS STEEL APPLICATION

And as a last resort.......

✓ Request "cut out sample" of problem areas

NDT in Canada 2019 | June 18-20 | Edmonton, AB

VERIFICATION

Sample piece was provided with damage.

Inspection began with the normal process

Longitudinal (0 degree) Wave

- ✓ Lower Frequency,
- ✓ Larger Aperture (diameter)

Sampling of similar thickness/material calibration block revealed:

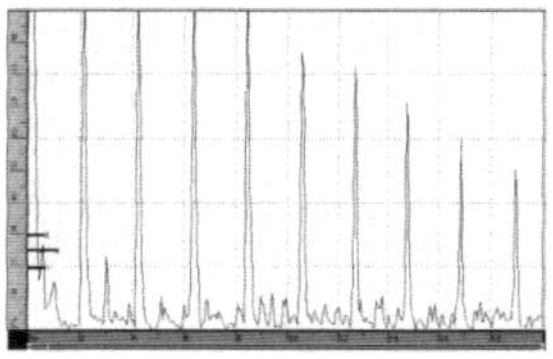

Actual piece revealed this:

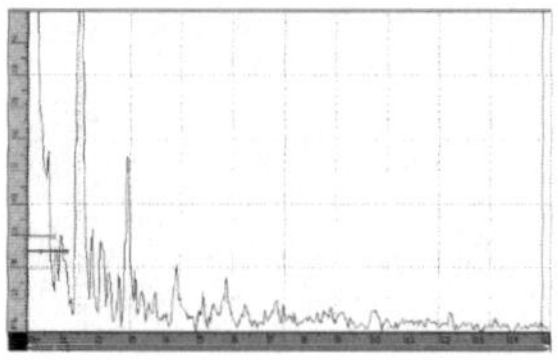

NDT in Canada 2019 | June 18-20 | Edmonton, AB

VERIFICATION

Sample piece was provided with damage.
Inspection began with the normal process.

Shear Wave

- ✓ Lower Frequency,
- ✓ Larger Aperture (diameter)

Quick Screening with Shear wave results revealed the following:

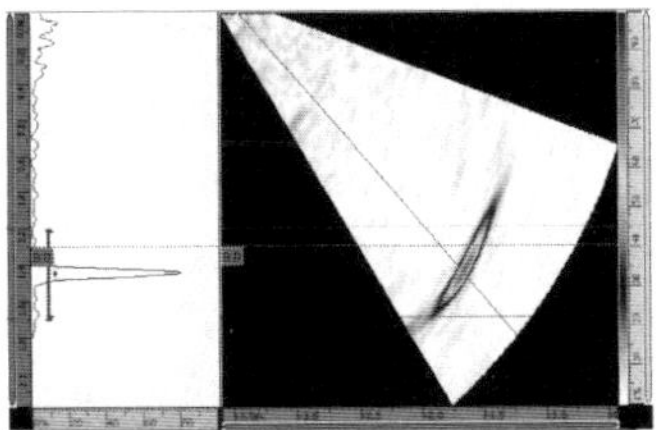

Reflection from Corner of Calibration block

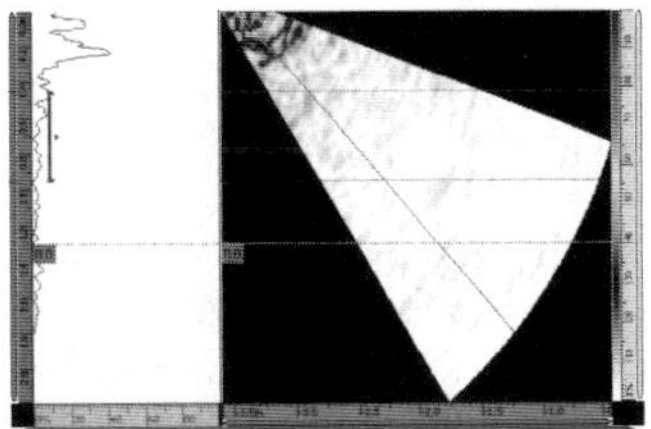

Reflection from Corner of Sample

VERIFICATION

Taking one step further, SDHs were introduced into the sample piece.

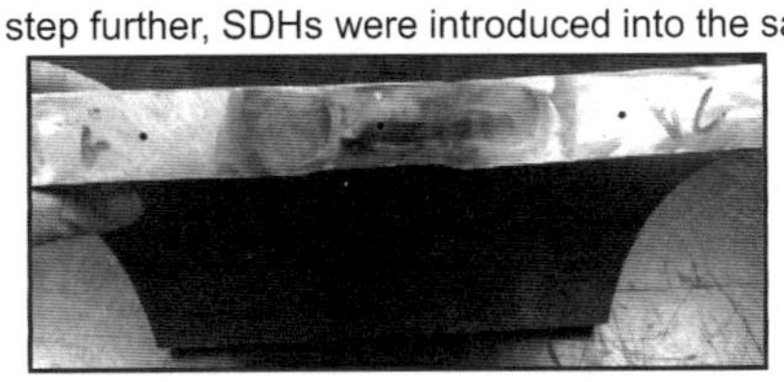

Quick Screening of the holes with Shearwave revealed the following:

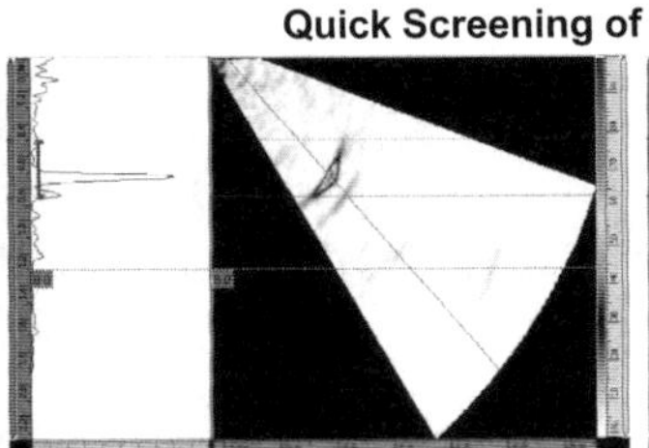

Reflection from SDH Calibration block

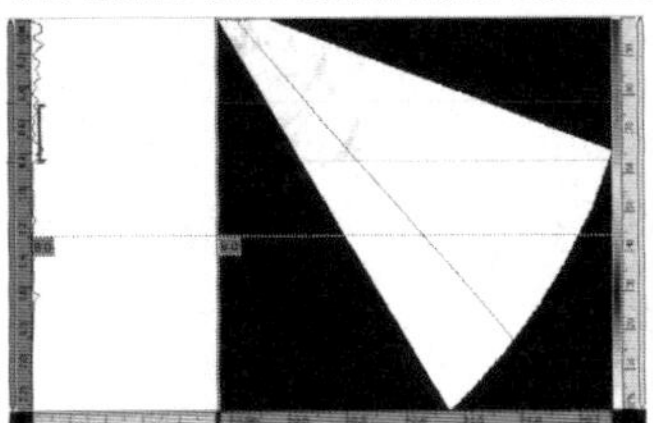

Reflection from SDH Base material (Sample)

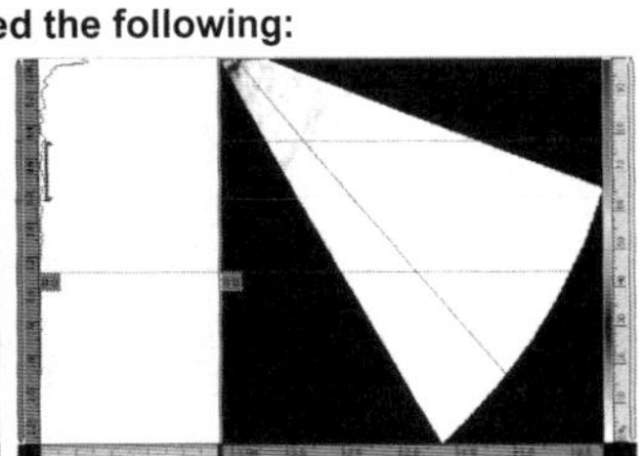

Reflection from SDH In weld (Sample)

GRAIN SIZE DISCREPANCY

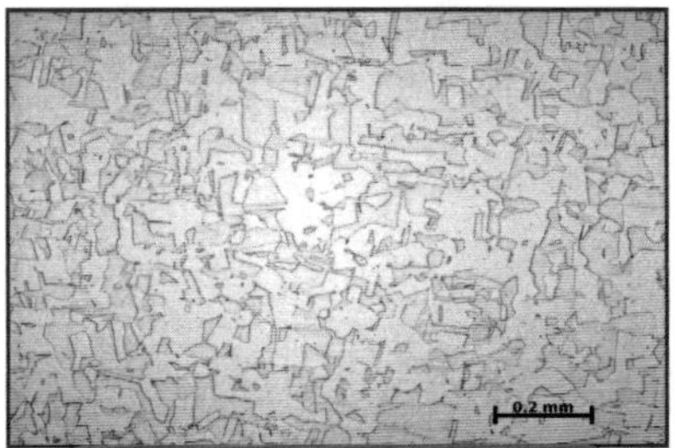

Typical 304SS Grain Size

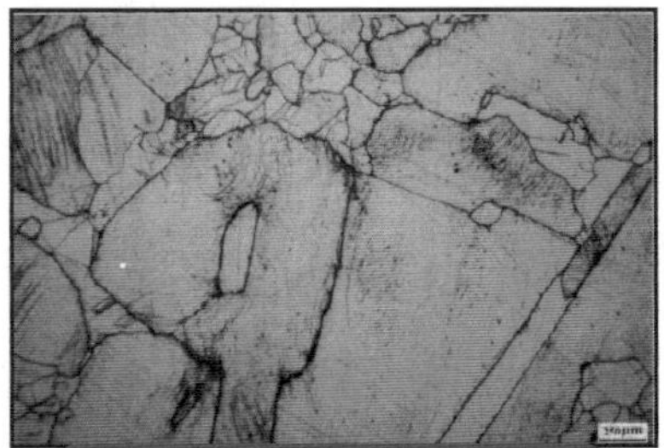

Typical 304SS Grain Size

Microstructural Analysis revealed that grain sizes were off the microscopic grain chart in ASTM E112.

It is believed that this was due to extreme temperatures (2000 -2200 F°).

GRAIN SIZE VISIBLE

Grain Size is actually visible to the NAKED EYE!

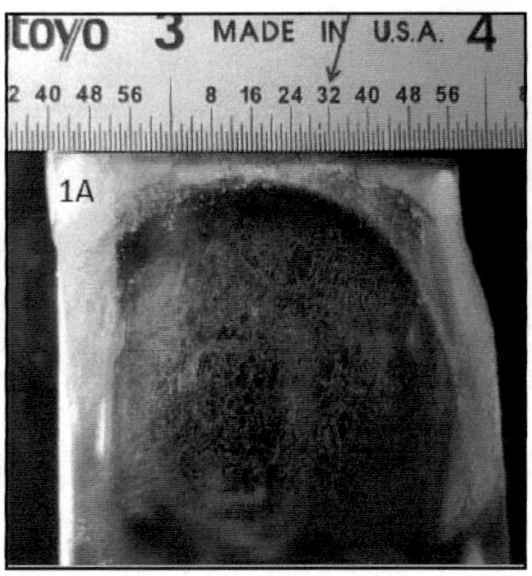

NDT in Canada 2019 | June 18-20 | Edmonton, AB

SELECTION

- ✓ Due to previous steps, it was clear that a different selection of probes was necessary.
- ✓ Shear waves were not offering a good solution to the problem.
- ✓ Next step was to evaluate DMA (Dual Matrix Array) probes with varying frequencies/sizes to compare effectiveness and come up with an inspection plan. These probes operate by producing longitudinal waves, but in a T/R Mode.

Benefits of these types probes are typically:

- ✓ Ability to penetrate coarse grain structures
- ✓ Higher signal to noise ratio
- ✓ Near Surface Resolution

Potential Drawbacks:

- ✓ 1st Leg information only
- ✓ Mode converted signals can be difficult to interpret
- ✓ Oversizing/Distortion due to low frequency and large aperture size.

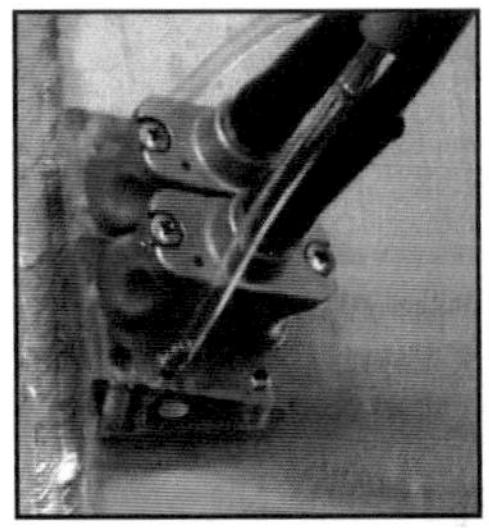

DIFFERENCE BETWEEN

DMA's versus Shear wave on large coarse grain materials

SENSITIVITY COMPARISON IMAGES

To better understand probe performance on different sections of the block, we gathered data and comparison of each, some from corners, SDHs, and OD notches.

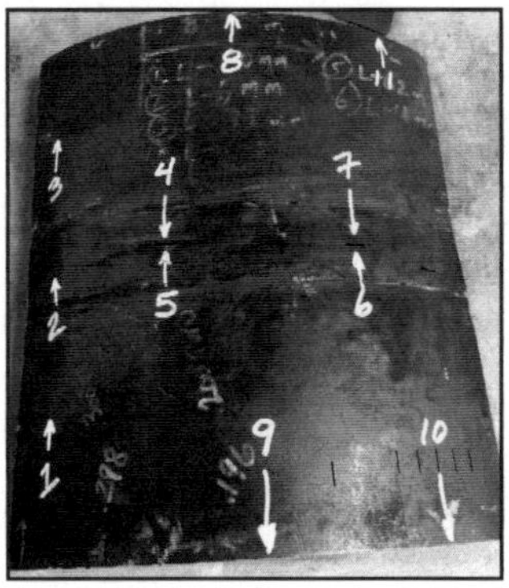

SENSITIVITY COMPARISON IMAGES (CORNER)

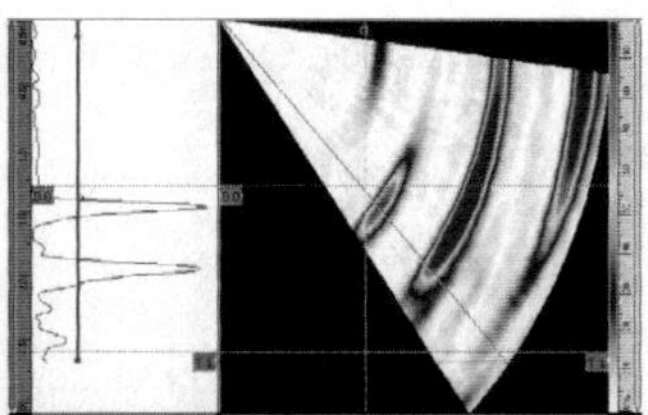

DMA Position 10

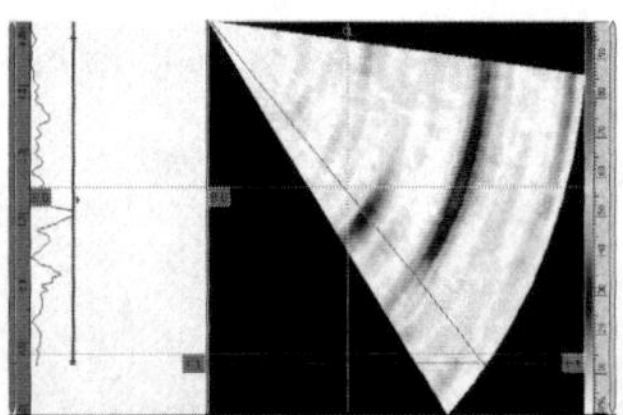

DMA Position 8

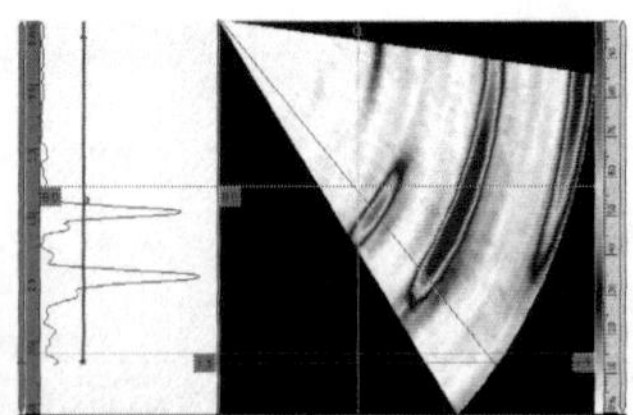

DMA Position 11

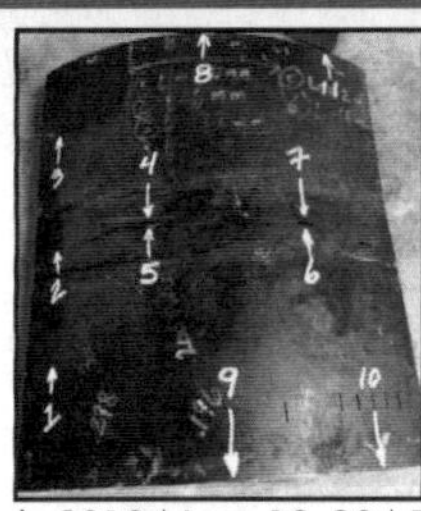

SENSITIVITY COMPARISON IMAGES (SDH)

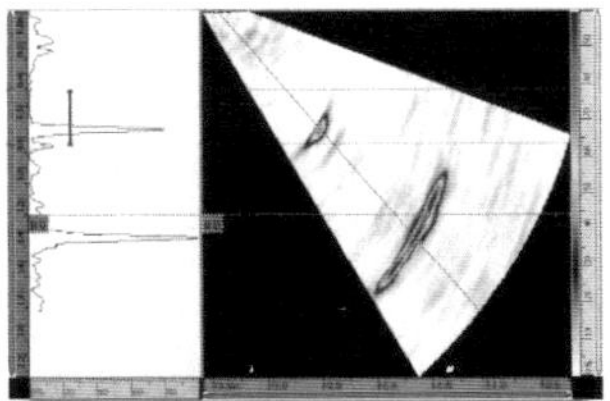
Shear Ref

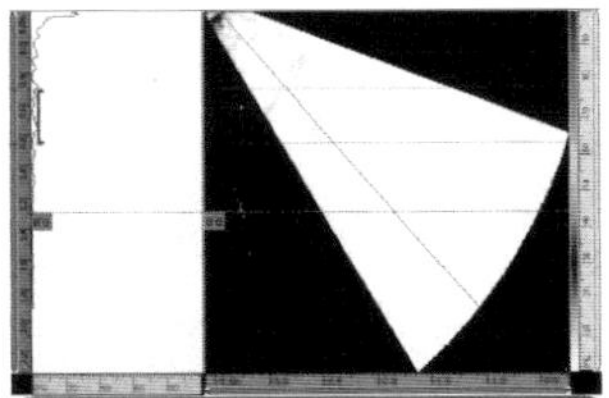
Position 2

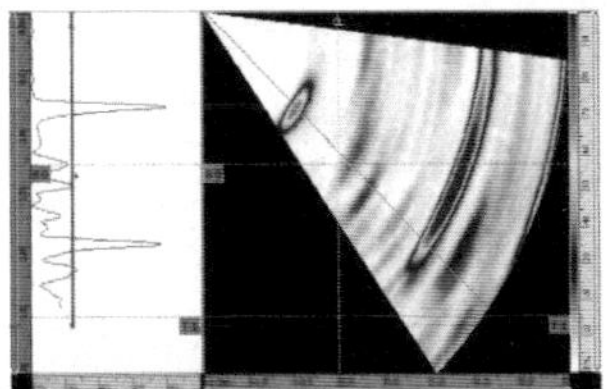
DMA Ref

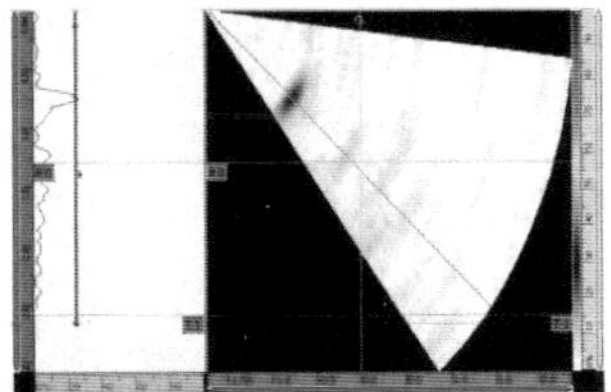
Position 2

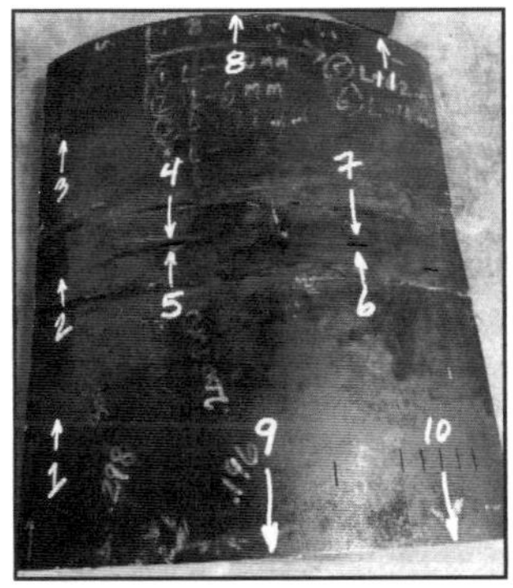

RESULTS

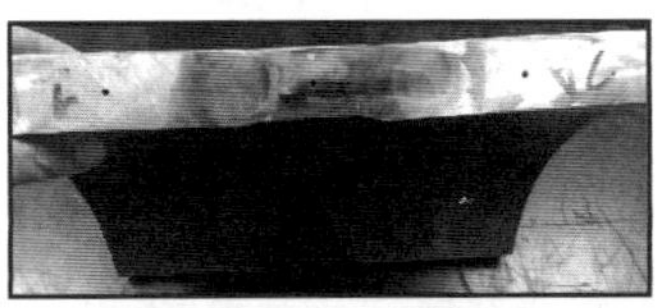

Reference sample=2" T 304 SS, Field Sample=1.5" T 304 SS	Shear wave 2.25Mhz x 1/2" Dia	DMA-1 1.5Mhz x 0.75" x .050"	DMA-2 4Mhz x 0.625" x 0.238"	Pulse Echo PA 2Mhz x 0.75" x 0.80"	0 Degree base material 2Mhz x 1.0" Dia
Commercially produced Reference Weld Sample, 4% Notches/ 0.125"SDH	0	0	0	0	0
Sample Side A SDH (BM)	-5.4 db	-5.9 db	-7.4 db	-4.3 db	-15 db
Sample Side B SDH (BM)	-23 db	-13.3 db	-16.2 db	-20.8 db	-20 db
Weld SDH	-21 db	-9.7 db	-12 db	-18.7 db	-
Sample Corner A (BM)	-11 db	-3.8 db	-6.4 db	-4.8 db	-
Sample Corner B (BM)	-39 db	-14.5 db	-17.7 db	-24.8 db	-

POTENTIAL OUTCOME

We need to open up our thought process and ask ourselves:

- ✓ Have similar inspections been conducted, in which nothing was found?
  - ✓ How do we ensure that nothing means nothing?
- ✓ Have all the proper processes been followed to ensure we didn't mistake a good inspection for a "bad" one?

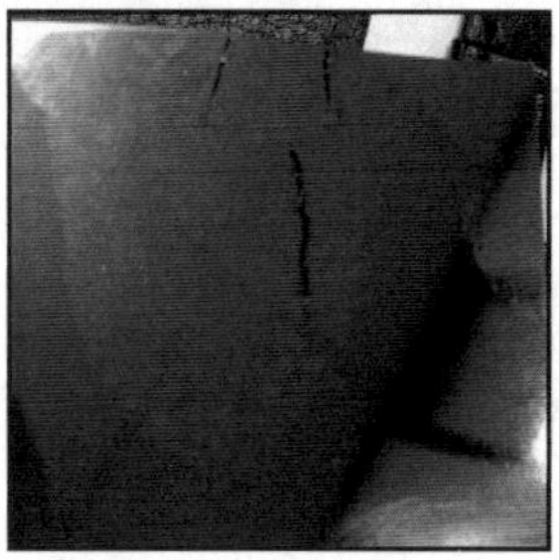

NDT in Canada 2019 | June 18-20 | Edmonton, AB

TAKEAWAYS

- ✓ Inspection results might not reflect reality
- ✓ Expectations on time to complete inspections need to be understood better
  - ✓ Focus on the result, rather then a result
  - ✓ Don't forget that results are feeding remaining life calculations or inspection intervals.
- ✓ There are ways to validate materials to raise level of awareness including:
  - ✓ Attenuation
  - ✓ Graininess
  - ✓ Beam Redirection
  - ✓ Velocities

NDT in Canada 2019 | June 18-20 | Edmonton, AB

Thank You!

Questions or Comments?

NDT in Canada 2019 | June 18-20 | Edmonton, AB

HATCH

STRUCTURAL HEALTH ASSESSMENT IN HEAVY INDUSTRIES

Dr. Afshin Sadri
Hatch

NDT in Canada
NDTiC 2019
Inform - Engage - Advance
June 18 - 20
River Cree Resort & Casino
Edmonton, Alberta

CINDE
CANADIAN INSTITUTE FOR NON-DESTRUCTIVE EVALUATION

HATCH

Content

- Purpose of SHM
- Methods of SHM
- Case Study on Industrial Application
- Conclusion

NDT in Canada 2019 | June 18-20 | Edmonton, AB

Purpose of SHM

HATCH

- Detect deterioration of structures due to corrosion, weathering, fatigue, chemical attack
- Locate damage zones
- Provide structural assessment for maintenance and rebuild plans

NDT in Canada 2019 | June 18-20 | Edmonton, AB

Methods of SHM

HATCH

- Periodic techniques
  - Visual Inspection
  - Ground Penetrating Radar
  - Impact Echo
- Continuous techniques
  - Strain gauges
  - AE monitoring
  - Fiber Optic Sensing

NDT in Canada 2019 | June 18-20 | Edmonton, AB

Case Study of Industrial Application

HATCH

- Six Cylindrical Concrete Silo (1957)
- One Steel Silo (1959)
- Subject to Typical Canadian Freeze and Thaw Condition for >60 years

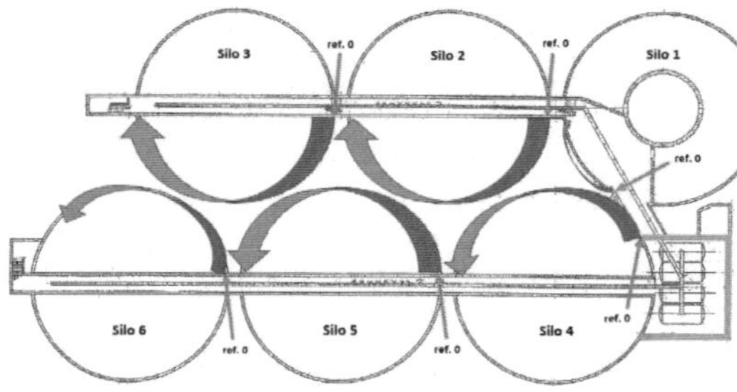

NDT in Canada 2019 | June 18-20 | Edmonton, AB

Case Study of Industrial Application

HATCH

- Assess Structural Health suitable for Service Life Extension?

Comprehensive NDT program:

- Visual Inspection
- Ground Penetrating Radar
- Impact Echo
- Schmidt Hammer Rebound Test
- Ultrasound Test

NDT in Canada 2019 | June 18-20 | Edmonton, AB

Visual Inspection

HATCH

- Locate Visible Damages and Main Structural Elements for further NDT Inspection
- Reduce Cost of NDT Inspection

Exterior Damages on Wall

Interior Damaged Wall

GPR

HATCH

Machine Room Roof 2D GPR Results

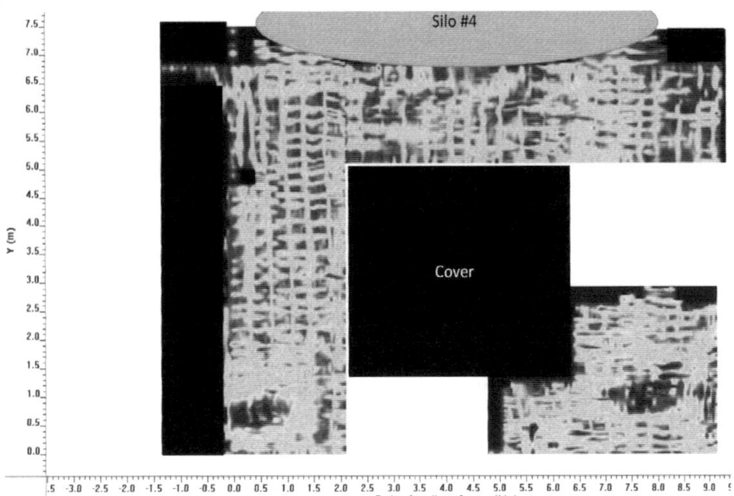

GPR Results of Delamination Map

HATCH

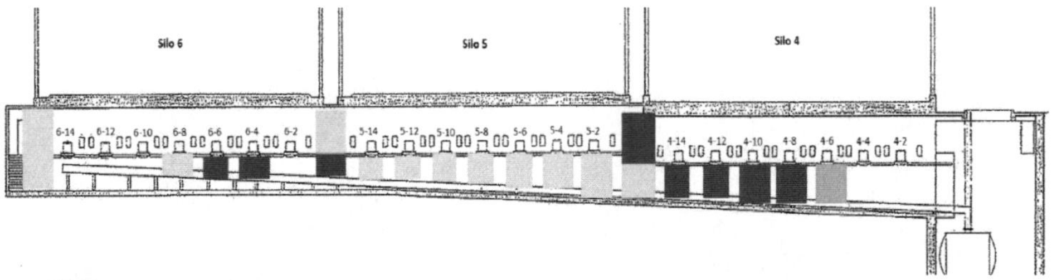

Severely Delaminated

Moderately Delaminated

Consolidated Concrete

Impact-Echo (IE)

HATCH

- Determine Cracks or Voids

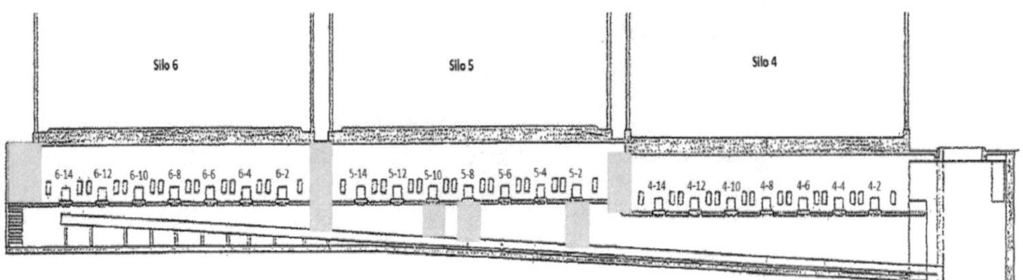

Anomalies Location

NDT in Canada 2019 | June 18-20 | Edmonton, AB

Schmidt Hammer

HATCH

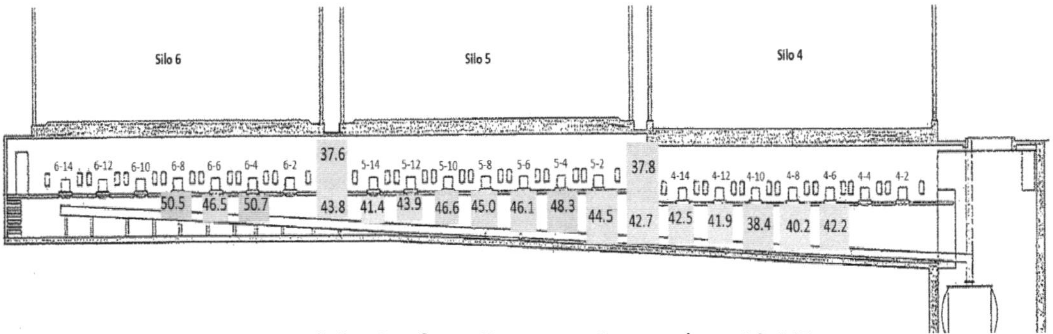

Most of the Surface Concrete Strength > 40 MPa

NDT in Canada 2019 | June 18-20 | Edmonton, AB

Ultrasound Test

HATCH

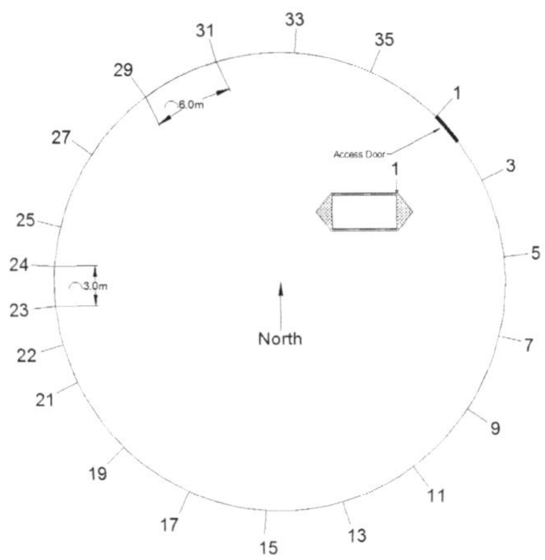

543 UT Measurements
Average Thickness Loss is 0.4 mm

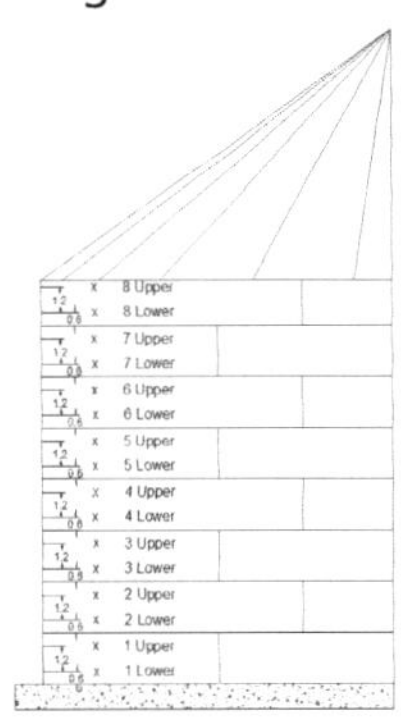

Corrosion and Deteriorated Membrane

NDT in Canada 2019 | June 18-20 | Edmonton, AB

Conclusion

HATCH

Comprehensive SHM Program:

- Prolong Campaign Life
- Plan for Maintenance → Timely Mitigation Measures
- Localize the Extent of Repair → Reduce cost

NDT in Canada 2019 | June 18-20 | Edmonton, AB

+

For more information,
please visit www.hatch.com

Dr. Afshin Sadri, P. Eng.
Global Director of NDT Technologies
Hatch Ltd.
Mississauga, Ontario, Canada

HATCH

Comparison between Two Methods of Non-Destructive Evaluation of Standing Trees

Shadabeh FATHI[1], Ferenc DIVOS [2] Laszlo BEJO[3]

[1]Institute of Wood Based Products and Technologies, University of Sopron, Hungary
E-mail: shadabe.fathi@gmail.com
[2]Head, FAKOPP Enterprise, Sopron, Hungary
E-mail: divos@fakopp.hu
[3]Institute of Wood Based Products and Technologies, University of Sopron, Hungary
E-mail: bejo.laszlo@uni-sopron.hu

Abstract

The main topic of this investigation is the root safety factor (SF) of forest trees, which was measured by dynamic methods. For this purpose, 5 decayed Ash trees were selected from Donaudorf regia in Vienna, Austria. The root stability was measured by the Pulling test and the DynaRoot system. The pulling test involves applying a bending load on the trunk via a cable attached to the tree. The method can be used to assess the uprooting stability of the tree. The dynamic root stability determination (DynaRoot), however, is based on real life wind loads, and therefore more appropriate for assessing real life risks. The results from the two methods were compared to each other. To achieve the best result, the crown surface was estimated from pictures using the ArborSonic 3D software and the results added to pulling test data for error minimization. The results from the two types of equipment correlated relatively well This means that both pieces of equipment are useful and safe for measuring the root stability of trees.

Keywords: Pulling Test, DynaRoot system, Safety Factor, ArborSonic, Ash trees

1. Introduction

Dynamic analysis is a term used in software engineering to describe the testing of the dynamic behavior of code [7]. Dynamic analysis may also refers to the examination of the physical response from the system to variables that are not constant and change with time.
The sun provides light for plants. Plants provide food for animals that are, in turn, consumed by other animals. Therefore, destroying the environment would disrupt the food chain system. Ensuring the safety of trees can help protecting all of the valuable components of the ecosystem. The dynamic method offers an efficient way to achieve this goal. Forests play a key role in mitigating or preventing the impact of a natural hazards, including falling rocks, avalanches, erosion, landslide, debris flow or flooding, on people and their assets in mountainous areas. A protection forest generally covers the sloping area between a hazard potential (e.g. an unstable rock cliff or an avalanche release zone) and the endangered or exposed assets. In the Alps, protection forests are increasingly considered equal to engineered mitigation measures against natural hazards. Some European countries consider trees important in protecting soil and to preventing it from eroding or blowing away. Austrian law differentiates between a protected forest (Schutzwald) and a protection forest (Bannwald). A protected forest is largely protecting the natural features of the forest itself. Whereas a protection forest is seen as also having a protective function, it is specifically said to protect against a concrete threat that is specified in the Bann or protection order. [1] Nowadays, there are many ways for protect and predict the defects.
The dynamic evaluation is clearly safer and more functional than the static tests. An excellent literature review by Kent et al. [5] summarizes research activity in dynamic tree evaluation These models rely on physical models or finite element modeling and account for many of the species, crown

area, age, pruning, site conditions, climate and seasonal factors. However, taking all of these factors in consideration is not possible from a practical point of view. [4, 5, 6] The behavior of a tree in the wind is highly sensitive to the initial conditions. Small differences in initial conditions yield widely diverging outcomes for such dynamic systems (a situation popularly referred to as "the butterfly effect"). It is impossible to predict the behavior of such systems on the long run. A definite relationship exists between the wind velocity and the inclination of the tree, but it is a complex relationship, not one of immediate cause and effect. One way to evaluate such systems is using statistical parameters of the data taken over longer intervals. There is no immediate relationship between the wind velocity and the inclination at any given moment, but there is a relationship between their averages and other statistical parameters observed over a longer interval. Our goal was to evaluate the root stability of trees using both static and dynamic tests, and compare the results.

2. Pulling test [3]

The pulling test (see (figure 1) consists of:

- Cable and winch: our system contains a 20-meter length of high capacity metal cable with a 1.6 metric ton manually operated winch. The winch has a ratchet mechanism that multiplies the force of the operator to exert sufficient tension on the cable. The cable and the winch were equipped with safety hooks and two soft belts for fitting it around the tree trunk and the anchor point.
- Load cell.: calibrated cable-mounted load cell, 5T capacity with a sampling rate of 1 Hz.
- Inclinometer: biaxial inclinometer sensor, ST-015 mounted on the tree collar with a measurement range of ± 2 degrees, and a Resolution of 0.001 degree.

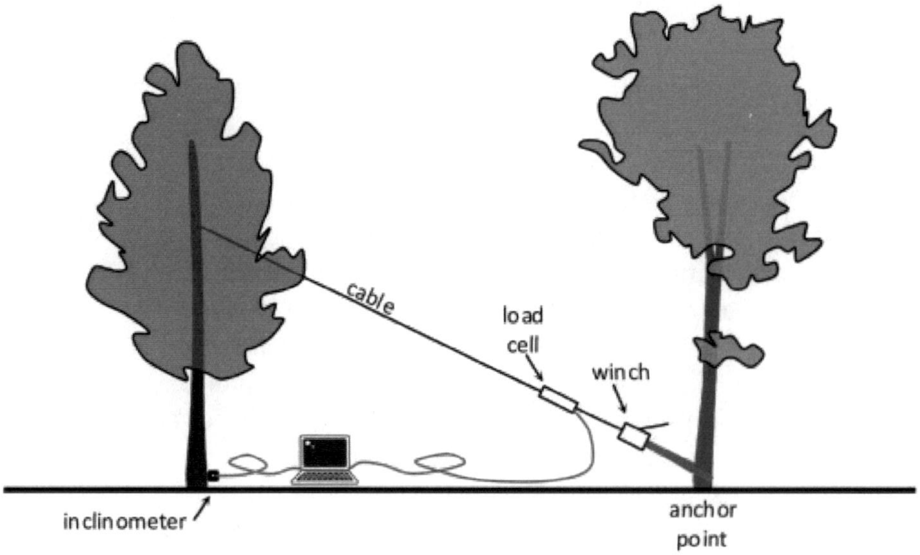

Figure 1- Schematic view of Pulling test

The pulling test is based on affixing a cable at approximately mid-height to the tree to be evaluated, and applying a moderate load, while measuring the inclination at the base of the trunk. The induced inclination is slight (less than .25 degrees), to make sure that the test itself does not damage or start uprooting the tree. The cable is attached to the tree at approximately mid-height in the crown. This typically requires a ladder, or climbing the tree to the appropriate height. A metal cable of appropriate loading capacity is attached to the trunk. A soft belt is typically used for this to avoid damaging the

tree. The other end of the cable enters a winch, which is affixed to an anchor point. The anchor point can be any object that is safely secured to the ground, most often a stump or the bottom of another tree. If another tree is used, care should be taken that the bark is not damaged. The winch applies tension to the cable. A load cell attached to the cable measures the tensile load. Since the cable is at an angle, the horizontal component of the load is calculated and used for the evaluation.
Load and inclination is continually measured throughout the measurement and sent to a computer for recording and evaluation. The instruments send the data is to a computer where it is evaluated real time and also recorded. The recorded load and inclination data provides the inclination curve, which can be approximated using a special tangential function. The maximum load – and from that, the maximum torque – required for uprooting the tree can be estimated by extrapolating this curve. From this value, a so-called safety factor (SF) can be calculated, as explained in detail in [3]. SF is always calculated with reference to a given wind velocity, taking the crown surface area and other parameters into account. A SF below one means that the reference wind velocity is likely to uproot the tree. When SF above 1.5 are considered safe, and in-between these two values there is a region of uncertainty [2].

3. DynaRoot [3]

The DynaRoot system consists of three components (see Figure 2):

- Anemometer: an instrument for measuring wind velocity at or near the tree to be evaluated. The closer to the better, but, depending on wind velocity DynaRoot may provide reliable data even with measurements taken several kilometers away. Ideally the anemometer should be clear of buildings or other objects that may obstruct the wind, at a height of at least 10 m.
- Inclinometer (the same as used in the pulling test)
- Evaluation software: a PC software for evaluating wind velocity, x and y inclination. The data, recorded over a period of several hours, are transferred from the anemometer and inclinometer on memory cards or wirelessly. The software breaks the data down into batches based on intervals of several minutes, and calculates statistical parameters for each batch. These statistical parameters are used for the tree stability evaluation.

The Safety Factor calculation is similar to that in the pulling test, except, in this case, wind pressure is used instead of force, and statistical parameters are used, instead of the momentary wind pressure and inclination values. There is a tangential relationship between the wind pressure and the inclination of the tree, and the critical wind pressure can be calculated from the curves. This critical value is used for calculating the SF, which is interpreted the same way as the one calculated from the static pulling test.
The following conditions were observed during the measurements:

- The distance between the measured trees and the anemometer, was less than 1 km.
- Wind velocity, was between 40 and 50 km/h
- Wind load and inclination data was statistically evaluated based on 10 min intervals.
- All of the trees were measured on the same day, and the wind direction was 45 degrees off east (NE).

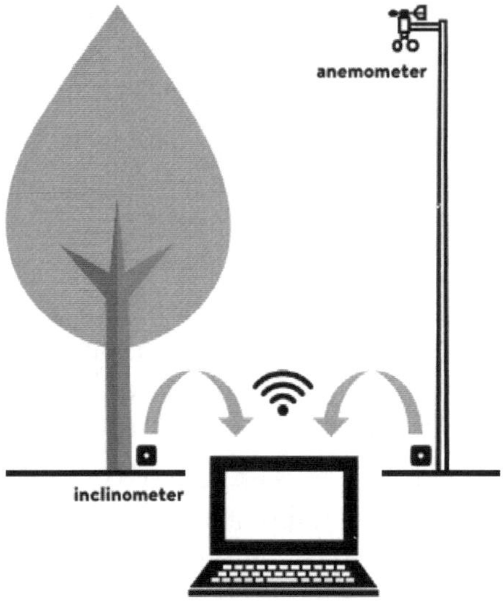

Figure 2- Schematic view of DynaRoot system

The reference wind velocity for the safety factor calculation was 33 m/s (approx. 120 km/h)

In this study, 5 severely decayed ash trees were evaluated whose SF was likely to be low at the reference wind velocity. Before the analysis of the pulling test data, the crown surface and crown center height values were calculated using the ArborSonic 3D software, which can calculate the crown surface and crown center height of a tree based on photographs. The crown surface was added to the results of pulling test, and then results obtained from DynaRoot and the Pulling test were compared to each other.

4. Result and discussion

Figure 3 shows sample output information from the Arborsonic software (crown surface determination), the pulling test and the dynaroot system.
After completing the pulling test and Dynaroot system, the safety factors were calculated based on both for each tree, and compared to each other. The results are plotted in Figure 4. The diagram shows relatively good agreement between the two sets of results, although the safety factors calculated from the pulling test were generally somewhat higher than those based on DynaRoot. There are several possible reasons for the discrapancies:

- Pulling direction was not exactly the same as the wind direction. It was not always possible to find an anchorage point in exactly the prevalent wind direction.
- Dynamic testing is affected by nearby objects (other trees) that may provide some shading from the effect of the wind, while the pulling test does not take these into consideration
- Because dynamic testing results show a greater degree of variability, SF calculation is based on the lower confidence limit of the results, which accounts for the lower SF values calculated by this method.
- Finally, the different nature of the two measurements will inevitably lead to some differences in the results.

A)

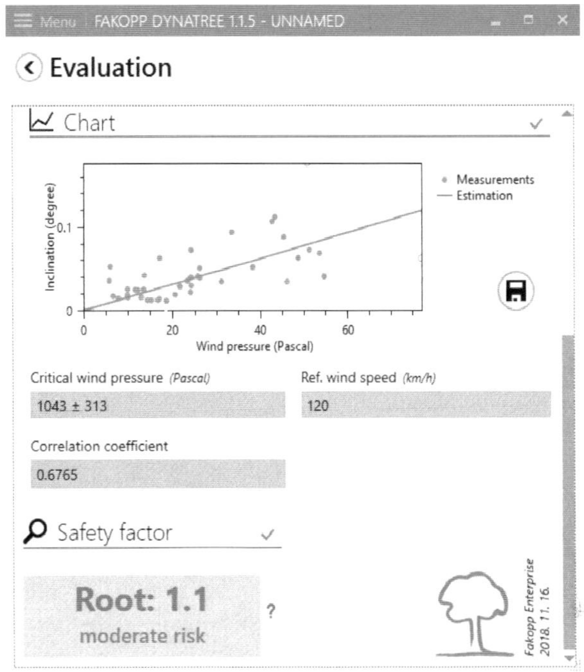

B)

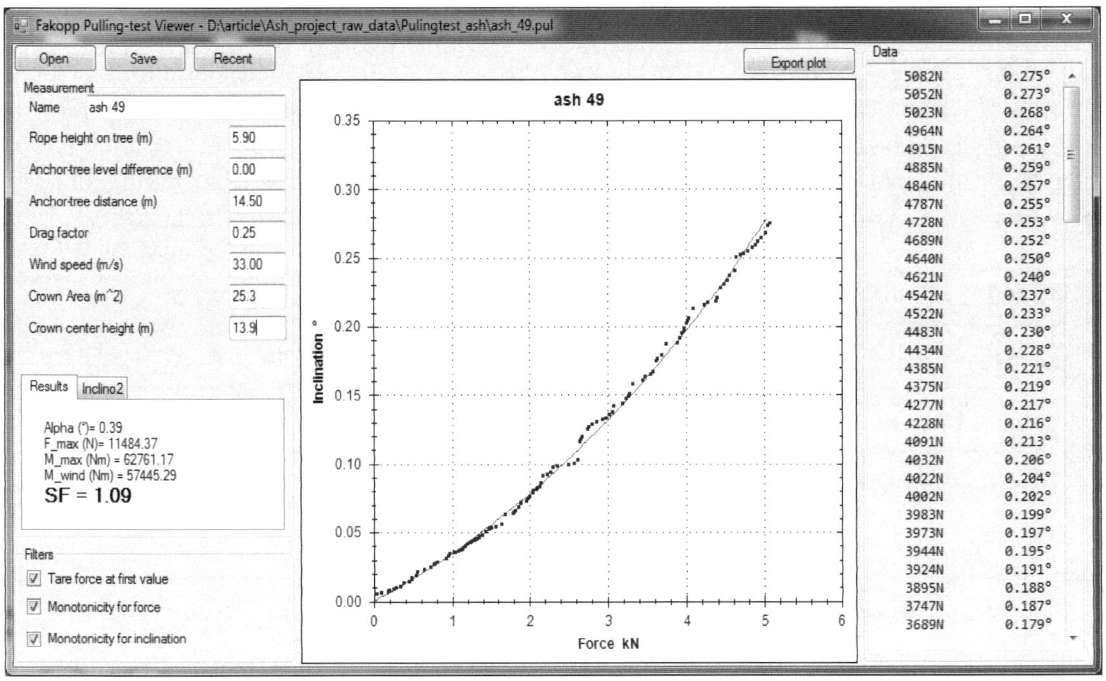

C)

Figure 3 The result of tree Nr. 1. A) Crown Surface and crown midpoint height is calculated using the Arborsonic software B) DynaRoot results analysis C) Pulling test curve analysis

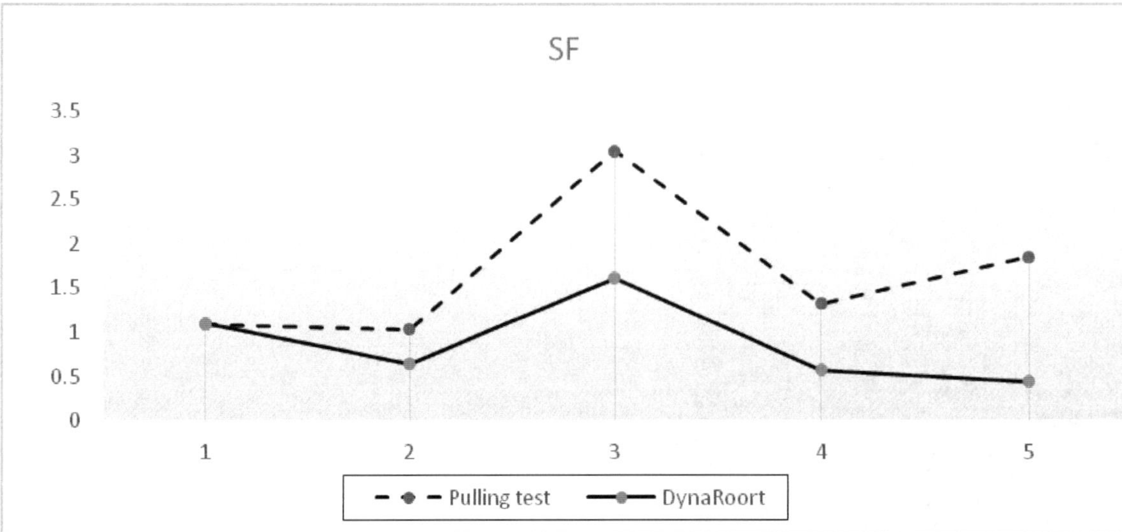

Figure 4 - Comparing the SFs obtained from Pulling test and DynaRoot system

Despite the discrepancies between the two systems, the SF values in Figure 4 show a relatively good correlation. Without actually uprooting the trees, it is difficult to assess which method is more accurate. However, Dynaroot provides a more conservative estimate, and the low SF values are likely to be closer to reality, as the examined trees were severely decayed.

Another way of comparing the pulling test and the DynaRoot system is by calculating the critical torque values, at which the tree is predicted to be uprooted. The critical torque is actually part of the output of the pulling test analysis. It may also be calculated from the DynaRoot results, based on the critical pressure, using the crown surface area and midpoint height, also taking the so-called aerodynamic drag factor into consideration. Table 1 shows the results of the analysis.

Table 1- Detailed results obtained by the Pulling test and the DynaRoot system

Tree Nr.		1	2	3	4	5
ArborSonic	A (m^2)	25.3	27.2	21.3	13	9.4
	cch (m)	13.9	14.1	11	11	8.7
Pulling test	F_{max} (N)	11484	13245	22113	5889	4817
	M_{max} (Nm)	62761	64401	116732	30815	24695
DynaRoot	P_{crit} (Pa)	1043 ±313	572 ±144	1460 ±383	476 ±94	1329 ±1038
	F_{crit} (N)	6596	3889	7774	1547	3123
	M_{crit} (Nm)	91697	54843	85519	17017	27171

A – crown surface area, cch – crown centerpoint height F_{max}, M_{max} = Predicted uprooting force and torque, pulling test; P_{crit}, F_{crit}, M_{crit} = Predicted uprooting pressure, force and torque, DynaRoot

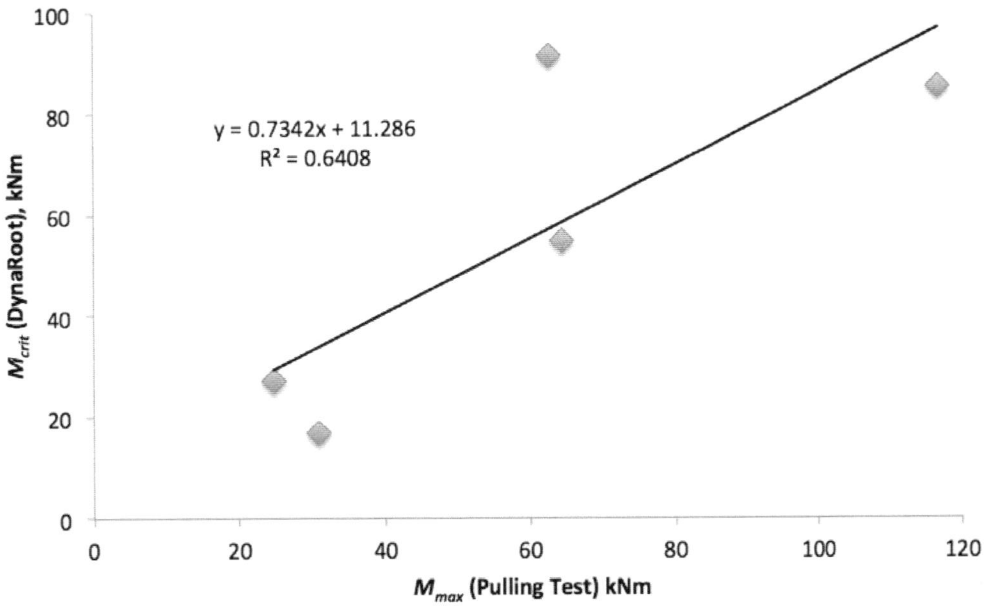

Figure 5 - the compare between the Torque of Dynaroot system and Torque of pulling test

The critical torque values determined based on the two stability evaluation methods showed very good correlation (Figure 5). The correlation coefficient was 0.64, which is acceptable when dealing with natural materials with a high degrees of variabilty. In most cases Dynaroot provided similar or lower uprooting torque values, which further confirms that Dynaroot provides a more conservative estimate for the stability of trees.

Conclusions

5 severely decayed Ash (*Fraxinus Excelsior*) trees were evaluated for stability using the traditional pulling test and an innovative dynamic tree stability evaluation method. The comparision of the two methods yielded the following conclusions:

- The two methods yielded different, but comparable Safety Factor (SF) values at the same reference wind velocity level. Differences were due to differences in measurement direction, the effect of possible shielding on dynamic measurements, and differences inhetent in the two methods.
- The dynamic method provides more conservative SF values.
- Uprooting torque estimates correlate reasonably well between the two methods (R^2 = 0.64) The dynamic method, again, typically, but not consistenty, provided more conservative estimates.

Further investigations, including the measurement of actual uprooting force, will provide more of the necessary evidence for the reliability of the two methods.

Acknowledgments
The research described in this article/study was carried out as part of the project GINOP-2.3.3-15-2016-00038, "Further processing of wood and wood products based on green chemistry and technology, through creating modern research infrastructure" in the framework of the Széchenyi2020 Programme. The implementation of this project is supported by the European Union, co-financed by the European Regional Development Fund.
Instruments, including Arborsonic and DynaRoot, were supplied by FAKOPP enterprise

References

1. Bundesgezetz vom 3. Juli 1975, mit dem das Forstwesen geregelt wird (Forstgezetz 1975) (Austrian Forest Law)
2. Bejó, L., F. Divós, S. Fathi. 2017. Dynamic root stability assessment – basics and practical examples. In: Proc. 20th International Nondestructive Testing and Evaluation of Wood Symposium. Madison (WI), USA, 2017.09.12-15. 7pp.
3. Fakopp Ltd., DynaRoot Dynamic Root Evaluation System, User's manual http://fakopp.com/docs/products/dynaroot/UserManualDynaRoot.pdf
4. Baker, C.J., and H.J. Bell. 1992. Aerodynamics of urban trees. Journal of Wind Engineering and Industrial Aerodynamics 44:2655–2666.
5. James, Kent.R., N. Haritos, and P.K. Addes. 2006. Mechanical stability of trees under dynamic loads. American Journal of Botany 93(10):1361–1369.
6. Moore, J.R., and D.A. Maguire. 2008. Simulating the dynamic behavior of Douglas-fir trees under applied loads by the finite element method. Tree Physiology 28:75–83
7. Gergely T, Balogh G, Horvath F, Vancsics B, Beszedes A, Tibor Gyimothy, 2018 Differences between a static and a dynamic test-to-code traceability recovery method, Software Quality Journal https://doi.org/10.1007/s11219-018-9430-x , Springe

Le génie pour l'industrie

Evaluation of the dead zones in the welded joints of hydraulic turbine runners

By : Mohammad Ebrahim Bajgholi

Département de Génie Mécanique, École de Technologie Supérieure (ETS)

CANADIAN INSTITUTE FOR NON-DESTRUCTIVE EVALUATION

NDT in Canada
NDTC 2019
Inform - Engage - Advance
June 18 - 20
River Cree Resort & Casino
Edmonton, Alberta

Outline

- Problem Statement
- Introduction
- Results and discussion
- Conclusions

NDT in Canada 2019 | June 18-20 | Edmonton, AB

Problem Statement

- In Quebec province, the vast majority of electrical power is produced by hydraulic turbines.
- In an attempt to insure equipment reliability and to control production costs, IREQ has developed a life estimation model which is based on the distribution and the size of structural discontinuities in turbines.
  - Undetected flaws may lead to unplanned failure.
  - Unplanned failure results in high repair costs and long down time.

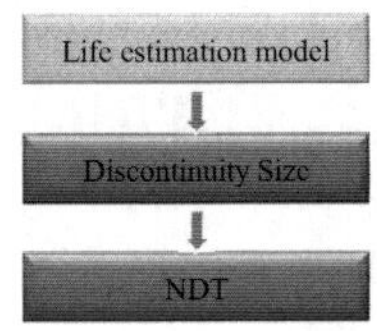

- NDT methods are used to detect, locate, and size flaws.
- Reliability of a specific NDT method depends on flaw type, size, and location and it is evaluated through POD analysis.
- The current project aims at providing IREQ with reliable flaw data (both measured and simulated) for their life estimation model.

Introduction

Francis Turbine Runner

- The Francis turbine runners of Beauharnois hydroelectric power station will be the subject of current study.
- Runners are the heart of any turbine
- Runners: multivariable section components.
  - 15 to 90 mm blade thickness
- The turbine runner weld is a series of Y-type joints.
  - 60° to 120° angle

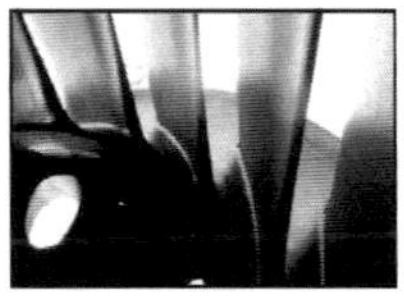

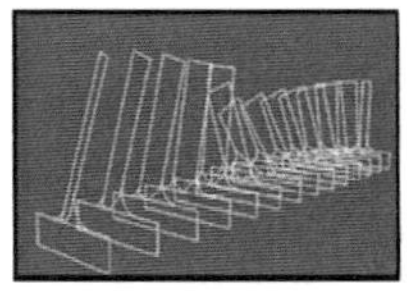

Welding Flaws

- The multipass Flux-cored arc welding (FCAW) used for turbine runner joints, generates some discontinuities.
  - Thermal cycles
  - Solidification
  - Cooling distortion
  - Residual stresses
- Welding discontinuities
  - Volumetric (Porosities, Slag inclusions)
  - Planar (LOF, LOP, Cold and Hot cracks)
- Slag inclusions and LOF are the most important discontinuities
- Welding discontinuity
- Dynamic stress concentration

  → Failure probability
- The critical weld region is approximately 300 mm from the trailing edge for both blade-to-crown and blade-to-web welds.

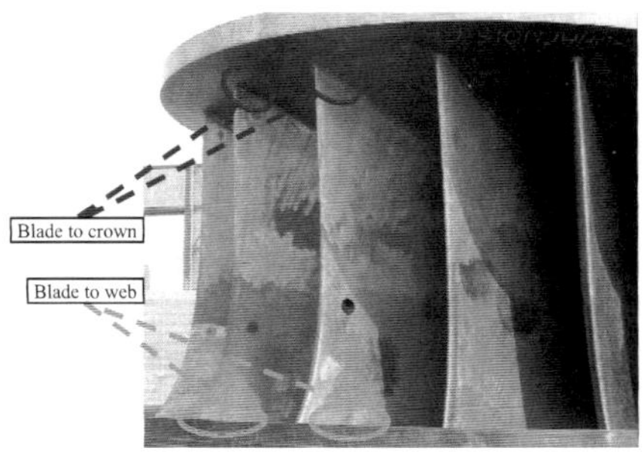

NDT Methods

Non-destructive Testing (NDT)

Technical method to examine materials or components in order to detect, locate, measure and evaluate flaws without causing damage.

Limited inspection space }
Complex geometry } Ultrasonic Testing → Lower signal quality → Low Probability of Detection (POD)

- Phased array (PA) has attracted the experts' attention
  - Capabilities in the inspection of complex shape structures
  - High detectability of randomly oriented flaws
  - Speed of testing

Scan whole thickness of the weld at one time

Advanced NDT Methods

- Different techniques with phased array in advanced ultrasonic inspection:

❖ Time-of-Flight Diffraction (TOFD):

TOFD uses the difference in the time of received waves, instead of their amplitude for defect.

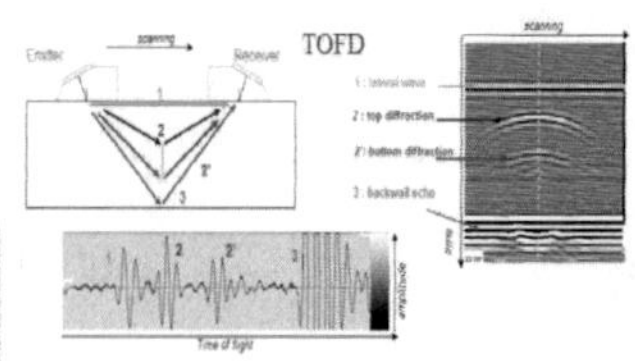

❖ Dynamic depth focusing (DDF):
Scanning is performed with different focal depths and provides higher sensitivity.

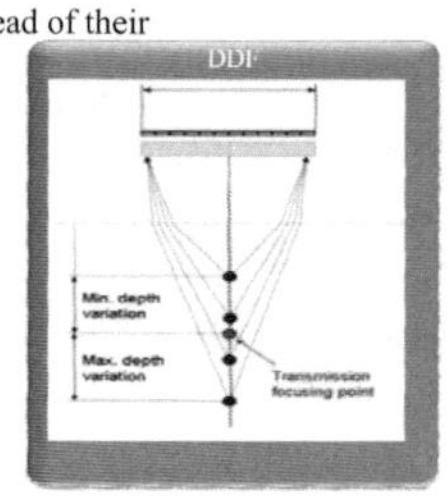

❖ Full Matrix Capture (FMC) and Total Focusing Method (TFM):
FMC is a UT data collection process and TFM is data processing to produce an image which is focused at every specified point.

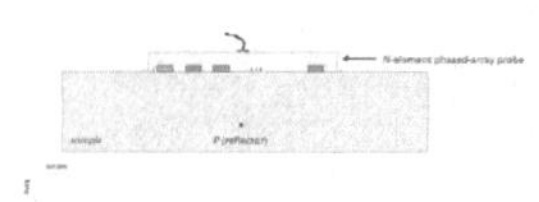

TFM

♦Problem ♦Introduction ♦objective ♦Methodology ♦Research Planning

Probability of Detection (POD)

- POD is a statistical estimation of the probability to detect, with a given inspection system, a discontinuity of a certain size with a certain level of confidence.

- POD curves can be built for two forms of data. Binary response where the information is in form of hit / miss and signal response where data is in a quantitative form to build a more detailed scenario.

Hit / Miss

- Defect is
  - Detected – Hit- 1
  - Not detected – Miss- 0
- More data required for POD
- Larger uncertainty intervals compare to signal response analysis
- ASTM E2862-12: Standard practice for probability of detection analysis for hit/miss data

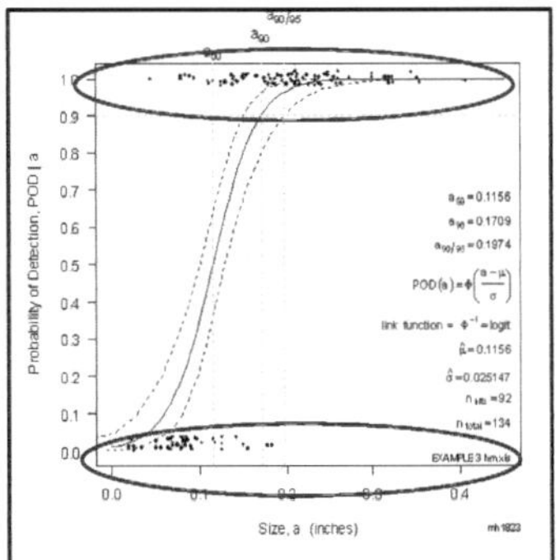

POD curves for hit/miss data (Annis,2009)

Probability of Detection (POD)

Signal response ($\hat{a}$) vs. flaw size (a)

- It has more information compare to hit or miss
- Less data points required for POD
- Maximum likelihood relationship between signal response ($\hat{a}$) and flaw size (a)
- A critical point in signal response data is the definition of the decision threshold (signal above threshold: defect, signal below threshold: noise)
- ASTM E3023-15: Standard practices for probability of detection analysis for â versus a data

Different results will be obtained when the two POD analysis methods are applied to the same data set.

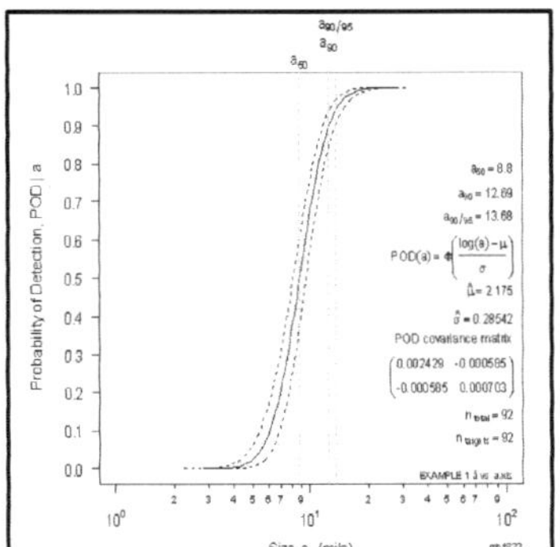

POD curves for signal response data (Annis,2009)

Probability of Detection (POD)

- Not all parameters can be considered in POD studies
- POD actual = POD intrinsic - g(AP) - h(HF)

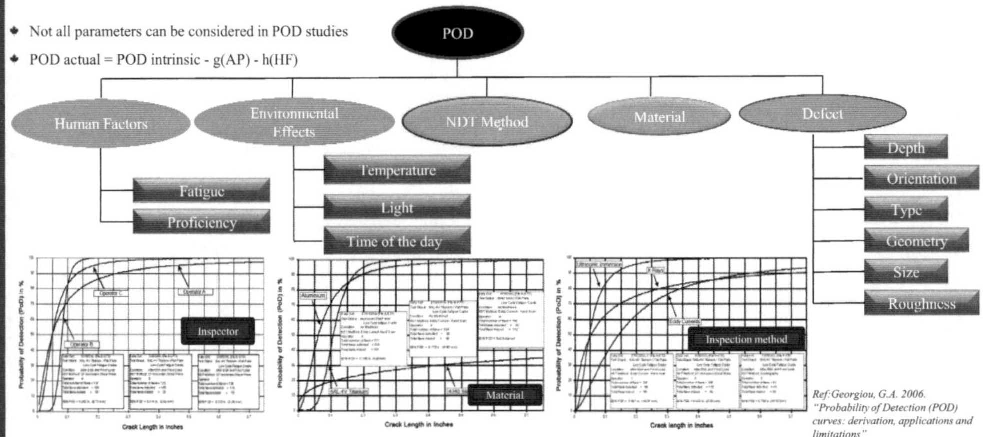

Ref:Georgiou, G.A. 2006. "Probability of Detection (POD) curves: derivation, applications and limitations"

Simulation

- CIVA can simulate the most common inspection methods and it is composed of simulation, imaging and analysis modules.
- Probes : Single element, dual element, TOFD, immersion, phased array.
- UT simulation tools in CIVA Software includes
  - **"Beam computation":** Beam propagation simulation
  - **"Inspection Simulation":** Beam interaction with flaws and specimen

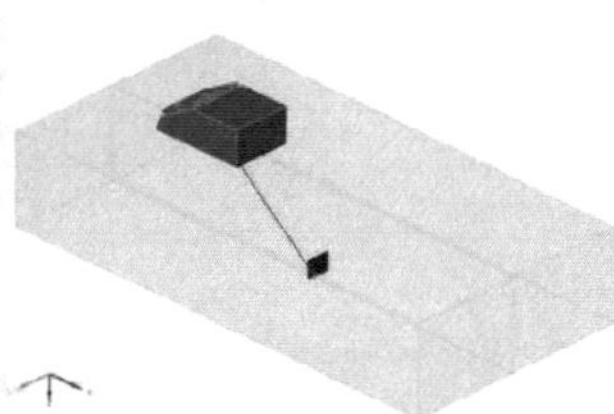

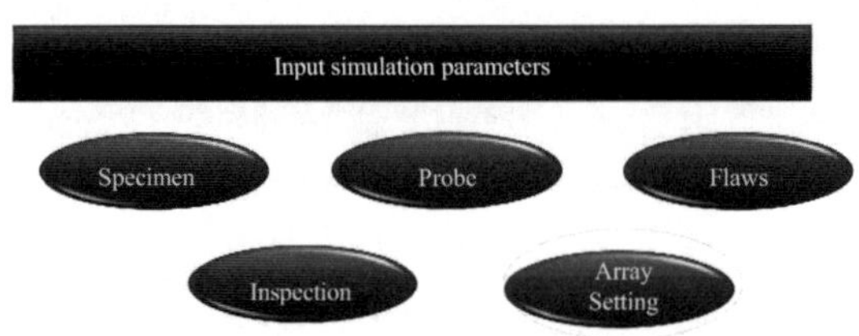

Evaluation of the Dead Zones

NDT in Canada 2019 | June 18-20 | Edmonton, AB

13

Evaluation of the Dead Zones

- Geometry of welding joint was applied to the turbine runners similar to the T-joint configuration.
- To evaluate the dead zone, T-joint configuration of martensitic stainless steel (UNS S41500) was considered.
- The 64-elements transducer probe with a frequency of 5MHz was considered in the simulation.
- The ray tracing method was performed in two steps which are called first half skip and second half skip.
- It was found that the dead zone is smaller in the first half skip.

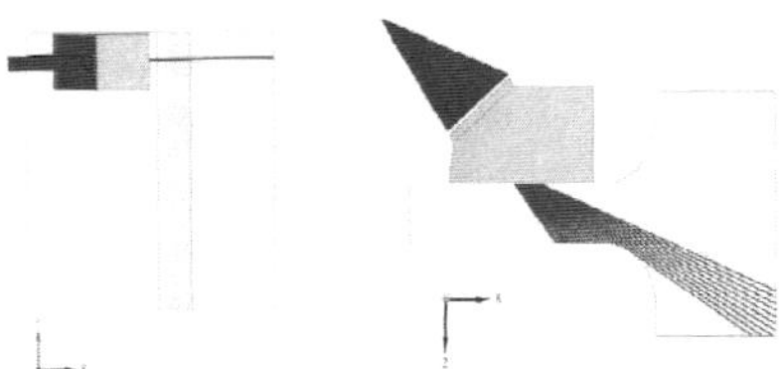

(a) (b)
First-half skips with considering of HAZ; (a) top view, (b) cross-section view

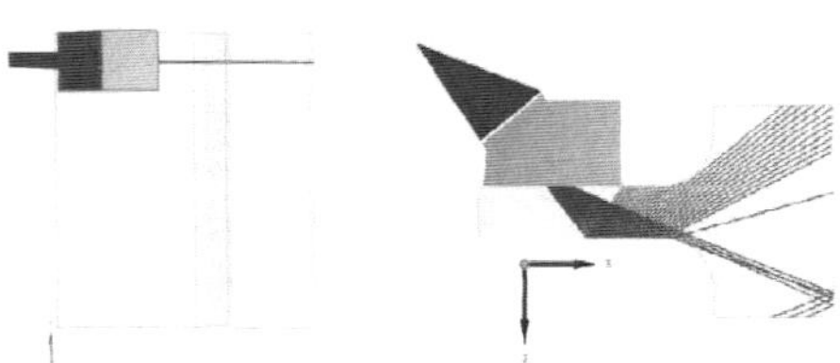

(a) (b)
Second half skip (a) top view, (b) cross-section view

NDT in Canada 2019 | June 18-20 | Edmonton, AB

14

Area 1-Dead Zones

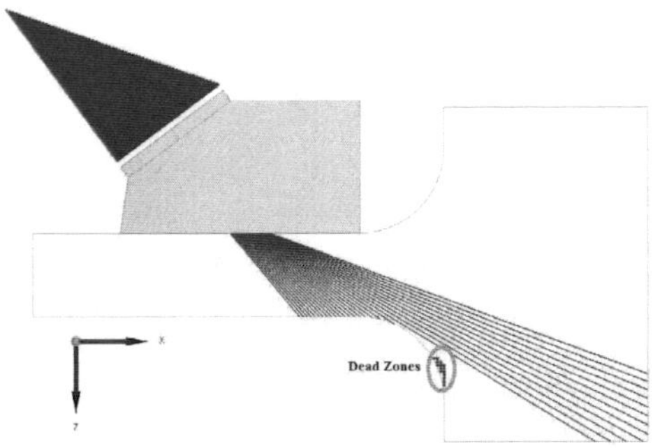

Defect Number	Positioning		
	X	Y	Z
1	-62	40	33
2	-62	40	32
3	-62	40	31
4	-62	40	30
5	-62	40	29
6	-62	40	28
7	-62	40	27
8	-63	40	29
9	-63	40	28
10	-63	40	27
11	-63	40	26
12	-64	40	27
13	-64	40	26
14	-64	40	25
15	-65	40	25

I put 15 defect on the dead zone and CIVA does not show the result in the dead zone.

Area 3-high POD

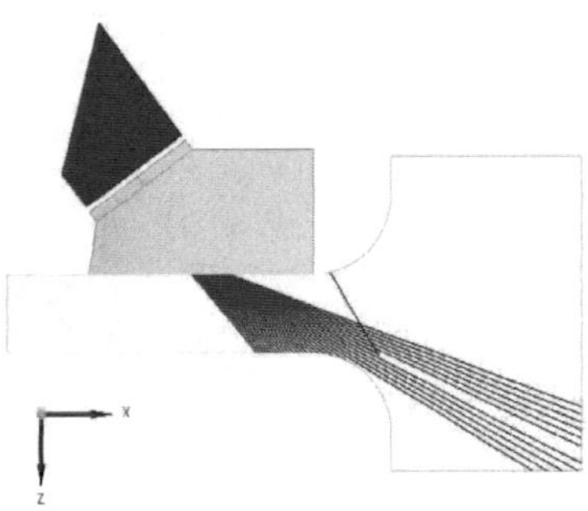

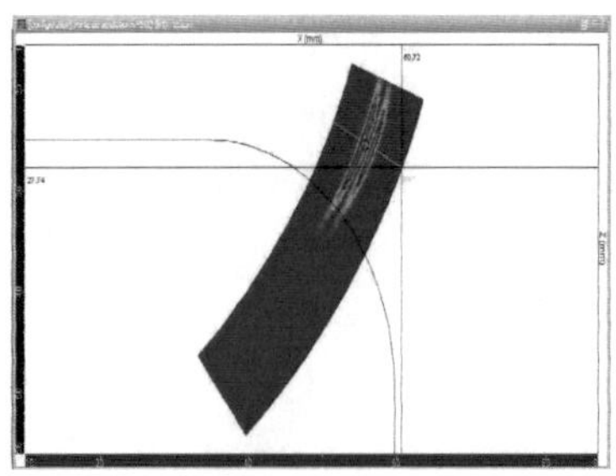

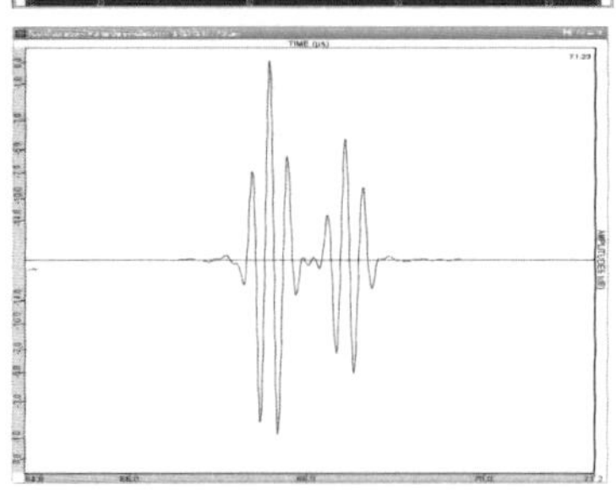

NDT in Canada 2019 | June 18-20 | Edmonton, AB

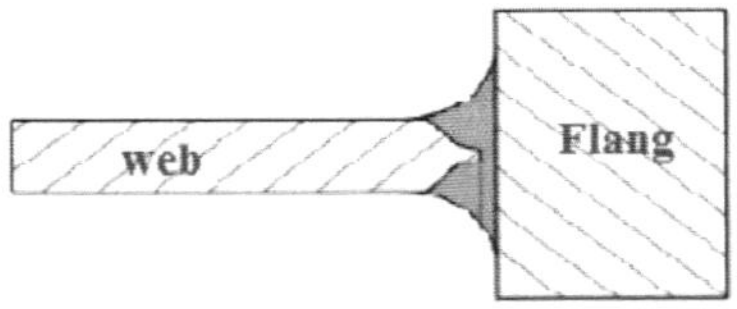

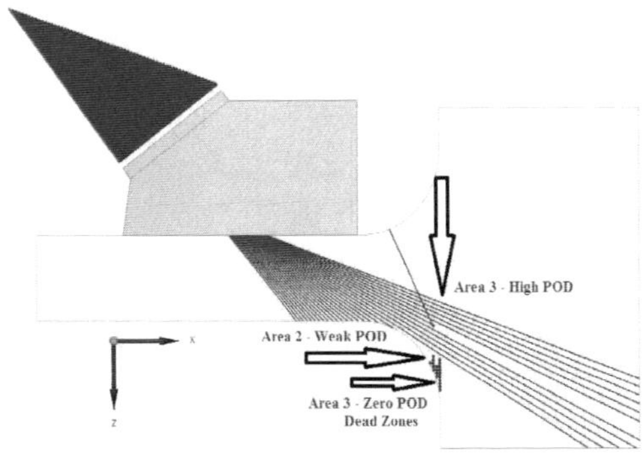

NDT in Canada 2019 | June 18-20 | Edmonton, AB

- Although, dead zones could be roughly estimated using simple ray tracing method, more accurate and realistic estimation is provided using beam computation method.

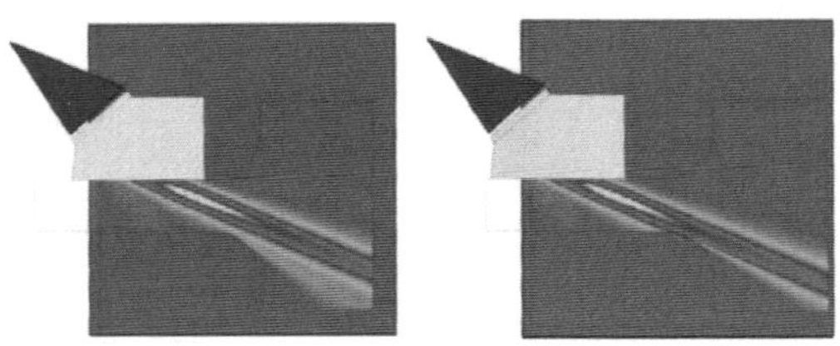

(a) (b)

Beam Computation; (a) first half skip, (b) second half skip

NDT in Canada 2019 | June 18-20 | Edmonton, AB

18

Conclusions

- The existence of dead zones is a challenging issue in the assessment of the integrity of a welded joint when these dead zones are in highly stressed areas.

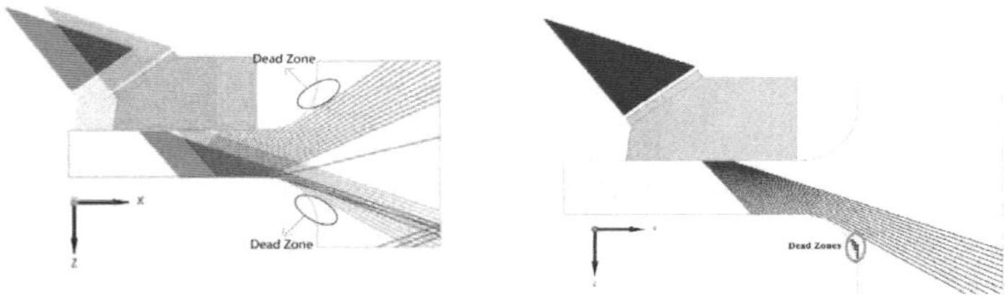

(a) Dead zone locations, (b) CIVA does not show the result in the dead zone with 15 defects

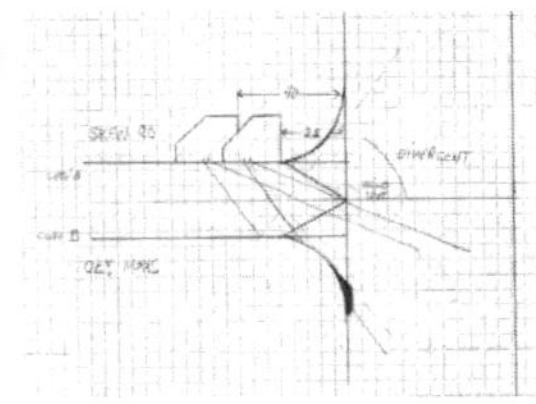

- Due to runner geometry and inspection limitations, some areas cannot be covered by the ultrasound (the tip of the blade is the most problematic area).
- CIVA does not show the result in the dead zone
- Complementary NDT methods will be proposed to avoid this issue.

NDT in Canada 2019 | June 18-20 | Edmonton, AB

A Comparative Study Between Different Guided Waves

Zeki Gokce

CINDE
CANADIAN INSTITUTE FOR NON-DESTRUCTIVE EVALUATION

NDT in Canada
NDTiC 2019
Inform - Engage - Advance

June 18 - 20
River Cree Resort & Casino
Edmonton, Alberta

A Comparative Study Between Different Guided Waves

GUIDED WAVES ON PIPELINES

- Low frequency Ultrasound (20-100kHz)
- Developed for Corrosion Under Insulation
- First introduced into the market in 1997 by TWI Ltd
- Established method for screening pipelines
- Inspection of inaccessible areas

A Comparative Study Between Different Guided Waves

CONVENTIONAL UT VERSUS LRUT

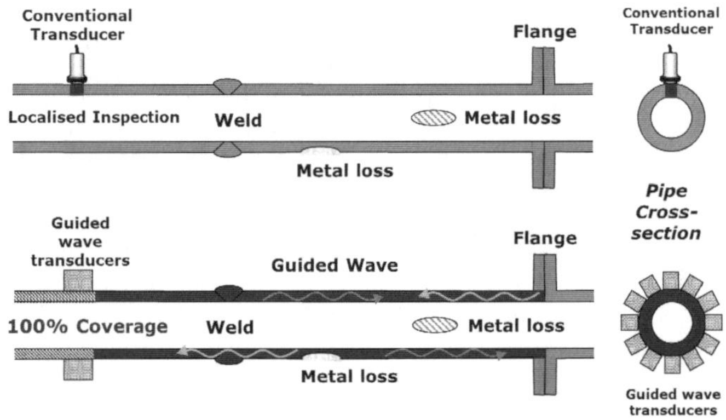

A Comparative Study Between Different Guided Waves

WAVE MODES IN PIPES

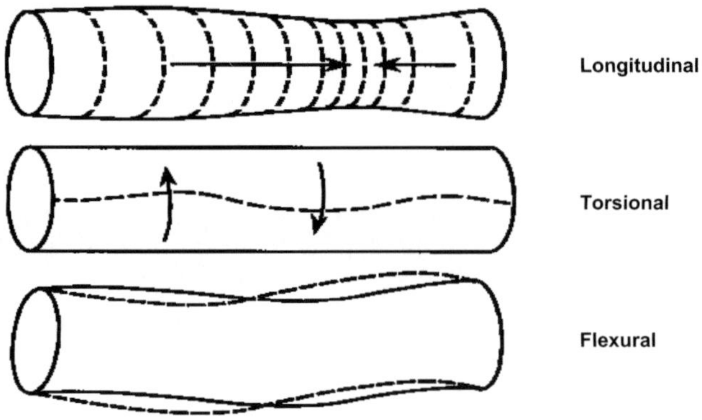

A Comparative Study Between Different Guided Waves

PIPE WALL DISPLACEMENTS

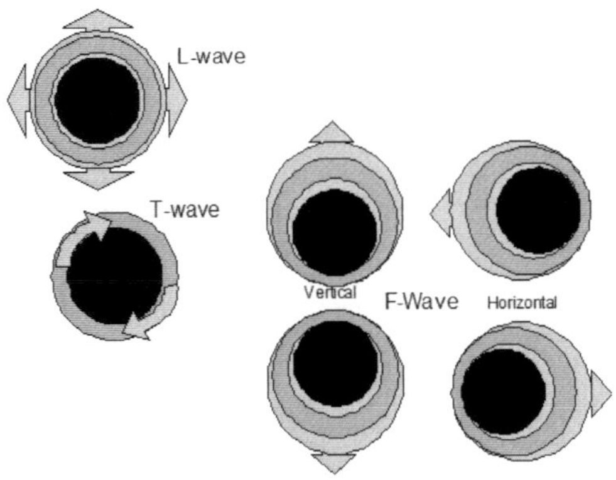

A Comparative Study Between Different Guided Waves

DISPERSION CURVES - PIPES

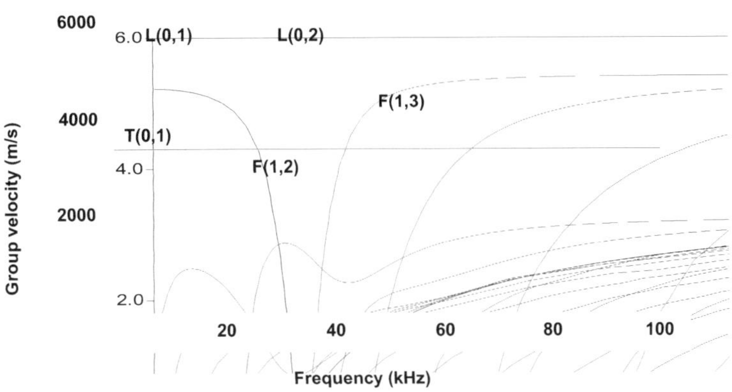

A Comparative Study Between Different Guided Waves

GUIDED WAVE RESPONSES

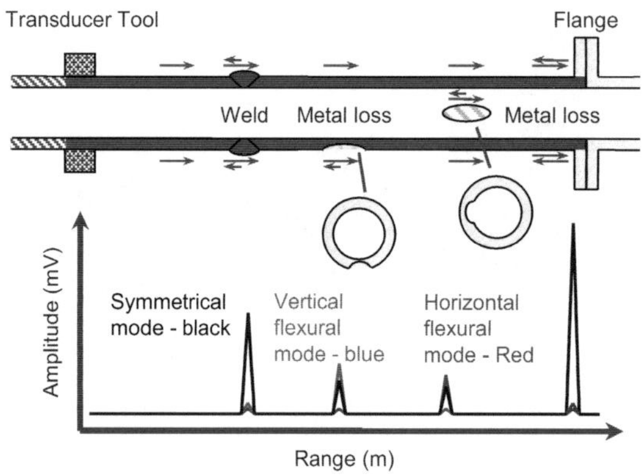

Symmetrical mode

A Comparative Study Between Different Guided Waves

DIFFERENT REFLECTED WAVES

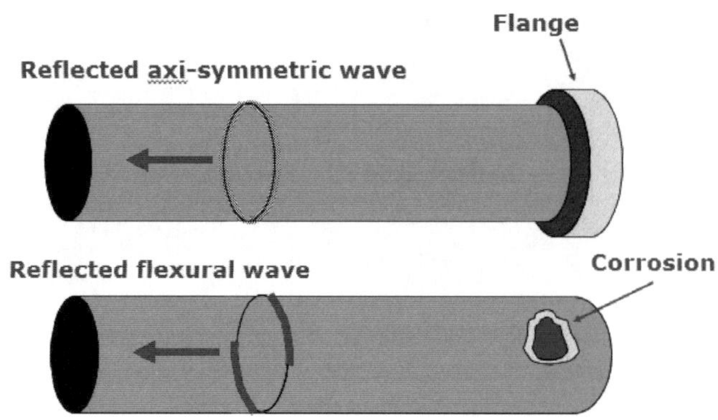

A Comparative Study Between Different Guided Waves

PROPAGATION OF AXI-SYMMETRIC WAVES

- Axi-symmetric waves, Longitudinal or Torsional, will propagate in both directions from a single ring.
- To prevent the wave from propagating in both directions a second ring is introduced.
  This ring will be phase delayed with respect to the first one.
- When using the Longitudinal wave mode it is necessary to dampen the unwanted L(0,1) wave mode.
  This necessitates a third ring.
- For optimum cancellation of L(0,1) the spacing between the rings must be equal to the L(0,1) wavelength, so the test frequency needs to be adjusted accordingly.
- The delay between the signals applied to the rings is equal to the spacing divided by the wave velocity.

A Comparative Study Between Different Guided Waves

TORSIONAL ONLY MODULES

- 3 Ring torsional only modules
  - 3rd Ring Provides 66% greater sound energy
- 5 transducers at 2 spacings
  - 30mm
  - 45mm
- Specifically for lines unsuitable for Longitudinal
  - Liquid filled lines
  - Thick pipes
- Proven better penetration over 2-ring

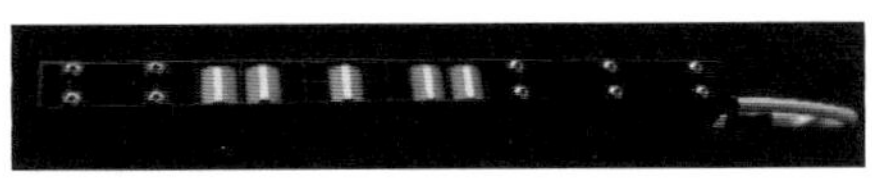

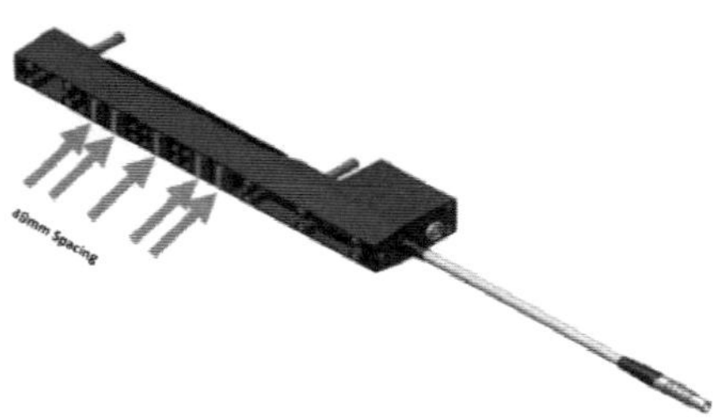

Eddyfi Technologies

A Comparative Study Between Different Guided Waves

5-RING TORSIONAL

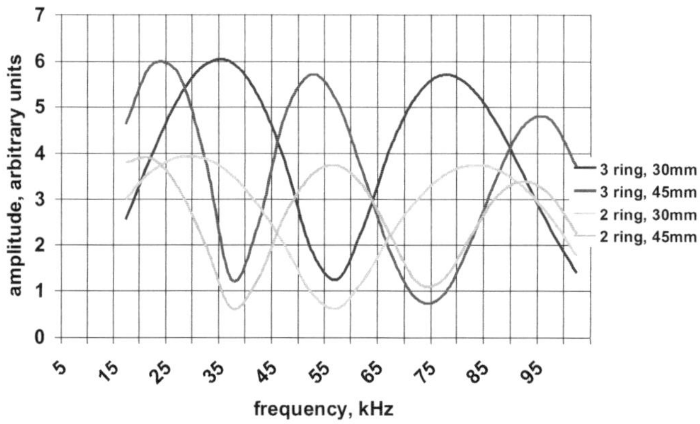

A Comparative Study Between Different Guided Waves

GOOD DATA QUALITY – 180M 590' IN EITHER DIRECTION

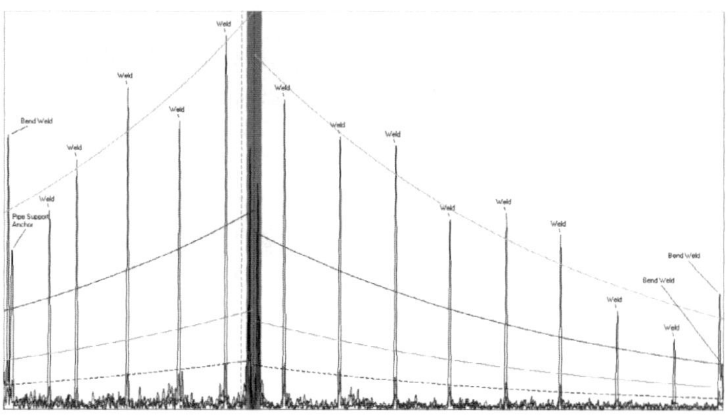

Reflectivity with reflector Cross Section – constant depth

[illegible] of the reflector also increases.

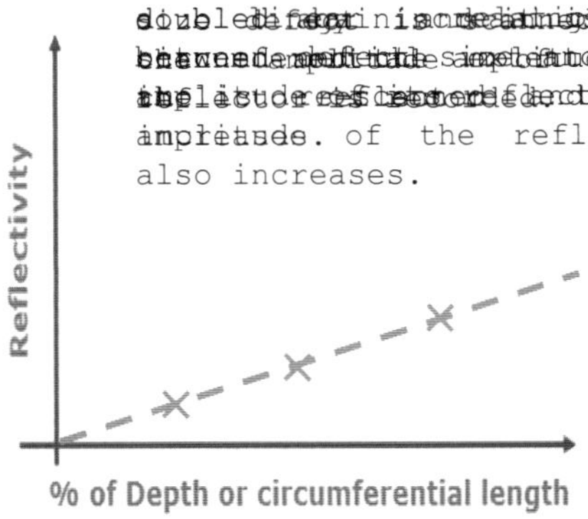

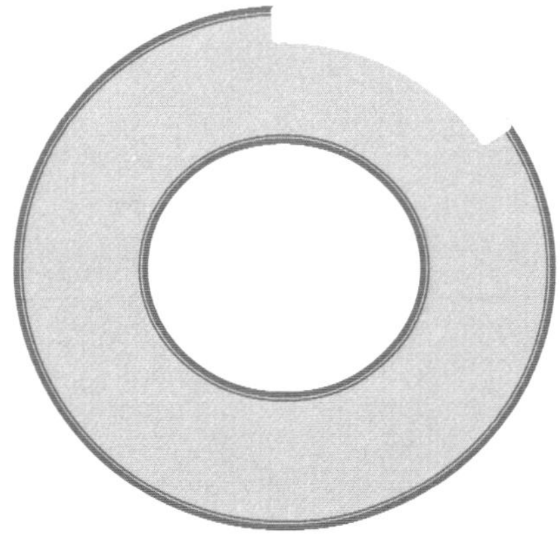

Reflectivity with reflector Cross Section – constant circum extent

[illegible] instead. between defect size and the amplitude of its reflection.

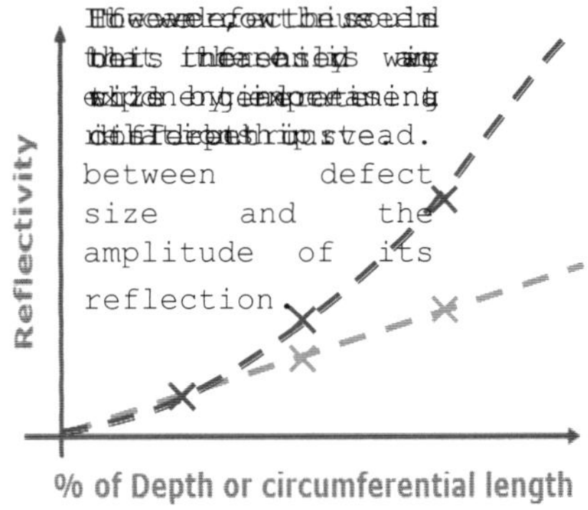

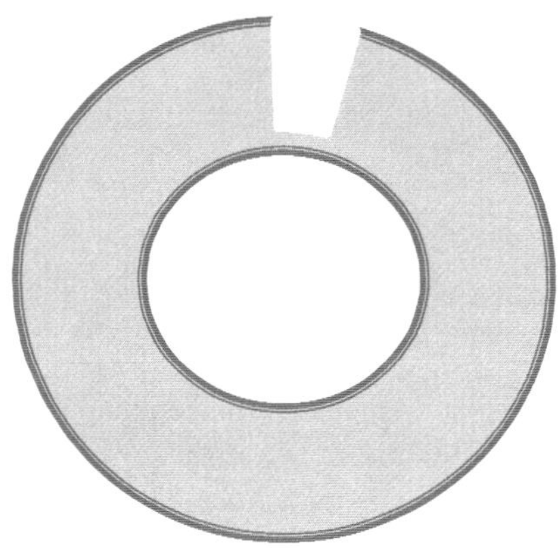

Reflectivity with reflector Cross Section – constant depth

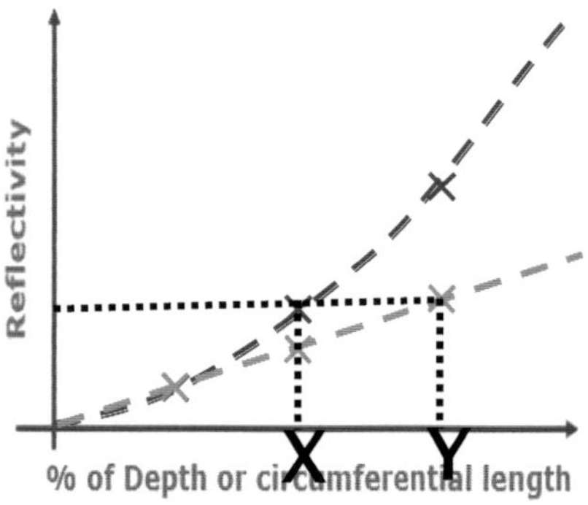

The relationship between a defect of known size and its amplitude can be ascertained.

The problem is that in reality the amplitude of the defect is known and we are trying to ascertain its size.

As information about the defects aspect ratio is not know, an error is introduced into the estimation of the defects true size.

The real size of the defect could be anywhere between X and Y, depending on whether it is shallow and wide or deep and narrow.

A Comparative Study Between Different Guided Waves

C-SCAN TO COMPLEMENT A-SCANS

- Also called synthetic focussing
- Single wave mode transmitted
- Pipe features cause mode conversion
- The collection of reflected modes is analysed
- The inferred location and extent of features is presented on a map

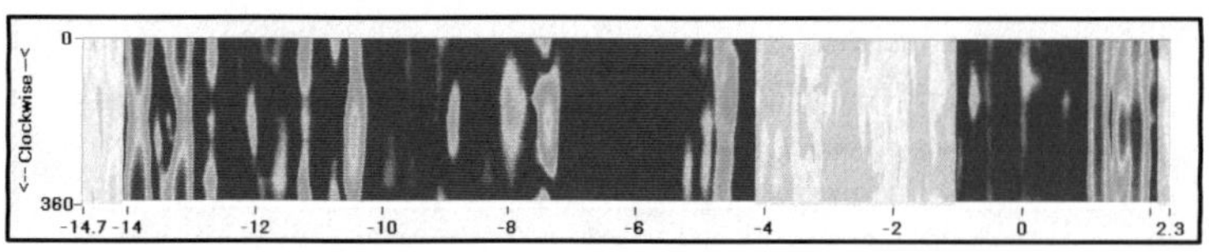

A Comparative Study Between Different Guided Waves

SECONDARY FOCUSING

- Focusing allows the energy to be concentrated where the defect is, increasing sensitivity and giving position and size information
- Sound energy concentrated in one region
- Focus results link directly to Report Manager
- Rotation 8 times around pipe
- 4 times greater sensitivity
- Multi defect focus capability

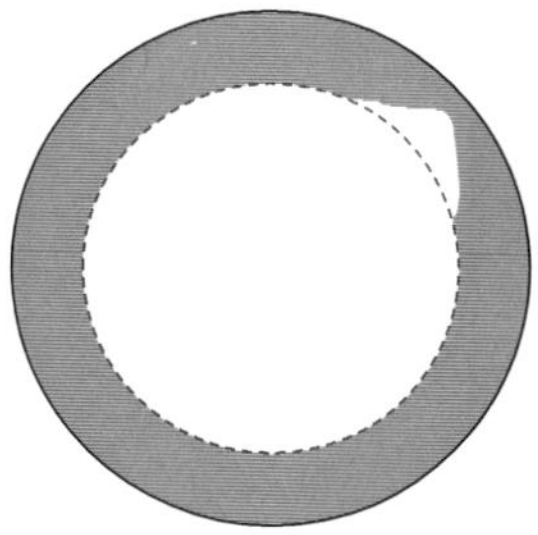

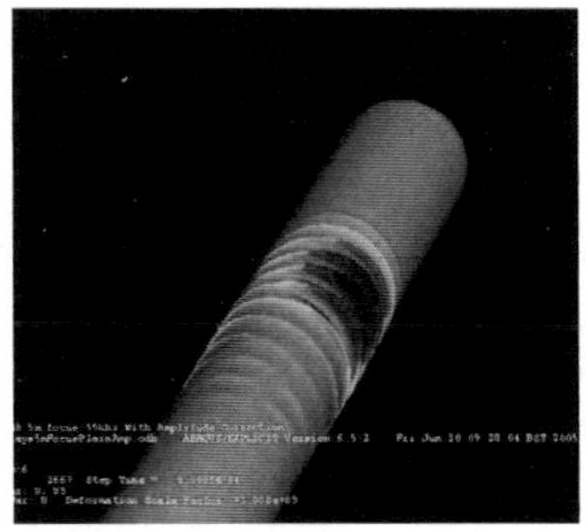

A Comparative Study Between Different Guided Waves

FOCUSSING – INFORMATION ABOUT THE CIRCUMFERENTIAL EXTENTS

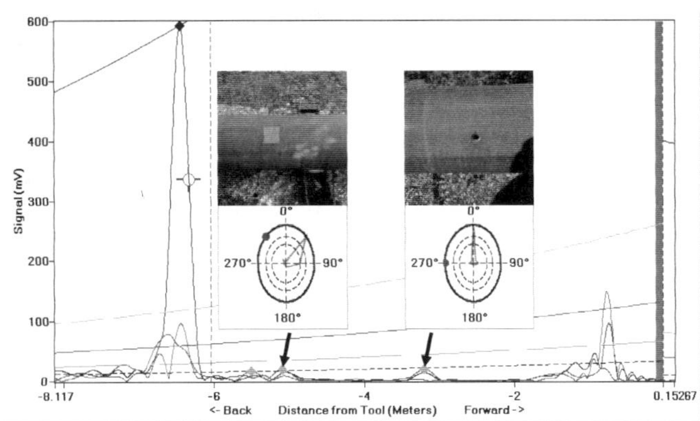

A Comparative Study Between Different Guided Waves

APPLICATIONS

Roads
Crossings

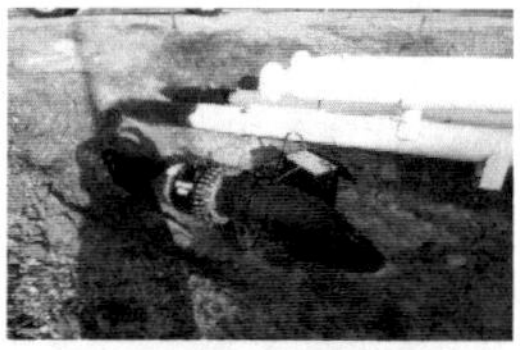

Buried
Pipe

Jetty
Lines

Gas
Pipelines

A Comparative Study Between Different Guided Waves

TANK FARM PIPEWORK

Ideal for application of guided waves:

- Long lengths of pipes
- Insulated line
- Link lines
- Jetty line inspection
- Bund wall penetrations
- Culvert Inspection
- Road crossings

A Comparative Study Between Different Guided Waves

REFINERY PIPEWORK

- Corrosion under Insulation
- Corrosion at simple Pipe-supports
- Hot pipe inspection – max 350°C
- Inspection of elevated pipe
- Flare line inspection
- Jetty pipe work

A Comparative Study Between Different Guided Waves

OFFSHORE APPLICATIONS

- Corrosion under insulation
- Riser inspection
- Deck penetrations
- Splash zone inspection
- Fretting on Caissons
- Caisson inspection
- Top side pipework
- Seals for deck hatches and fire seals

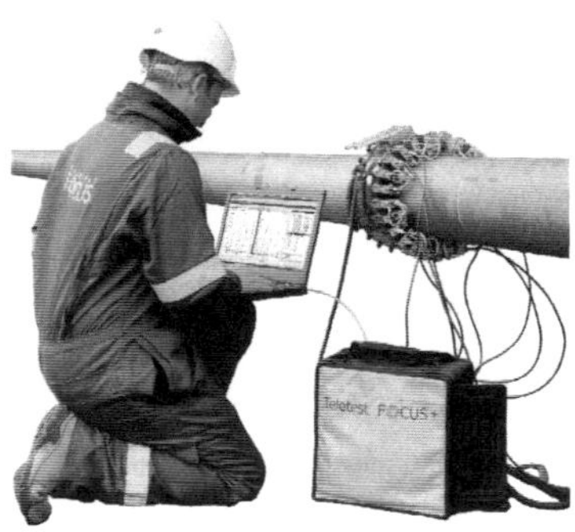

A Comparative Study Between Different Guided Waves

OTHER INSPECTIONS

- Road Crossings
- River Crossings
- Transmission lines
- Unpiggable pipelines
- Buried pipelines
- Insulated Sphere legs
- Air-soil interface

A Comparative Study Between Different Guided Waves

MULTIMODE MODULES

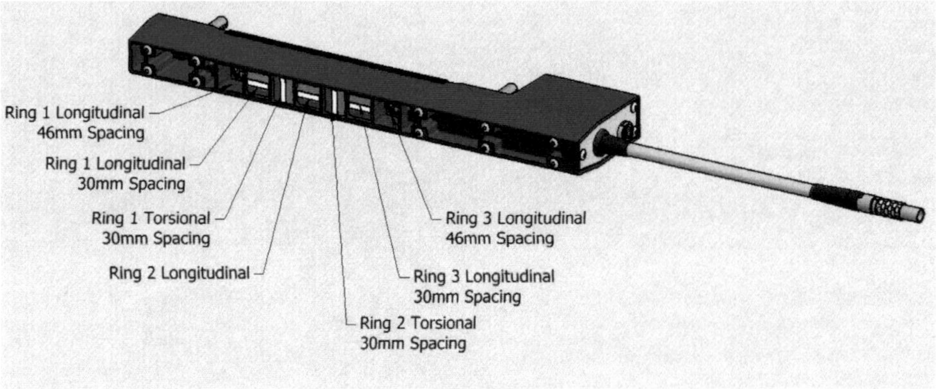

A Comparative Study Between Different Guided Waves

INSPECTION OF FURNACE TUBES

- Difficult to inspect using conventional techniques
- Cut the U-bends off for access and an internal tool
- Expensive in cost and time
- Guided Waves can screen pipes quickly during shut-down
- Access is difficult for conventional tooling

A Comparative Study Between Different Guided Waves

INSPECTION OF FINNED TUBES

- Torsional: nothing could be seen using Torsional mode
- Longitudinal: both the pipe end and the defect could be identified
- Only Teletest capable of Longitudinal

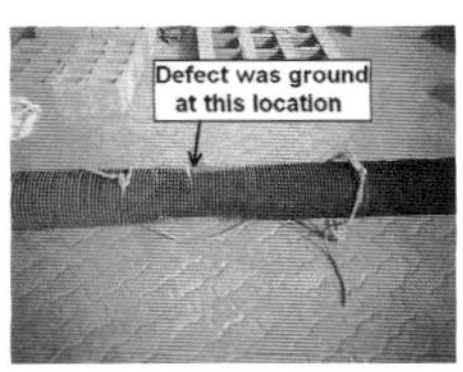

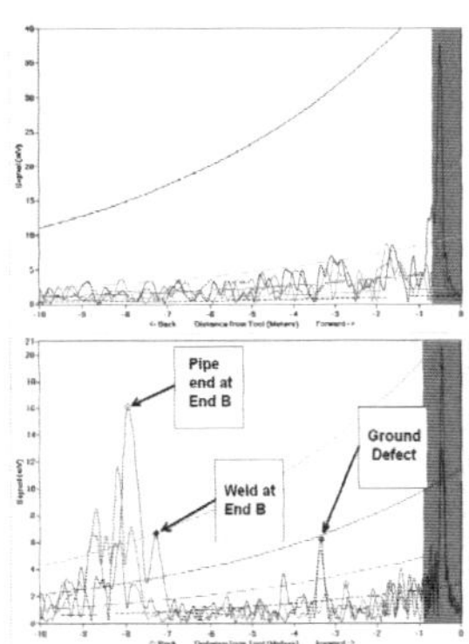

A Comparative Study Between Different Guided Waves

LONGITUDINALLY WELDED PIPE SUPPORTS

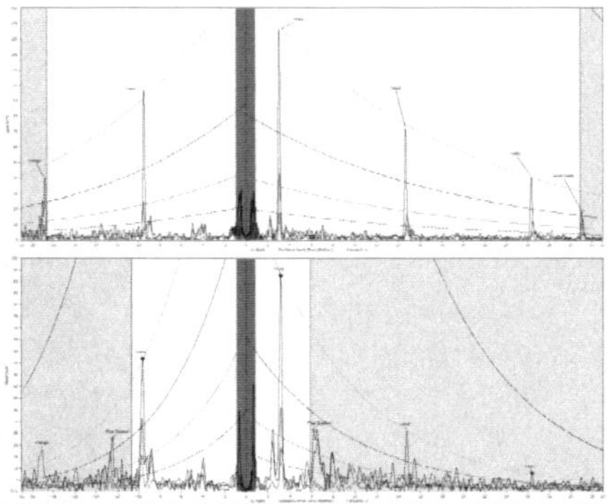

A Comparative Study Between Different Guided Waves

GOOD DATA QUALITY – 180M 590' IN EITHER DIRECTION

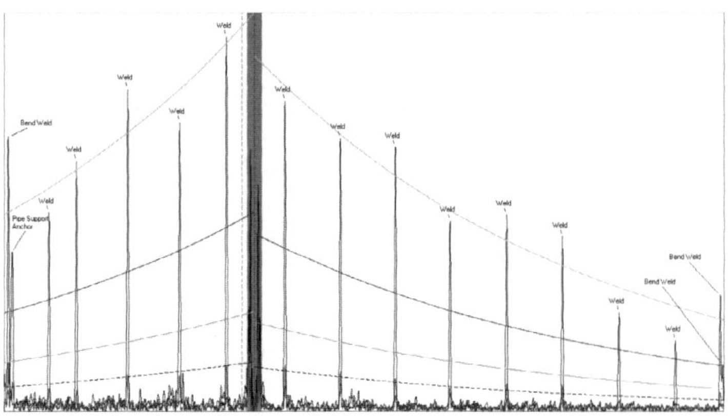

A Comparative Study Between Different Guided Waves

REVERBERATIONS – LIQUID FILLED LINES

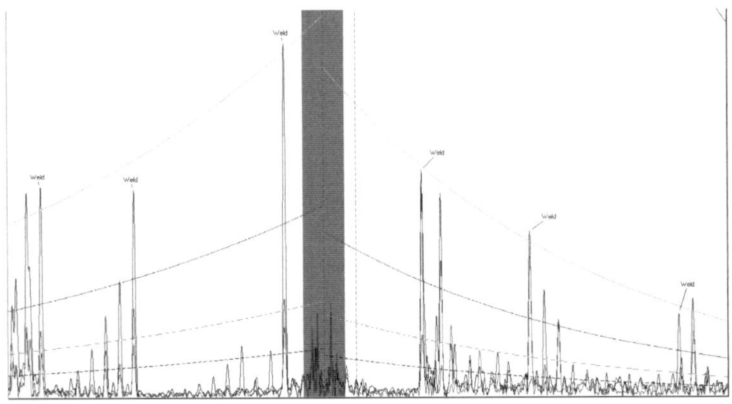

A Comparative Study Between Different Guided Waves

REVERBERATIONS

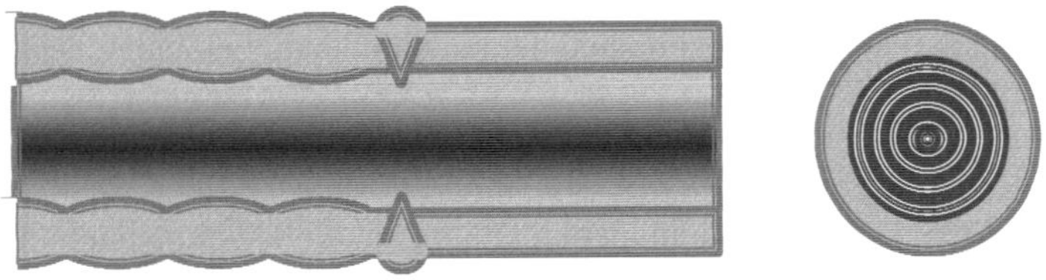

[illegible] wave.
[illegible] the fluid in the pipe.

A Comparative Study Between Different Guided Waves

SUMMARY

Guided Wave Testing can inspect 10s of meters of pipe from one location with good probability of detection.

A screening tool the allows operators to prioritize localized inspection where needed.

Extremely valuable for inspection of inaccessible areas.

A cost effective method for inspection of non-piggable pipe

An important tool for inspection of pipe

Consider LONGITUDINAL

Non-destructive evaluation of impact damage in carbon fibre reinforced polymer using infrared thermography, shearography and ultrasonic testing

Ali Khademi Far[1], Martin Viens[1], Louis-Daniel Théroux[2], Olivier Arés[2]

[1] École de technologie supérieure; Montréal, Canada
Phone: +1 514 396 8800, Fax: +1 514 396 8950; E-mail: ali.khademi-far.1@ens.etsmtl.ca, martin.viens@etsmtl.ca
[2] Centre technologique en aérospatiale; St-Hubert, Canada; E-mail: louis-daniel.theroux@cegepmontpetit.ca, olivier.ares@cegepmontpetit.ca

Keywords: infrared thermography, shearography, CFRP, impact damage, non-destructive testing

Abstract
Low-velocity impact in carbon fibre reinforced polymer (CFRP) is a known cause of failure in aerostructures. Nondestructive evaluation (NDE) is used to detect defects in CFRPs. The present paper aims to compare the features and capabilities of different NDT methods to detect and quantify impact damages in CFRP laminates. To evaluate the impact damages, ultrasonic testing (UT), pulse phased thermography (PPT), and shearography are examined. These methods are conducted to evaluate damages for different impact energy ($10\,J$, $20\,J$ and $25\,J$) on laminate composite. The extents of the damages are quantified using ultrasonic testing. Base on the ultrasonic results, the defect sizing capability of infrared thermography techniques and shearography are evaluated. The experiments show that infrared thermography and shearography are effectively able to quantify the defects. Moreover, shearography presents a qualitative correlation between fringe density and level of impact energy.

Keywords: infrared thermography technique, ultrasonic testing, shearography, CFRP, impact damage, non-destructive testing

1. Introduction

CFRP has been increasingly used in aerospace structures due to their specific mechanical properties such as high strength to weight ratio [1]. Typical damage scenario in CFRP is the low-velocity impact (LVI) which is susceptible to cause transverse cracks, fibre breakages, and multiple interlaminar delaminations. It is therefore, a known cause of premature failure in aero structures [3]. To reveal and evaluate this damage, adequate NDE methods are needed [4]-[5] at different stage of fabrication, maintenance and service [2].

Among available NDE methods, visual testing (VT) is, at the same time, the simplest and the fastest one. However, VT has an inherent disadvantage in the characterization of impact damages: in some cases, internal failure mechanism absorbs impact energy without showing any external sign of flaw [6]. UT has an excellent ability in sizing and revealing the location of flaws such as matrix crack and delamination in CFRP [7]. However, UT is time-consuming, and requires a contact medium [8]. Infrared thermography (IRT) techniques are reported as a practical method to evaluate the nature and the extent of defect in the laminar composites [9]. As an IRT technique, pulse thermography (PT) is particularly effective because a large section of the structure can be imaged at a time. PT is however sensitive to the variation of surface geometry, environmental noises and uneven heating [10]. For these reasons, pulse phased thermography (PPT) is conducted to increase the quality of the thermal image. However, PPT needs image processing, such as phase transformation and spatial filtering for high-quality phasegram. [11]. Under loading condition, structural defects in CFRP cause strain concentration that could be detected by shearography which is sensitive to strain anomalies caused by such weaknesses. A disadvantage of shearography is that defect types other than delamination are hard to detect [6].

Due to the complexity of damage mechanisms in impacted laminate composites, evaluation methods can overestimate or underestimate the defect. The current work is an experimental study which evaluates the features and abilities of different NDE methods: namely UT, IRT and shearography. These methods are used to detect and characterize LVI damages in composites. Their advantages and drawbacks are then presented.

2. Inspection Technique

In this section, a description of the selected methods and experimental procedures, are presented. In the experiments, the extents of the damages are quantified using ultrasonic testing. Based on the ultrasonic results, the defect sizing capability of infrared thermography techniques and shearography are evaluated.

2.1. Ultrasonic

A pulse-echo UT technique is performed to detect impact damage on the laminates. This method is used for flaw sizing and thickness measurements. However, traditional pulse-echo has some limitations for damage detection in laminated composites. C-scan 2D images reveal the size, and depth of the defects. In the experiment, the ultrasonic C-scan is performed from the impacted side. The inspection is conducted using an immersion tank with two 13-step millimetre encoders using the Zetec Topaz system, with a 64-element 5MHz probe configurated with an active aperture of 18 elements.

2.2. Shearography

Shearography is an optical technique based on the interference of laser speckle patterns. The principle of the speckle interferometry involves a comparison of the recorded speckle patterns before and after stimulating the specimen by a thermal or mechanical load. Measuring the variation of the speckle patterns leads to a phased shearogram. Change in the phase map correspond to the spatial derivative of the out-of-plane displacement [12]. For the shearography test, a DANTEC Q-800 portable shearography system with eight lasers of 120 mW at a wavelength of 660 nm is used. The data is analysed by ISTRA4D X 86 software. Thermal stress is used to induce deformation for making the shearograms [13]. A heating lamp of 750 W applies a thermal load for 10 seconds, and the images are recorded for 10 seconds during cooling by natural convection.

2.3. Infrared thermography

For infrared thermography, PPT techniques are conducted for the inspection. In pulse thermography, a specimen is heated with a short thermal pulse. While the thermal pulse travels through the material, an infrared camera records the temperature of the surface at regular interval time. The captured sequence form a 3D matrix of pixels, and their value corresponds to the temperature [14]. In PPT, the sequence is processed using Fast Fourier Transform (FFT) [11].

In this study, the acquisition is performed using cooled Telops camera (640×512 pixel array) and two Hensel flash lamps each generates 3 kJ heat pulse in 4 ms. The frames are recorded for 25 seconds in the cooling process with a frequency of 50 Hz. The lamps were placed at an angle of 45° and a distance of 0,4 m. Matlab with IR-view [15] is used to calculate phase and amplitude images.

3. Inspected parts

The sample studied in this work are three 30 cm×30 cm CFRP test coupons which are composed of 16 plies of unidirectional prepreg tapes (Cycom 5276) with a quasi-isotropic layup $[45/0/-45/90]_{2s}$. The thickness of the samples is 2.3 mm, and they are impacted at energy levels of 10J, 15J and 25J using a drop-weight impact system. The impactor has a mass of 5.5 kg and the striker tip has a diameter of 16 mm according to ASTM D7136/D7136M-15 standard [16]. A summary of the three samples is shown in table 1.

Table 1 Impact intensity

Coupons	Impact Energy
E2	10J
E3	20J
E4	25J

4. NDE results

In the following section, the test results are presented and discussed in detail. The test indications are shown, and sizing ability of the methods are compared.

4.1. Ultrasonic

Results from the UT inspection are shown in Fig.1. The depth of the indications is color coded blue for the near-surface signals and red for the back wall indications.
The results show that multiple delaminations are formed around the impacted area at different depths. The centre of the damaged zone, coded in blue, corresponds to surface deformation induced by the impact. For these pixels, the surface deformation causes local loss of the signal due to the non-normal incidence of the ultrasonic beam.

Table 2 presents the C-scan estimation of the damaged areas for the three impact values. The results show that higher impact energy induces larger damage extent. The measured defective areas are considered as a reference to evaluate the differences of the other methods in terms of defect sizing.

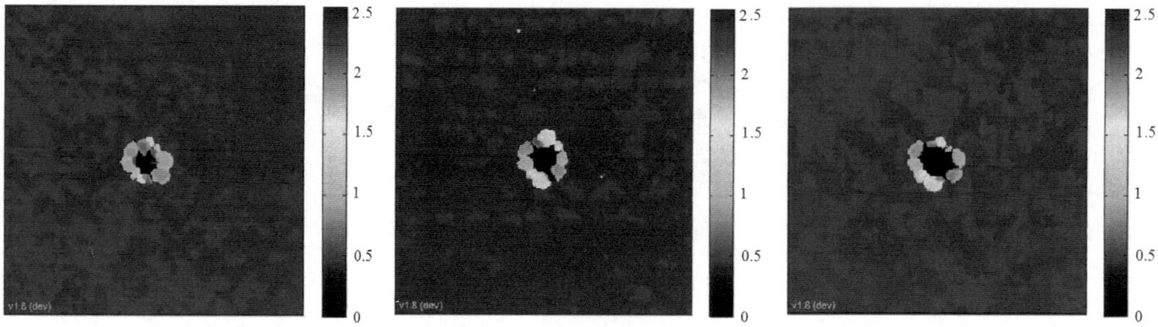

(a) (b) (c)

Figure 1 C-scan time of flight image of the damages (a) E2 sample impacted by 10J, (b) E3 sample impacted by 20J and (c) E4 sample impacted by 25J

4.2. Shearography

The shearography results show that all the impact damages are detectable by this method. (table 2). The size of the defects is measured by the full-width half amplitude for the defective area in the shearing direction.

The indication sizes are 310 mm² for 10J, 394 mm² for 20J and 487 mm² for 25J impact energy for 2,4 mm shearing distance. The shearograms illustrate that the size of the defects is proportional to the value of the impact level. Furthermore, the wrapped phase shearograms (Fig. 2) demonstrate that the intensity of the impacts is corresponding to the fringe density. In that way, the severer impact damage is identified by an indication with a higher fringe density.

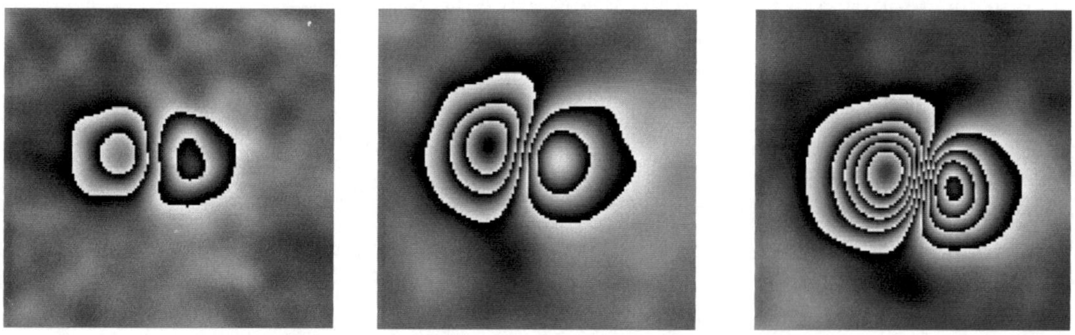

(a) (b) (c)

Figure 2 The shearography images of the damages after the thermal load, (a) E2 sample impacted by 10J, (b) E3 sample impacted by 20J and (c) E4 sample impacted by 25J, for the pixel size of 0.27 mm

4.3. Thermography

For PPT, the phasegrams was captured in the minimum available frequency of 0.03Hz. To increase the phasegram quality, a cold image subtraction (CIS) and recursive median filter (RMF) are applied by IR-view software [17]. In the phase images, all the defects are clearly visible, Fig. 3. Since the irregular boundaries of the multiple delaminations produce complex heat diffusion through the thickness [10], It causes underestimate damage sizing in PPT result. Table 2 shows the size of the indications, while the flaw sizing is conducted according to ASTM E2582-07 [18].

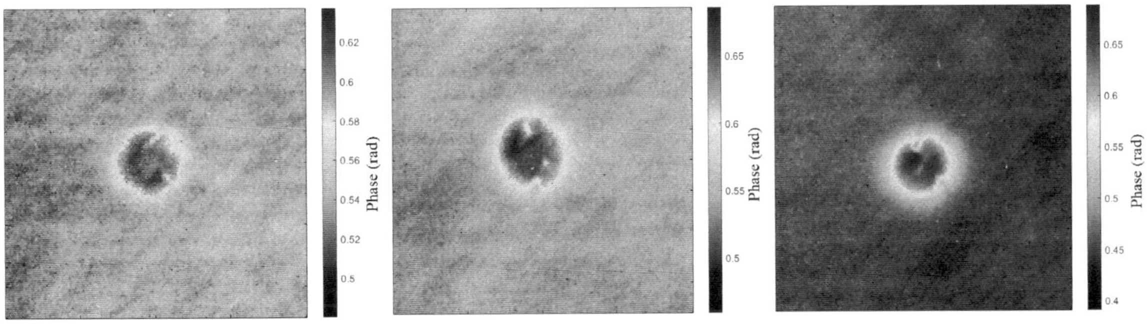

(a) (b) (c)

Figure 3 The phasegrams for (a) E2 sample impacted by 10J, (b) E3 sample impacted by 20J and (c) E4 sample impacted

Table 2 Comparison of the measured areas with different NDT techniques, the differences are calculated with considering the UT sizing as the reference

Coupons	Method				
	C-scan	Shearography		PPT	
	Area(mm²)	Area(mm²)	difference (%)	Area(mm²)	difference (%)
E2	441	310	29.7	265	40
E3	522	394	24.5	352	32.5
E4	547	487	11.5	408	25.4

5. Discussion and conclusion

In this work, three NDE methods were applied to study their capability in terms of detecting impact damages in CFRP. Therefore, the three coupons with different levels of impact damage were used. The size and shape of the indication for each method were compared.

The conclusion is, the applied techniques can be used as a reliable method to detect these level of impact damages in the CFRP laminate. PPT gave the high-resolution image for all the coupons in a relatively fast process. For shearography, the results showed, with the same size of impactor, the fringe density of the indications are proportional to the value of impact energy. This capability is a promising feature of shearography to evaluate impact damage. Concerning the shape and the damage size, UT results have more visual detail about the shape of the defect. PPT and shearography underestimated damage sizes with half of amplitude sizing method in comparison to UT. Although shearography sizing was closer to UT results, in comparison to PPT, it seems that the full-width half amplitude is not a precise method to size the impact damage for the two methods. Therefore, to address the issue, other sizing criteria should be implemented.

6. Acknowledgement

This study is supported by the Centre Technologique en Aérospatiale (CTA) through the framework of oN DuTy!–NSERC–CREATE programme in NDT.

References

[1] F. Ciampa, P. Mahmoodi, F. Pinto, and M. Meo, "Recent advances in active infrared thermography for non-destructive testing of aerospace components," *Sensors (Switzerland)*, vol. 18, no. 2, 2018.

[2] J. Baaran, "Visual Inspection of Composite Structures EASA-Research Project/2007/3 Final Report," 2009.

[3] Y. Aoki, H. Suemasu, and T. Ishikawa, "Damage propagation in CFRP laminates subjected to low velocity impact and static indentation," *Adv. Compos. Mater. Off. J. Japan Soc. Compos. Mater.*, vol. 16, no. 1, pp. 45–61, 2007.

[4] J. Gryzagoridis and D. Findeis, "Impact damage detection on composites using optical NDT techniques," *Insight Non-Destructive Test. Cond. Monit.*, vol. 52, no. 5, pp. 248–251, 2010.

[5] V. Vavilov, A. Chulkov, A. Smotrov, S. Smotrova, and A. Moskovchenko, "Characterizing impact damage in GFRP/CFRP composites by determining thermal effusivity/diffusivity," *Meas. Sci. Technol.*, vol. 30, no. 3, p. ab018e, 2019.

[6] S.Gholizadeh, "A review of non-destructive testing of composite materials," *NDT Int.*, vol. 15, no. 2, pp. 75–86, 2016.

[7] Q. Shen, M. Omar, and S. Dongri, "Ultrasonic NDE Techniques for Impact Damage Inspection on CFRP Laminates," *J. Mater. Sci. Res.*, vol. 1, no. 1, pp. 1–16, 2011.

[8] M. H. Sherafat, N. Quaegebeur, P. Hubert, L. Lessard, and P. Masson, "Experimental Model of Impact Damage for Guided Wave-Based Inspection of Composites," *J. Nondestruct. Eval. Diagnostics Progn. Eng. Syst.*, vol. 1, no. 4, p. 040801, 2018.

[9] and X. M. L.-D. Théroux, J. Dumoulin, "Square Heating Applied to Shearography and Active Infrared Thermography Measurements Coupling: From Feasibility Test in Laboratory to Numerical Study of Pultruded CFRP Plates Glued on Concrete Specimen," *Strain*, vol. 50, no. 5, pp. 389–403, 2014.

[10] S. S. Pawar and K. Peters, "Through-the-thickness identification of impact damage in composite laminates through pulsed phase thermography," *Meas. Sci. Technol.*, vol. 24, no. 11, 2013.

[11] C. Ibarra-Castanedo and X. Maldague, "Pulsed phase thermography reviewed," *Quant. Infrared Thermogr. J.*, vol. 1, no. 1, pp. 47–70, 2007.

[12] Y. Y. Hung, L. X. Yang, and Y. H. Huang, "Non-destructive evaluation (NDE) of composites: Digital shearography," *Non-Destructive Eval. Polym. Matrix Compos. Tech. Appl.*, pp. 84–115, 2013.

[13] Y. H. Huang, S. P. Ng, L. Liu, C. L. Li, Y. S. Chen, and Y. Y. Hung, "NDT&E using shearography with impulsive thermal stressing and clustering phase extraction," *Opt. Lasers Eng.*, vol. 47, no. 7–8, pp. 774–781, 2009.

[14] X. P. Maldague, "Introduction to NDT by Active Infrared Thermography," *Mater. Eval.*, vol. 6, no. 9, pp. 1060–1073, 2002.

[15] M. T. Klein, C. Ibarra-Castanedo, X. P. V. Maldague, and A. Bendada, "A straightforward graphical user interface for basic and advanced signal processing of thermographic infrared sequences," *Thermosense XXX Proc. SPIE Vol. 6939*, vol. 6939, p. 9, 2008.

[16] S. Te, "Back to Composite Standards," *ASTM Int. Des. D*, vol. i, no. C, pp. 4–6, 2011.

[17] H. Zhang *et al.*, "Thermographic Non-Destructive Evaluation for Natural Fiber-

Reinforced Composite Laminates," *Appl. Sci.*, vol. 8, no. 2, p. 240, 2018.
[18] ASTM, "E2582-07 Standard Practice for Infrared Thermography of Composite Panels and Repair Patches used in Aerospace Applications," *ASTM Int.*, vol. i, no. July, pp. 1–6, 2014.

How latest sensor and communication technologies help to improve UCI (ultrasonic contact impedance) hardness testing method

Tom Ott,
Proceq USA

CINDE
CANADIAN INSTITUTE FOR NON-DESTRUCTIVE EVALUATION

NDT in Canada
NDTiC 2019
Inform - Engage - Advance

June 18 - 20
River Cree Resort & Casino
Edmonton, Alberta

UCI is one of the most popular portable hardness testing methods

Fast and simple	Small indentation	High accessibility

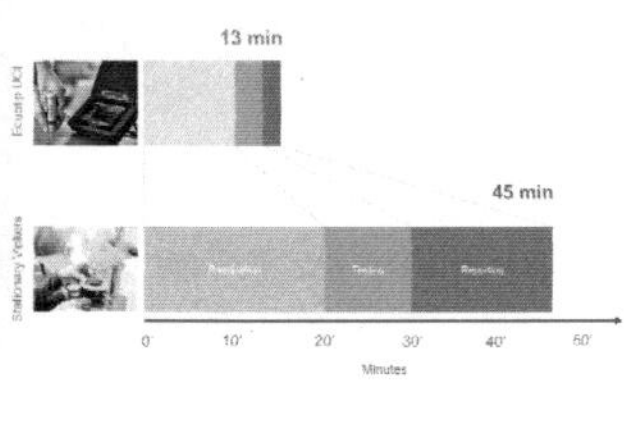

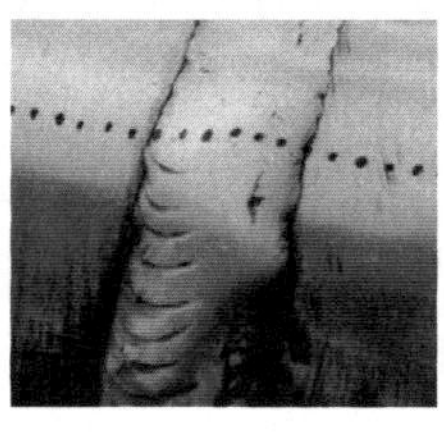

Practical benefits of utilizing Bluetooth and cloud connectivity with the adjustable-test-load Ultrasonic Contact Impedance (UCI) hardness testing method

Frequency shift is used to infer the hardness of the test piece

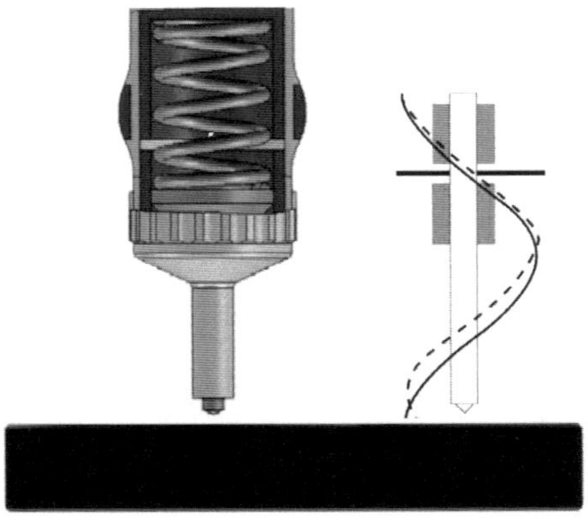

Longitudinal oscillation of resonating rod at ultrasonic frequency

Vickers indenter forced into test piece

Measured frequency shift converted to hardness

International standards ASTM A1038 and DIN 50159 specify UCI in detail

- Widely established
- Well standardized
- Diamond indenter in UCI: same **tip shape** as in stationary Vickers

Hardness scale: HV, indicated as "HV(UCI)"

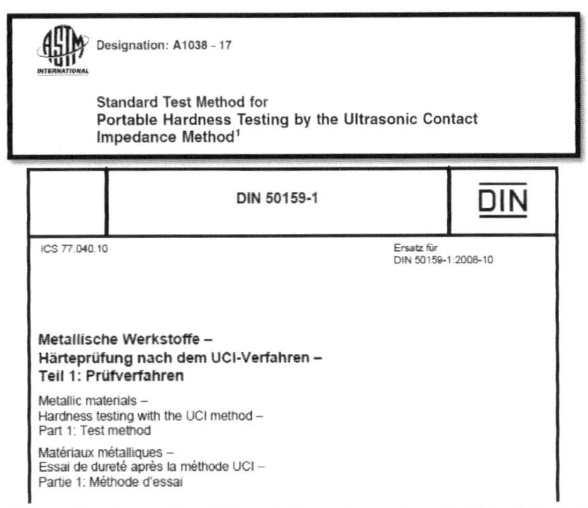

Designation: A1038 – 17

Standard Test Method for Portable Hardness Testing by the Ultrasonic Contact Impedance Method[1]

DIN 50159-1

DIN

ICS 77.040.10

Ersatz für DIN 50159-1:2008-10

Metallische Werkstoffe – Härteprüfung nach dem UCI-Verfahren – Teil 1: Prüfverfahren

Metallic materials – Hardness testing with the UCI method – Part 1: Test method

Matériaux métalliques – Essai de dureté après la méthode UCI – Partie 1: Méthode d'essai

Practical benefits of utilizing Bluetooth and cloud connectivity with the adjustable-test-load Ultrasonic Contact Impedance (UCI) hardness testing method

© Proceq 2018

Different applications require different UCI test loads

HV1 (1 kgf, ~10 N)	HV5 (5 kgf, ~50 N)	HV10 (10 kgf, ~98 N)
• Precisions parts • Thin coatings • Hardened layers	• Hardened material • Carburized material • Welds and HAZs	• Small forgings • Cast material • Welds and HAZs

Practical benefits of utilizing Bluetooth and cloud connectivity with the adjustable-test-load Ultrasonic Contact Impedance (UCI) hardness testing method

The mechanical working principle of conventional UCI probes only allows one fixed load per probe

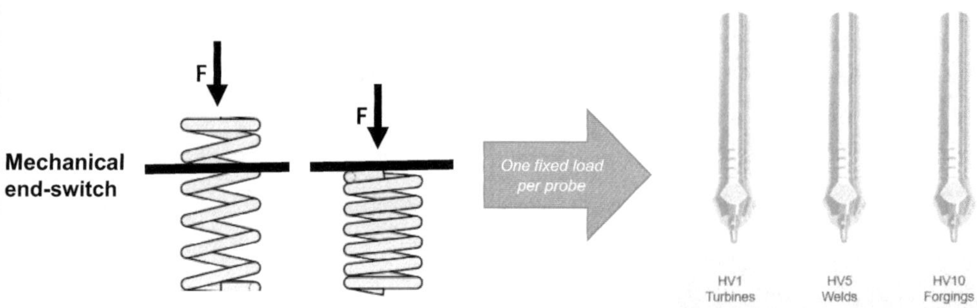

Practical benefits of utilizing Bluetooth and cloud connectivity with the adjustable-test-load Ultrasonic Contact Impedance (UCI) hardness testing method

The test results with conventional UCI probes are highly user-dependent

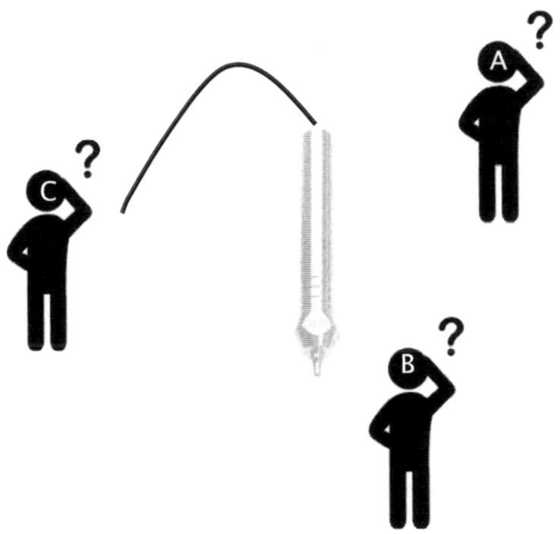

Why?

1. Constant and continuous load application is needed
2. Conventional UCI probes are **not able to** track the loading process
3. **No help provided** to reduce the dependency on user skill

Practical benefits of utilizing Bluetooth and cloud connectivity with the adjustable-test-load Ultrasonic Contact Impedance (UCI) hardness testing method

The end-to-end process with conventional UCI probes is rather "manual"

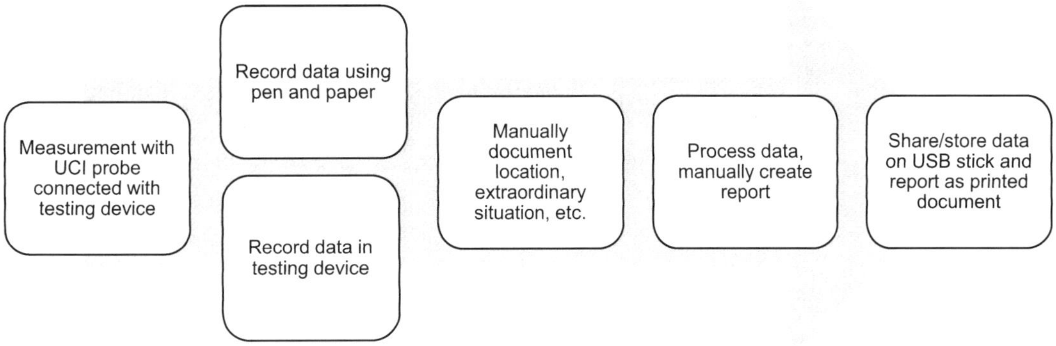

Practical benefits of utilizing Bluetooth and cloud connectivity with the adjustable-test-load Ultrasonic Contact Impedance (UCI) hardness testing method

In Lean Manufacturing, eight "wastes" make processes slow, ineffective and/or inefficient

- Transport
- Inventory
- Motion
- Waiting
- Over-production
- Over-processing
- Defects
- Skills

Practical benefits of utilizing Bluetooth and cloud connectivity with the adjustable-test-load Ultrasonic Contact Impedance (UCI) hardness testing method

5 of the 8 Lean wastes are evident during the traditional UCI testing process

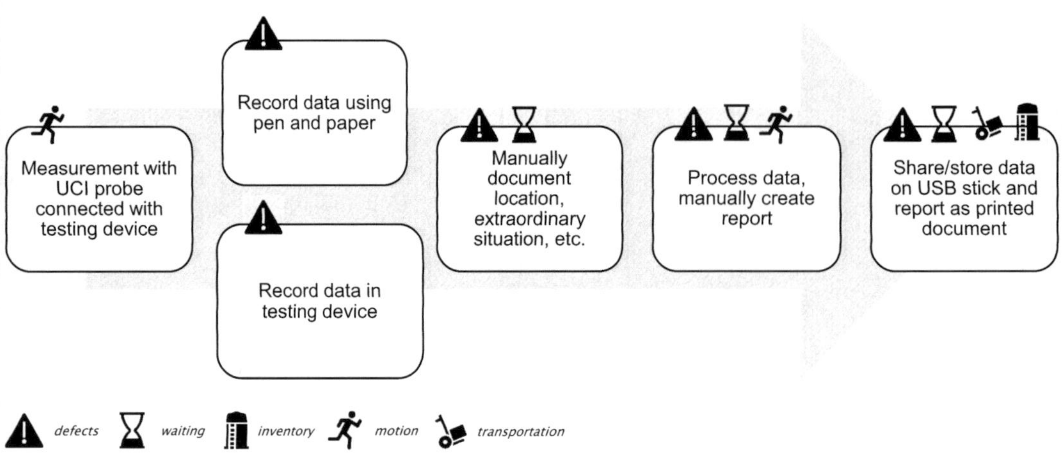

Practical benefits of utilizing Bluetooth and cloud connectivity with the adjustable-test-load Ultrasonic Contact Impedance (UCI) hardness testing method

Latest sensor and electronic technology allows to monitor and measure test load over a range

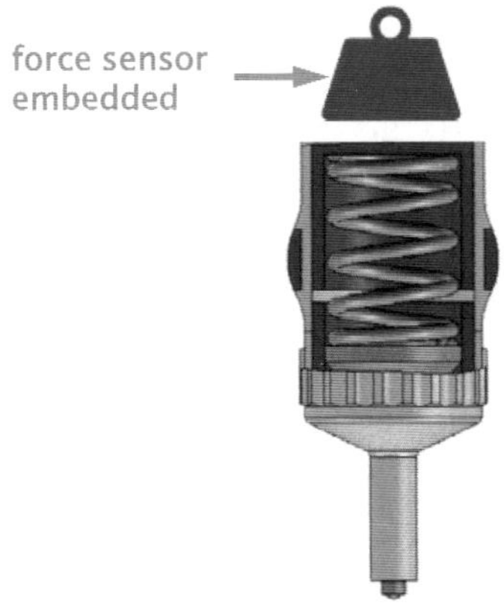

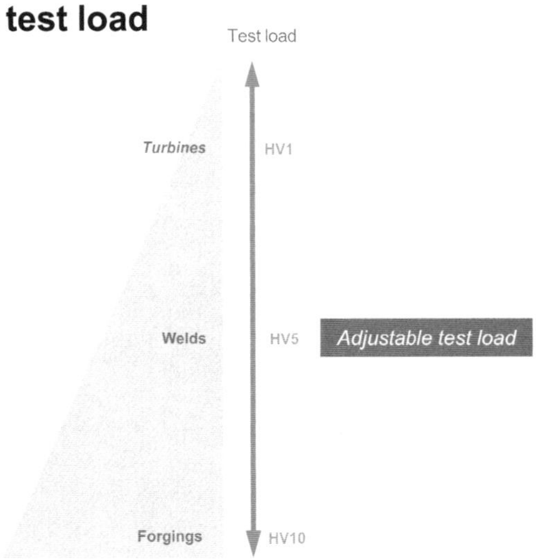

Practical benefits of utilizing Bluetooth and cloud connectivity with the adjustable-test-load Ultrasonic Contact Impedance (UCI) hardness testing method

A force sensor allows real-time monitoring and recording of the load

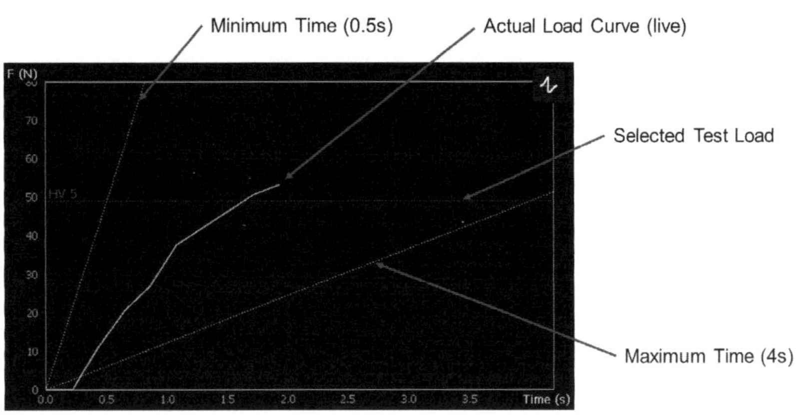

Source: Proceq

Practical benefits of utilizing Bluetooth and cloud connectivity with the adjustable-test-load Ultrasonic Contact Impedance (UCI) hardness testing method

User guidance can be shown based on the load monitoring

Example of user guidance

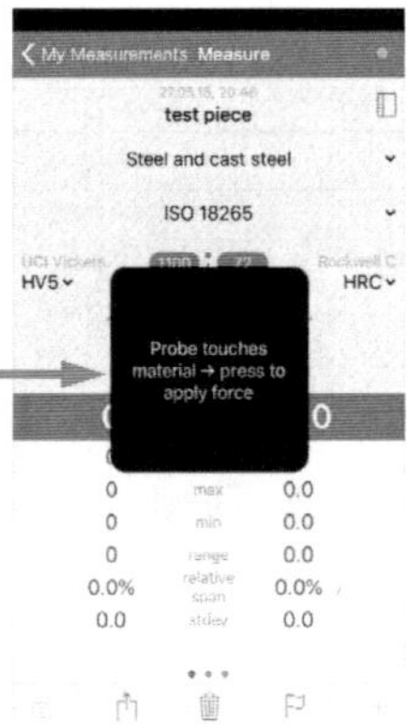

User guidance:

- Start to apply force
- Target force reached – lift probe
- Force applied too fast/slow

New communication technologies now improve the UCI hardness testing process

Smart devices	Bluetooth 4.0	Cloud connectivity

Bluetooth creates wireless connectivity reduces the waste of motion for data

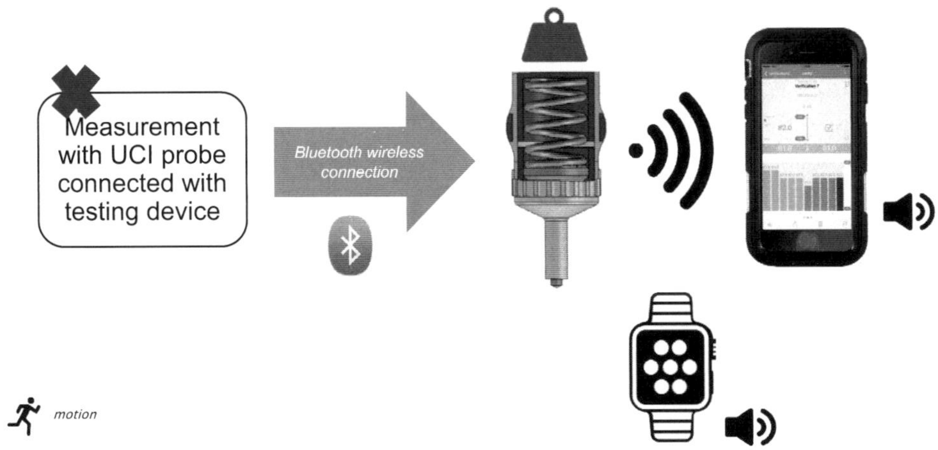

Cloud-based data eliminates worries of defective or lost data

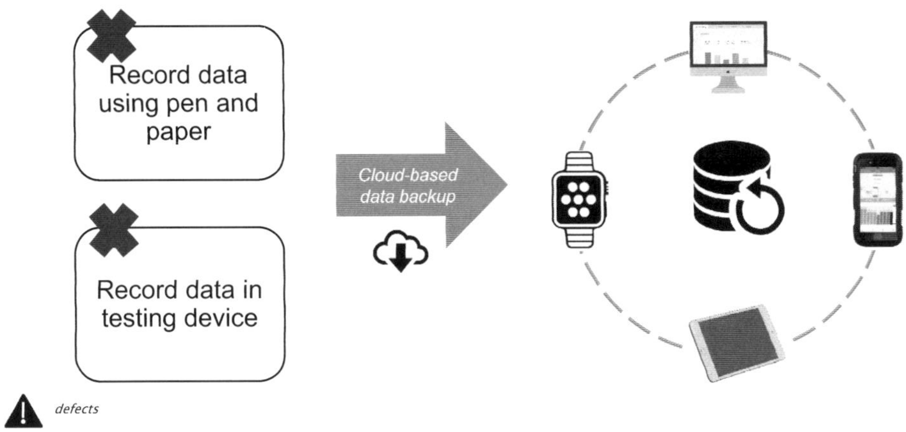

Full data traceability reduces waiting and defects in communicating results

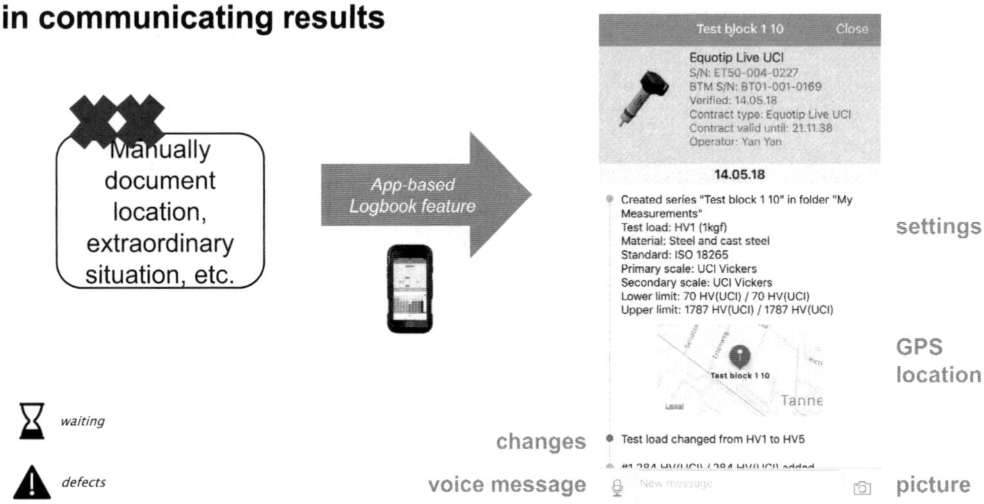

Practical benefits of utilizing Bluetooth and cloud connectivity with the adjustable-test-load Ultrasonic Contact Impedance (UCI) hardness testing method

Cloud-connected app enables on-the-spot data analysis and reporting

Practical benefits of utilizing Bluetooth and cloud connectivity with the adjustable-test-load Ultrasonic Contact Impedance (UCI) hardness testing method

Cloud connectivity enables real-time data and report sharing

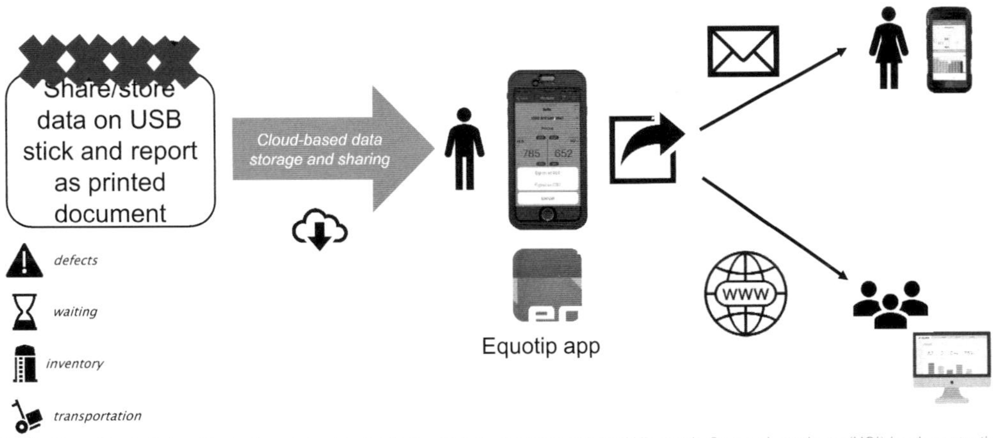

Practical benefits of utilizing Bluetooth and cloud connectivity with the adjustable-test-load Ultrasonic Contact Impedance (UCI) hardness testing method

New digital technology makes UCI testing overall more accurate, convenient and efficient

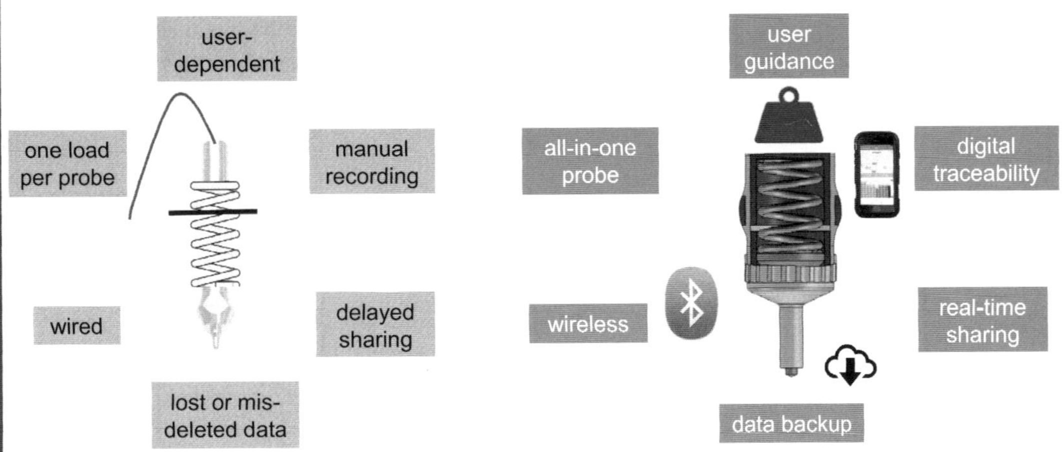

Practical benefits of utilizing Bluetooth and cloud connectivity with the adjustable-test-load Ultrasonic Contact Impedance (UCI) hardness testing method

Thank you for your attention.

Any Questions?

Tom Ott,
Proceq USA

CINDE
CANADIAN INSTITUTE FOR NON-DESTRUCTIVE EVALUATION

NDT in Canada
NDTiC 2019
Inform - Engage - Advance
June 18 - 20
River Cree Resort & Casino
Edmonton, Alberta

Investigation on the suitability of laser-excited thermography to detect cavities in metallic components

Malte MUND [1], Elisabeth STAMMEN [1], Klaus Dilger [1]

[1] Institut für Füge- und Schweißtechnik, Technische Universität Braunschweig; Braunschweig, Germany
Phone: +49 531 39195594, Fax: +49 531 39195599; e-mail: m.mund@tu-braunschweig.de, e.stammen@tu-braunschweig.de, k.dilger@tu-braunschweig.de

Abstract
Active thermography has become a widely used technique in non-destructive material testing and numerous excitation sources are applied. One field of application is the testing of joints. Especially the detection of defects in welds is a challenging tasks as two-dimensional defects like cracks can occur as well as volumetric inner defects as cavities.
Recently, the suitability of lasers as a contactless excitation source has been investigated and it was shown, that cracks can be detected. As the specimens are locally heated, a heat flow is initiated. Those heat flows can be used to detect crack-like defects that are perpendicular to the surface. However, the suitability of the laser-excited thermography to detect inner defects like voids and the influence of geometric as well as excitation parameters on the test result has not been fully investigated. Therefore, this study focus on the detection of inner defects in metallic components.
In a first step, thermographic measurements with varying excitation parameters were performed. The obtained results were than used as a basis for the numerical simulation of the testing process. A parametric FE-model has been implemented and the influence of excitation as well geometric parameters on the testing results were determined. Different approaches to evaluate the resulting surface temperatures were considered as well. The simulation results were than compared to the experimental data that was obtained by testing idealized specimens with cavities. Based on these results, the most relevant parameters are determined and limiting factors resulting from the experimental setup are discussed.

Keywords: Laser, Thermography, FE-Simulation, aluminum testing

1. Introduction

The non-destructive testing of welds is highly standardized. Beside the definition of welding defects that are defined in ISO 6520-1 for fusion welds and in 6520-2 for press welds, the corresponding quality levels are defined in ISO 5817 for steel, titanium and nickel and their alloys and in ISO 10042 for aluminum and its alloys. Some of the most common imperfections are cavities and their allowed size depends on the required quality level as well as the thickness of the weld. In order to assure the quality of these welds, the testing standard ISO 17635 defines non-destructive testing methods and their suitability for the evaluation of the weld. However, active thermographic testing methods are not considered within these standard although numerous investigations concerning the applicability of thermographic testing methods for welded joints have been investigated within recent years and some studies report the successful application of active thermographic methods for the detection of inner volumetric defects.
Several studies have been conducted to determine the detection boundaries of laser excited thermography. Most studies focus on the detection of cracks. Experimental tests [1, 2] as well as simulations [3, 4] were performed and it could be shown, that cracks within the low micrometer range can be detected. In order to detect void inner defects, exemplary results are shown in [5] for flaws in laser powder deposition components and in [6] for carbon fiber reinforced plastics.
In order to qualify laser based thermographic methods for the non-destructive testing of welds, a critical evaluation of the potential of these method to detect inner volumetric defects such as voids and their suitability to fulfil the existing standards for the quality levels in welds is required. Therefore, this investigation propose a numeric model to evaluate the influence of excitation parameters as well as some geometric properties on the resulting surface temperature

and evaluates the suitability of laser excited thermography for the testing of butt-welds according to the existing testing standards.

2. Methods and Materials

In order to determine the suitability of laser excited thermography for weld inspection, a simulation based approach was chosen to eliminate the influence of excitation and recording equipment. The outcome of thermographic measurements with laser excitation depend on the utilized excitation source and the excitation parameters, the thermal and geometrical properties of tested specimens as well as the applied IR-camera. As it was intended to make a general statement on the suitability of the testing method, in a first step the functionality of the measurement principle was demonstrated. In order to be able to define the influence of several parameters, a simulation based approach was chosen. Experimental investigations on the influence of excitation parameters on the heating behavior were performed as a basis for the simulations. The accuracy of the proposed simulation model was shown and finally the influences of the parameters were computed using finite element method. The experimental setup and the simulation model are described below.

2.1 Experimental setup

A sketch of the experimental setup used in this investigation is and is displayed in Figure 1. The test bench consist of a laser based excitation, the IR-camera as a recording device and an even specimen placed on a scissor lift as well as control units for and a device for data processing.

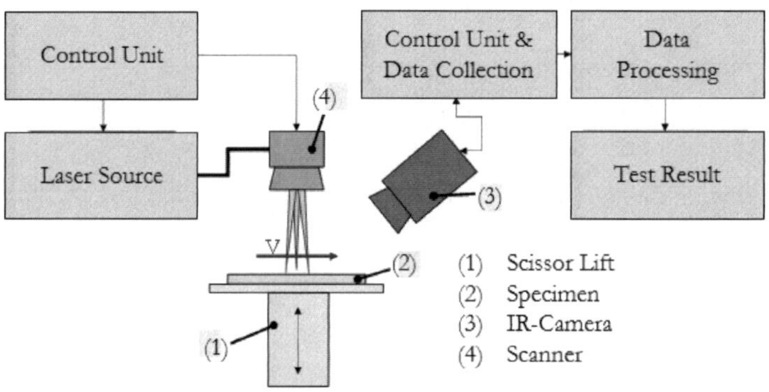

Figure 1: Sketch of the experimental setup used within the experiments

Within this study, a redEnergy G4 fiber laser by SPI Lasers UK Ltd. emitting radiation at a wavelength of 1062 nm with an average output power of 70 W was utilized in continuous mode. The laser source generates the monochromatic light and is connected to a laser scanner via glass fiber. The scanner is positioned perpendicular to the surface of the specimen to be heated. The scanner attached to the laser was an Axialscan-20/-30 by Raylase AG allowing the guidance of the laser beam across the surface of the specimen. The excitation parameters were defined using the control software weldMARK3 by Raylase AG. Within this study, the average output power as well as the scanning speed were varied. While the average output power was set in between 14 W and 70 W, the scanning varied between 10 mm/s and 100 mm/s. When the test is started, the laser beam travels over the surface of the specimen with a predefined average output power and scanning speed along a given path. Within the experiments, the path was chosen to be a straight line with a length of 50 mm.

The specimen consisted of the aluminium alloy EN AW 6082-T6. It hat a quadratic shape with a side length of 50 mm to avoid heat accumulation on the edges and a thickness of 8 mm. This was chosen, as the thinnest category of metal sheets within ISO 17635 is defined up to this

thickness. The surface of the specimen were blackened to suppress reflections and to allow an accurate temperature measurement. The specimen was placed on a scissor lift that allows the variation of the distance between the surface of the specimen and the scanner. Therefore, a variation of the spot size can be achieved. The deflection from the focal level of the laser beam was varied between ±25 mm.
An IR-camera records the heating process. The IR-camera was a VarioCAM® HD produced by Infratec GmbH. The framerate within the measurements was set to 240 fps. The associated image size was 640 × 120 px and one pixel represents a sixe of approximately 100 x 100 μm. The results of the measurements were proceeded using the software Irbis 3.1 bx Infratec GmbH, Germany.

2.2 Finite Element Simulation

The numerical simulations were performed using the commercially available software Simufact Welding 8.0 by simufact engineering GmbH. This software was developed for the simulation of welding processes and offers implemented heat sources that can be tailored to the application. Within this investigation, a circular surface heat source with varying diameter was applied. The material used in the simulations was EN AW 6082 and the properties were assigned to the model from the databank included in Simufact Welding 8.0.
Two different models were considered within the simulations. The models were prepared using the software Abaqus CAE 2018. On one hand, an even plate with dimension 50 x 50 x 8 mm was used to demonstrate the accuracy of the simulation compared to the experiments. The size of the models was chosen based on pre-tests that showed no heat accumulation appeared on the edges with this size of specimen. On the other hand, models with an artificial cuboidal cavity with varying dimensions were prepared. The size of the cavities was chosen in conformity with the requirements of the testing standards. Permitted cavities with sizes lower than the required diameters were as well considered as larger, inadmissible cavities. The sizes differed between 0.1 mm and 4.307 mm. The distance of the cavities was also varied between 0.1 mm and 4 mm. The cuboidal shape of the cavities was chosen as a constant heat introduction required a uniform distribution of surface nodes to suppress excitation caused irregularities from superimposing with the cavity caused temperature changes. Furthermore, the cuboidal shape allowed a consideration of the influence of the projected surface compared to the cavity volume. The mesh size within the excitation area was chosen to 50 μm and increased continuously from the surface.

3. Experimental Results

Within this chapter, first a measurement of a specimen with artificial cavities is show. Then, exemplary results of the experimental tests to characterize the laser heating process are shown. Therefore, the maximum temperature of each pixel was determined and the temperature increase compared to the initial temperature was calculated. This was done to eliminate variations of the initial temperatures within the experimental measurements. To display the differences in the maximum temperature increases, measurement lines were placed across the specimen perpendicular to the scanning line. In order to evaluate the accuracy of the simulations, three measurement lines were placed in a steady distance from each other within the experiments and a mean value was calculated out of three measurements.

3.1 Detection of artificial voids

In Figure 2, an example of a thermographic measurement and a μ-CT measurement are shown. The depth of the artificial voids was 1 mm. The path of the laser is illustrated by the red line in the μ-CT picture. The results indicate, that the larger voids with a diameter of 8 mm and 4 mm are detectable, while the smaller void with a diameter 2 mm cannot be detected. This is caused by the smaller size and the larger distance of the scanning line to the void. However, the measuring principle can be demonstrated.

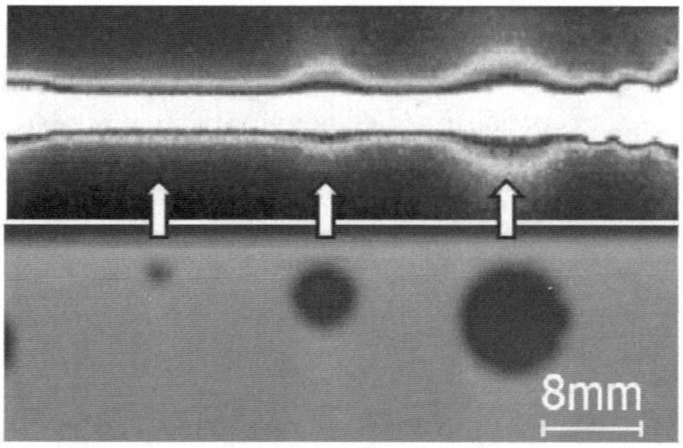

Figure 2: Comparison of thermographic measurement (scanning speed 10 mm/s spot size 1 mm and average output power 21 W) and µ-CT-measurement

3.2 Influence of Excitation Parameters on Heating Properties

As described, in a first step the influence of different excitation parameters on the heating of the specimens was determined. Exemplary results of these experiments are given in Figure 3. Figure 3a.) displays the influence of the average output power and the distance from the focal level (and therefore the spot diameter) on the maximum temperature increase ΔT_{max} while Figure 3b.) shows the influence of the scanning speed. While the influence of the scanning speed and the average output power can be measured over the whole measuring section, the influence of the varying spot diameter is only significant within the region of the heat introduction.

a.) b.)

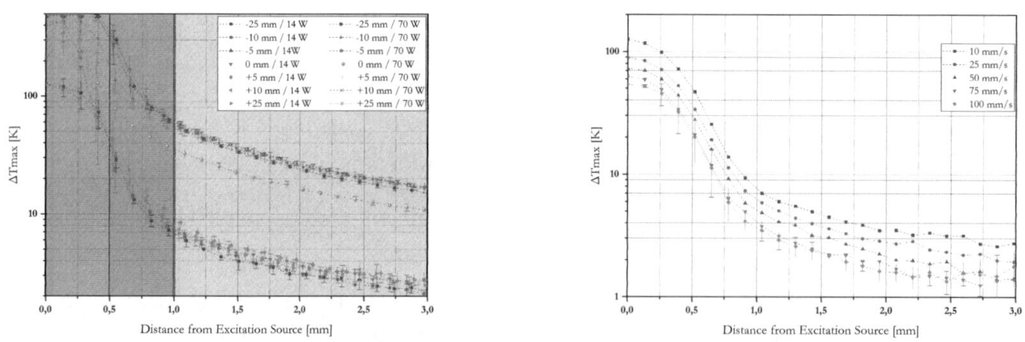

Figure 3: Influence of excitation parameters on maximum temperature increase: a.) Influence of distance to focal level and average output power (scanning speed: 10 mm/s) and b.) influence of scanning speed (focal level: 0 mm; average output power; 14 W)

It can be derived from the results, that an increase in the applied average output power results in a significant increase in energy input and therefore, the maximum temperature increase is obvious. In addition, an influence of the focal level can be observed within the heated area (red) as well as in a small zone next to it (orange). Due to the change in spot size resulting from the different focal levels, the heat introduction changes as the power per area increases. The closer the distance to the focal level is, the higher are the resulting temperatures. However, within the green area, which starts, independent from the applied power, in a distance of approximately 1 mm to the middle of the laser source, no influence of the focal level (and the spot diameter) can be observed. In addition, a decrease in temperature can be observed for one of the test series with an average output power of 70 W within the focal level can be observed. This is caused by ablation of the black paint and damages on the surface of the aluminum specimen as the energy input is too concentrated.

In comparison to the influence of the laser power, the influence of the scanning speed is rather small. However, an increased scanning speed results in a decrease in energy input and therefore lower temperature increases. This can be explained by the lower changes in energy per unit length.

4. Simulation Results

In a first step, the accuracy of the simulation model is demonstrated. Corresponding to the experimental test, simulation were computed and the results were compared to the experimental findings. Then, a short overview of different approaches for the processing of the date is given based on a exemplary simulation result. In a next step, the influence of different cavities is shown. On one hand, the size of the cavities was varied and on the other, the shape was modified. Finally, the influence of excitation parameters is shown.

4.1 Accuracy of the Simulation Models

As the average output power was identified as the main influence on the resulting maximum temperature differences of the specimen, a variation of this parameter was used to demonstrate the accuracy of the simulation model. In Figure 4, simulation and experimental results are compared. The focal plane in the experiments was kept constant at +25 mm which resulted in a spot diameter of approximately 1 mm. The experiments were conducted with an constant scanning speed of 10 mm/s. The simulation was performed using exactly the same excitation parameters and as can be seen in the diagram, the maximum temperature increase of the simulations and the experiments correlate sufficiently. However, within the area of the heat introduction, the experimentally measured temperatures exceed those of the simulations, but the overall energy introduction seems to be the same. The differences may be explained by a vaporization of the paint used to blacken the specimen that overlay the measurements without leaving significant marks on the surface. In a distance of 1 mm to the center of the excitation, the temperature increase in simulations and therefore, the simulation is considered accurate to examine the suitability of the laser based thermography to detect cavities in metallic components.

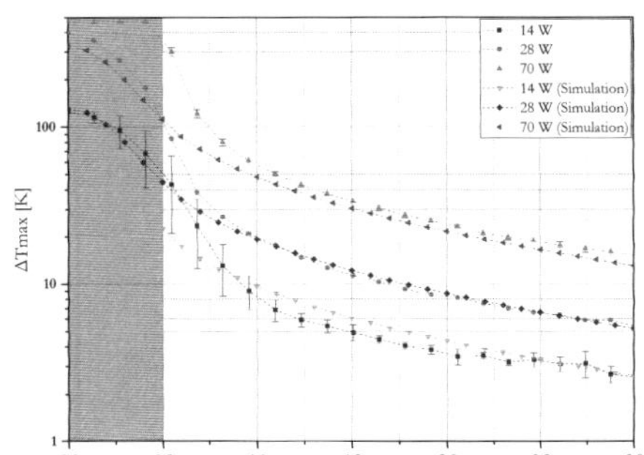

Figure 4: Comparison of experimental and simulation results depending on average output power (focal level: +25 mm; scanning speed: 10 mm/s)

4.2 Simulation Results – Result Processing

Different approaches for the visualization of the cavities have been investigated. Exemplary results are shown in Figure 5. As can be seen from the illustrations, the cavity (edge length of 0.1 mm and 0.1 mm under the surface) can be visualized using adequate data processing.

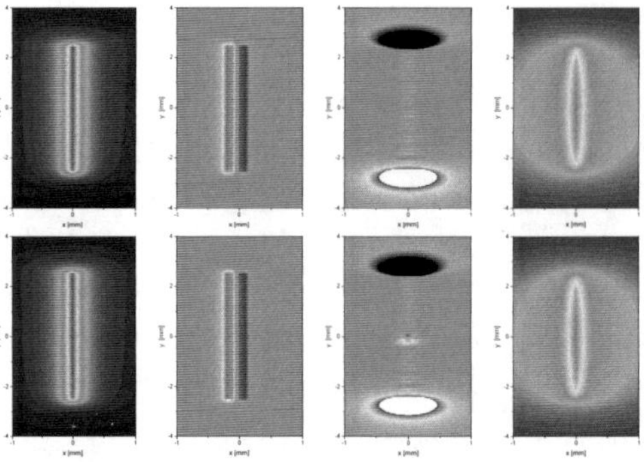

Figure 5: Approaches for cavity visualization – Reference samples (a.) T_{max}, (b.) d T_{max} /dx, (c.) d T_{max}/dy, (d.) mean temperature) and e.)-h.) Sample with cavity (e.) T_{max}, (f.) dT_{max}/dx, (g.) T_{max}/dy, (h.) mean temperature

While mean temperatures (d.) and h.)) for each measurement point suppress resulting temperature changes and therefore a detection of the cavity is impossible, a further processing of the maximum temperatures T_{max} by using derivations shows good potential to increase the signal. Especially results obtained by using the first derivation of the T_{max} in the scanning direction are quite significant (c.) and g.)). It has to be notices, that it is possible to detect the cavities by comparing the maximum temperatures as well, also the picture above indicate otherwise. This approach preferably used in the within this paper.

4.3 Simulation Results – Detection of Cavities Depending on excitation parameters

The proposed simulation model enables the variation of the spot diameter, the scanning speed and the applied power. The effect of these parameters on the detectability of the cavities is shown in Figure 6. The cubic cavity in this simulations had an edge length of 0.2 mm and was placed in a depth of 0.1 mm. The reference simulation were performed using a solid sample. The scanning speed for this simulations was 10 mm/s, the spot size was 400 µm and the power was 7 W.

a.) b.) c.)

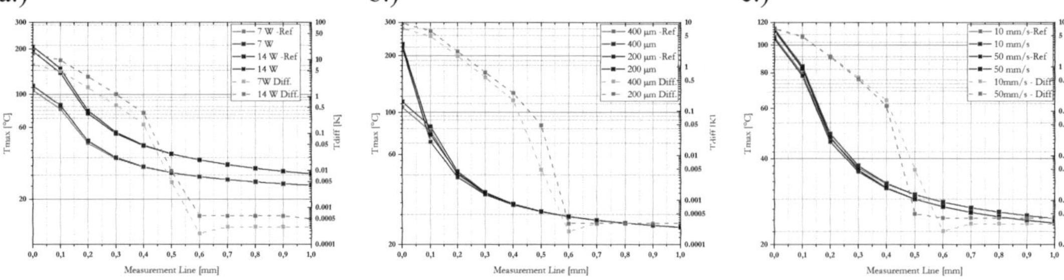

Figure 6: Influences of excitation parameters on detectability of cavities – a.) Influence of laser power; b.) Influence of spot size and c.) Influence of scanning speed

As the result show, the excitation parameters influence the heating of the specimen significantly. An increase of the laser power increase the resulting maximum temperatures and the maximum energy input, as well as a lower scanning speed. In contrast, the spot diameter influence only the maximum temperatures within the heated area as lower a lower spot diameter the intensity of the heating increase. However, it has to be noticed, that the excitation parameters do not affect the detectability of the void significantly. The example shown in the diagrams demonstrate, that the differences in maximum temperatures are significant (T_{diff}> 0.1 K) within a distance of 0.4 mm from the center of the excitation.

Therefore, it seems to be suitable to use lower excitation powers and larger spot diameters to improve the experimental results as the ablation can be avoided and the experimental test should be more accurate. In contrast, lower scanning speeds should be preferred to receive accurate date. The scanning speed should be chosen with respect to the framerate of the IR-Camera

4.4 Simulation Results – Influences of Geometric Properties on the Detection of Cavities

In Figure 7, the influence of two different geometric properties are shown In Figure 7a.) the size of a cavity placed in a depth of 0.1 mm is shown while Figure 7b.) displays the effect of the shape of a cavity with a constant volume of 1 mm³. All simulations were performed using a scanning speed of 10 mm/s, a spot diameter of 400 µm and a laser power of 7 W.

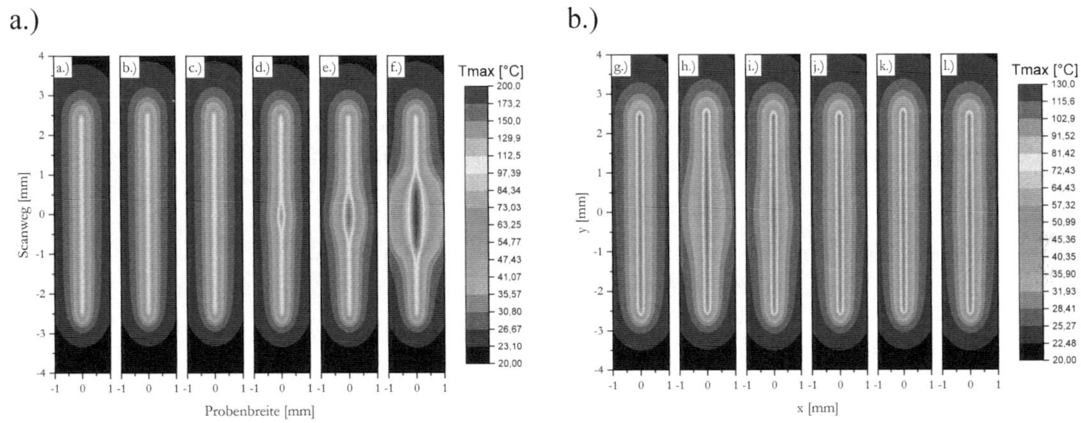

Figure 7: Influence of geometric properties –a.) Influence of size of cavity compared to a reference sample and b.) Influence of shape of cavity compared to a reference sample

As can be seen from the results, the size of the cavity has a mayor influence. While smaller cavities with edge length smaller 0.2 mm ((b.) and (c.)) can hardly be detected, larger cavities result in a local increase in temperature giving a clear indication of the cavity. However, not only the size, but also the shape influence the detectability. As can be seen from Figure 7b.), the projected area of the cavity is more important than the volume. The larger the projected area is, the larger is the probability of detection as the temperature differences increase. A void with a projected area of 10 mm² in a depth of 0.5 mm can be detected (h.), while a cavity with equal volume but smaller projected area of 1mm² cannot be identified (k.).

5. Conclusion

First of all, it can be shown that basic measurement principle is suitable to detect subsurface defects. The cavities shown in the example in chapter 3.1 are quite big and do not fulfil the requirements of the testing standards, but they are suitable to demonstrate the basic approach of the testing concept. The main influences of the excitation source were identified and the resulting effects of parameter variations are displayed. It was found, that the laser power has the largest impact on heating of the specimens. The effects of the scanning speed are much lower and it was found, that the influence of the spot diameter is negligible.
A simulation model was proposed and the accuracy was demonstrated by comparing simulation and experimental results. There a difference in the area of heat introduction that are suspected to be caused by an interaction of the paint used to improve the measurement results within the experiments and the laser radiation. In addition, it was found, that the basic properties of the heating process are well covered. Finally it was shown, that geometric have the largest effect on the detectability of the artificial defects while excitation parameters determine the heating of the specimen, but do not influence the detectability. Therefore, these parameters should be

chosen to minimize the possibility of ablation. It was found, that the shape of the defect is more important than the actual size of the cavities.
As a basic result it can be stated, that the measurement principle is suitable to detect small defects. The influence of the excitation source might be neglected and therefore further studies will focus on the influence of the recording device on the test results of experimental test to validate the simulation results.

References

[1] Almond, D. P.; Weekes, B.; Li, Teng; Pickering, S. G.; Kostson, E.; Wilson, J. ; Tian, G. Y.; Dixon, S.; Burrows, S.: Thermographic techniques for the detection of cracks in metallic components. In: Insight - Non-Destructive Testing and Condition Monitoring 53 (2011), Nr. 11, S. 614–620

[2] Burrows, S. E. ; Dixon, S. ; Pickering, S. G. ; Li, T. ; Almond, D. P.: Thermographic detection of surface breaking defects using a scanning laser source. In: NDT & E International 44 (2011), Nr. 7, S. 589–596

[3] Puthiyaveettil, N. ; Krishna, S.; Kidangan, R.; Unnikrishnakurup, S; Krishnamurthy, C. V. ; Zeigler, M; Myrach,; Balasubramaniam, K: In-line laser thermography for crack detection at elevated temperature: A Numerical modeling study. In: Proceedings of the 2016 International Conference on Quantitative InfraRed Thermography : QIRT Council, 4-8 July, 2016

[4] Li, T.; Almond, D. P.; Rees, D. A. S.: Crack imaging by scanning pulsed laser spot thermography. In: NDT & E International 44 (2011), Nr. 2, S. 216–225

[5] Santospirito, S. P. ; Łopatka, Rafał ; Cerniglia, Donatella ; Słyk, Kamil ; Luo, Bin ; Panggabean, Dorothée ; Rudlin, John: Defect detection in laser powder deposition components by laser thermography and laser ultrasonic inspections. In: Heisterkamp, Alexander; Herman, Peter R.; Meunier, Michel; Nolte, Stefan (Hrsg.): Frontiers in Ultrafast Optics: Biomedical, Scientific, and Industrial Applications XIII : SPIE, 2013 (SPIE Proceedings), 86111N

[6] Zhang, H.; Fernandes, H.; Djupkep Dizeu, F. B.; Hassler, U.; Fleuret, J.; Genest, M.; Ibarra-Castanedo, C.; Robitaille, F.; Joncas, S.; Maldague, X.: Pulsed micro-laser line thermography on submillimeter cavity in carbon fiber reinforced polymer composites: experimental and numerical analyses for the capability of detection. In: Applied optics 55 (2016), Nr. 34, D1-D10

Detection of Second Layer Cracks in Riveted Aluminum Structure

P. Ross Underhill, Mary Jo Aquilina, Thomas W. Krause

Department of Physics and Space Science
Royal Military College of Canada

CINDE
CANADIAN INSTITUTE FOR NON-DESTRUCTIVE EVALUATION

NDT in Canada
NDTiC 2019
Inform - Engage - Advance
June 18 - 20
River Cree Resort & Casino
Edmonton, Alberta

Outline

- Motivation
- Approach
- Results
- Summary
- Way FOrward
- Acknowledgements

Motivation

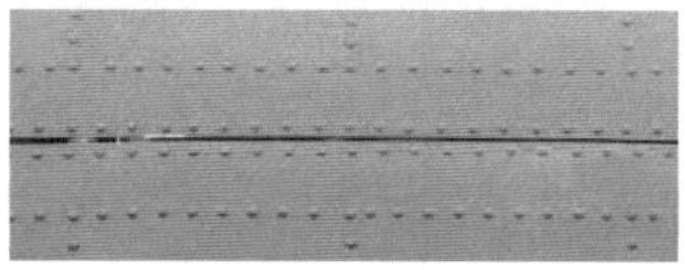

Approach - Probes I

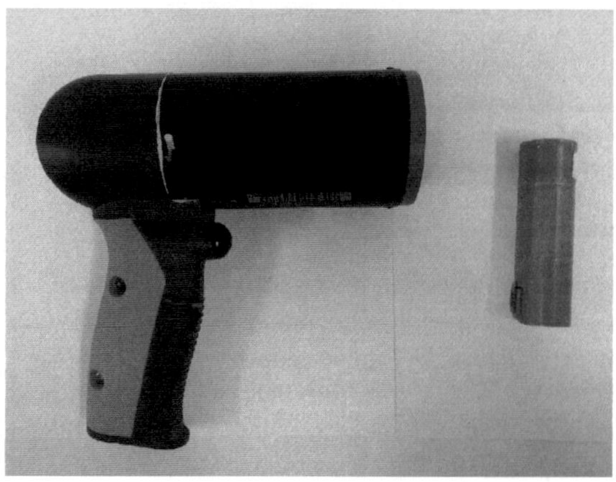

Approach - Probes II

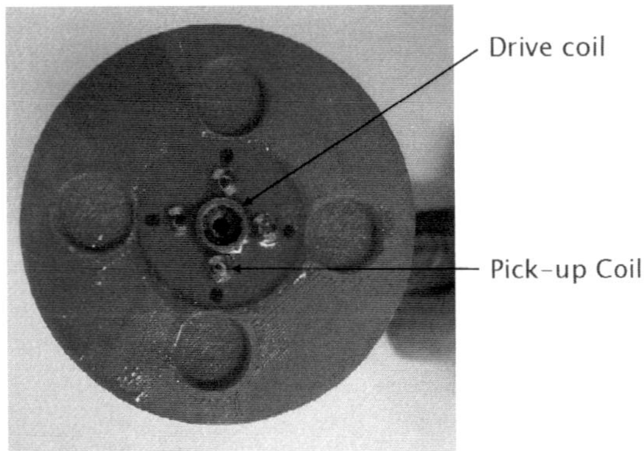

Sample

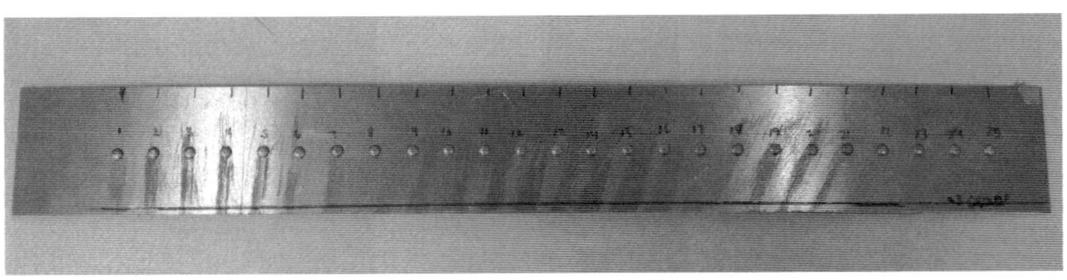

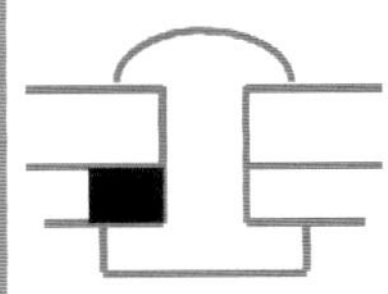

Approach - Analysis I

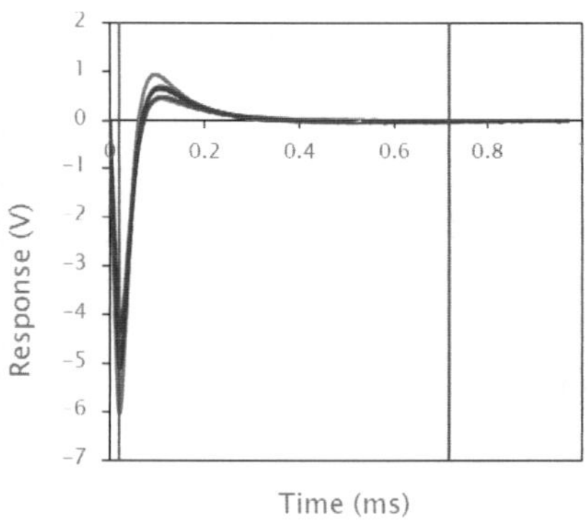

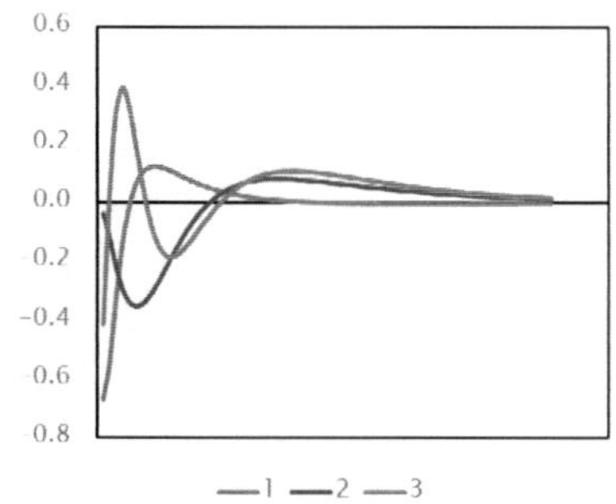

mPCA

-8.59	0.28	0.39
-7.55	-0.55	-0.20
-6.95	-1.02	0.06
-6.52	-0.86	-0.19
-7.64	-0.66	-0.08
-7.78	-0.49	-0.04
-6.34	-0.92	-0.13
-7.66	-0.69	-0.04

Approach - Analysis II

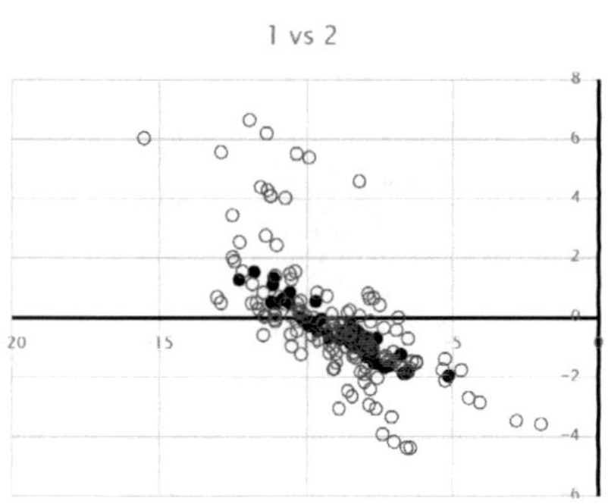

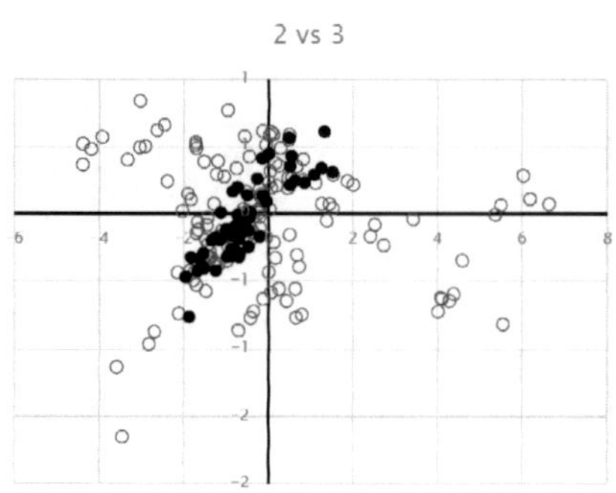

Approach – Analysis III

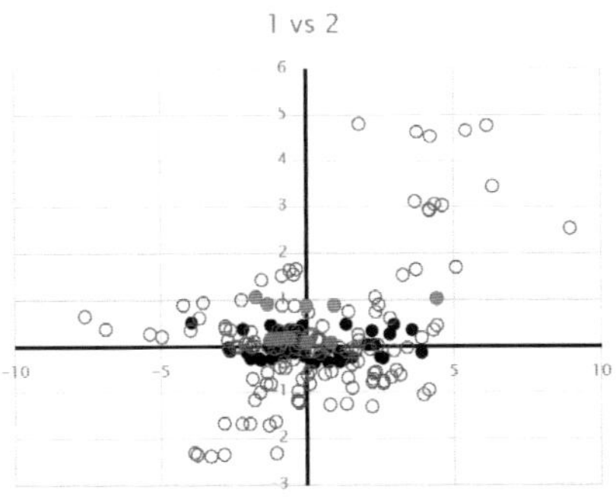

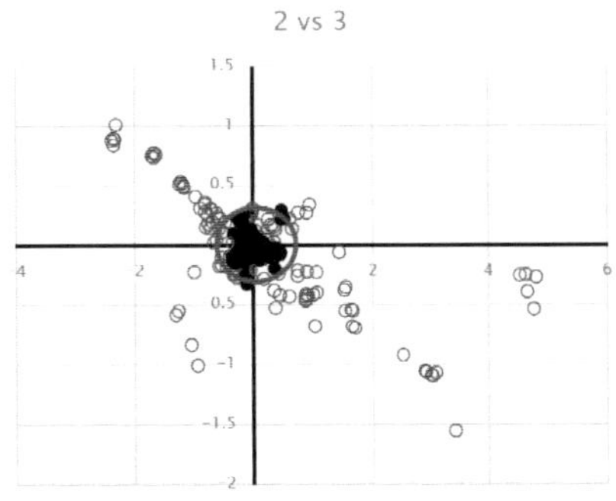

NDT in Canada 2019 | June 18-20 | Edmonton, AB 9

Results- 3/16 Rivet Longitudinal

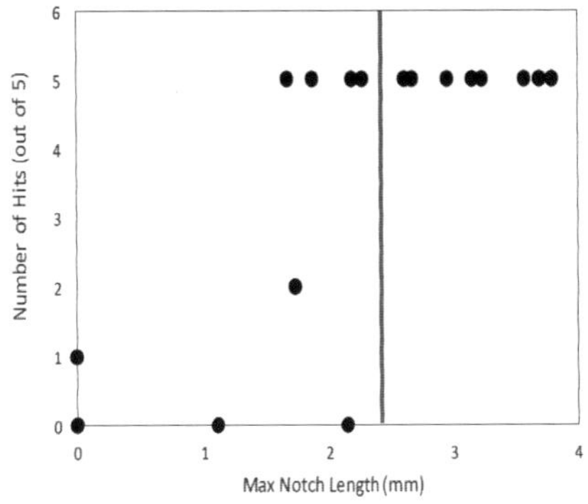

NDT in Canada 2019 | June 18-20 | Edmonton, AB 10

Results - 3/16 Rivet Transverse

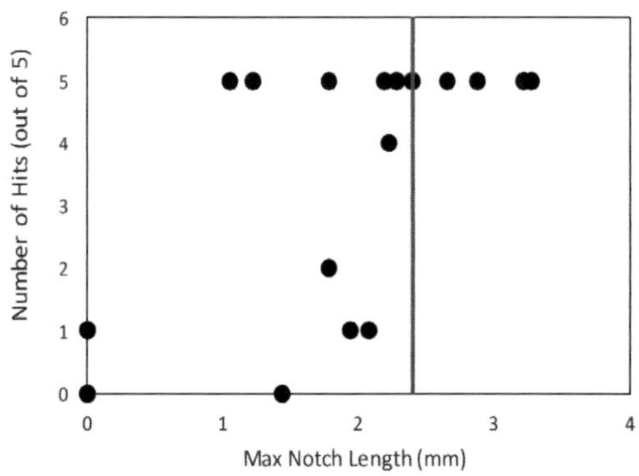

NDT in Canada 2019 | June 18-20 | Edmonton, AB

11

Results - Conventional EC

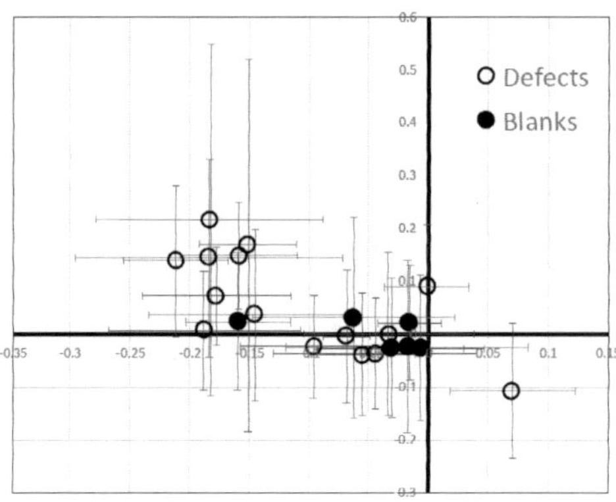

Average of 5 measurements

NDT in Canada 2019 | June 18-20 | Edmonton, AB

12

Results - 1/8" Rivet

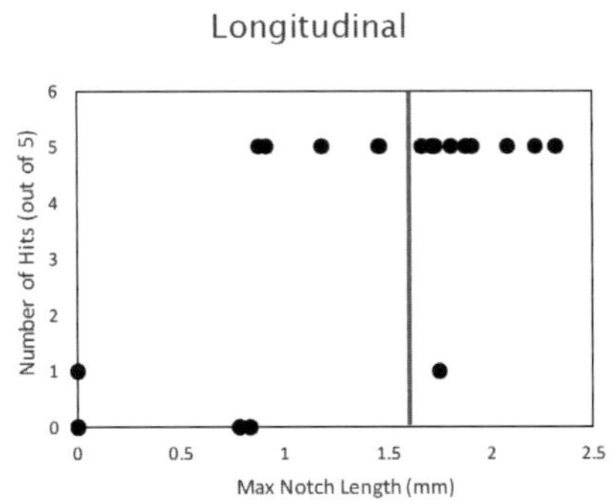

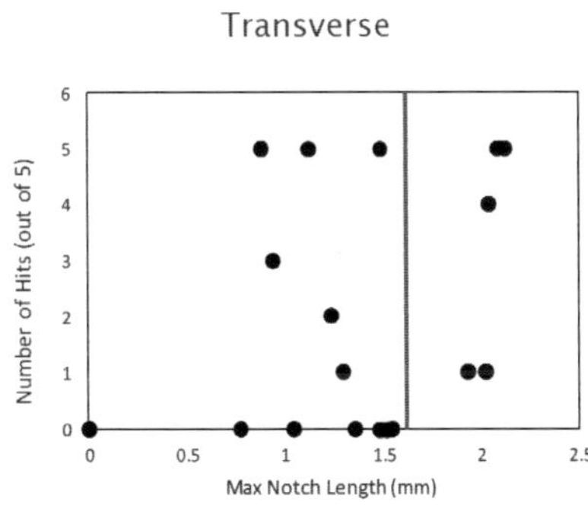

NDT in Canada 2019 | June 18-20 | Edmonton, AB 13

Side by Side Configuration

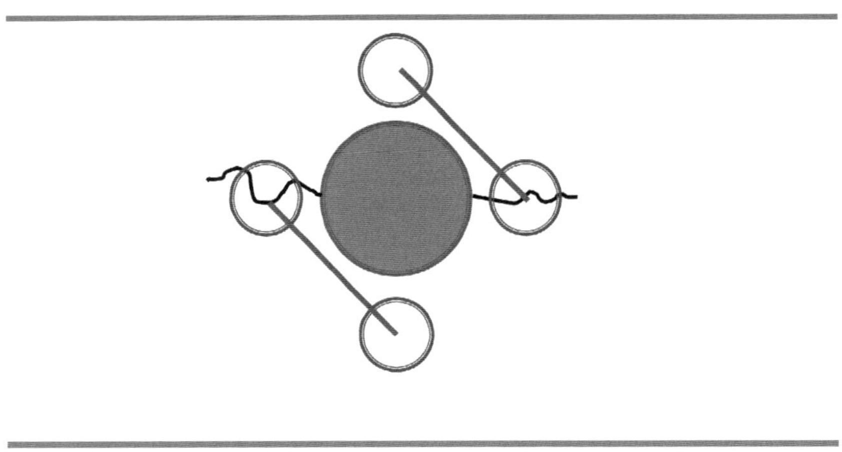

NDT in Canada 2019 | June 18-20 | Edmonton, AB 14

Side by Side Configuration

- Longitudinal
  - 100% detection
  - 0 out of 30 false calls
- Transverse
  - 100% detection
  - 2 out of 35 false calls

Summary

- Excellent detection rate with small number of false calls
- Side by side configuration appears to be superior

The way from here

- Confirm superior performance of side by side configuration with new probes
- Increase number of samples
  - Shape and position of 2nd layer
  - Hole position variability, especially with respect to edge

NDT in Canada 2019 | June 18-20 | Edmonton, AB 17

The way from here

- Simplify cluster detection

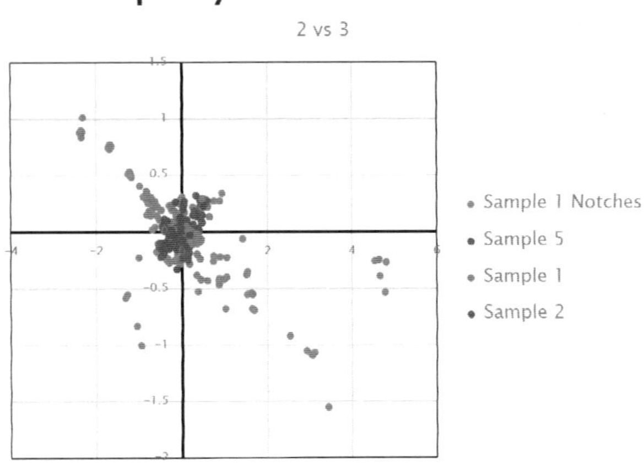

NDT in Canada 2019 | June 18-20 | Edmonton, AB 18

Acknowledgments

- AERAC committee of DTAES (DND)

NDT in Canada 2019 | June 18-20 | Edmonton, AB 19

Questions

NDT in Canada 2019 | June 18-20 | Edmonton, AB 20

Suitability of Quality Management Methods for Adhesive Bonding Processes

Malte MUND [1], Elisabeth STAMMEN [1], Klaus DILGER [1]

[1] Institut für Füge- und Schweißtechnik, Technische Universität Braunschweig; Braunschweig, Germany
Phone: +49 531 39195594, Fax: +49 531 39195599; e-mail: m.mund@tu-braunschweig.de, e.stammen@tu-braunschweig.de, k.dilger@tu-braunschweig.de

Abstract
Due to its ability to join different materials, adhesive bonding has established as one of the most important joining techniques especially for lightweight applications. In addition, it offers further advantages as a planar load transfer. However, there are also some challenges that limit the application of adhesive bonding and one of these challenges is the quality assurance. Adhesive bonding is a special process as the quality of an adhesive bond cannot be determined by non-destructive testing. Although numerous testing methods have been applied in recent years, none has proven to be able to verify the conformity of adhesive bonds reliable. Therefore, it is essential to implement a quality management systems.
Therefore, within this study, quality management methods are evaluated regarding their suitability to be used to implement a quality management system for adhesive bonding processes. Based on the evaluation, instructions for the implementation of a quality management systems are derived and a procedure for the introduction of a quality management systems for adhesive bonding processes is given.

Keywords: Adhesive bonding, quality assurance, quality management

1. Introduction

Adhesive bonding is a joining method that is widely used in industrial applications. This joining method uses an adhesive to join the adherends and offers some advantages compared to other joining methods like riveting or welding as it is not limited to certain materials and prevents tension peaks due to the planar load transfer. However, the application of adhesive bonding is complicated as it is a special process. That means, that the outcome of the process is influenced by many parameters on one hand and a non-destructive evaluation is not possible on the other hand.
In recent years, several non-destructive testing methods have been applied without satisfactory results. Thermographic testing methods [1], shearography [2] as well as ultrasonic testing methods [3, 4] have been applied and some relevant bonding defects like voids or a lack of adhesive can be detected. Other defects like kissing bonds and weak bonds cannot be detected with sufficient reliability. Therefore, quality management methods like Total Quality Management are applied [5].
As a result, the only way to achieve reliable adhesive bonds, quality management systems (QMS) have to be applied. In recent years, the introduction of QMS was specified in German standards concerning adhesive bonding. Within the standard DIN 2304-1 it is stated, that the requirements of the standard have to be integrate into a QMS for the bonding process. However, there are no guidelines how a QMS for adhesive bonding process should look like.
Therefore, within this study, a process for the planning of an adhesive bonding process as a key element of a QMS based on well-established methods for quality management is proposed and the suitability of the considered methods is evaluated. The presented experiences on the suitability of the methods are derived from two process chains that have been considered within an ongoing research project called "EcoQuality" that focus on the development of a model to implement QMS for adhesive bonding processes.

2. Planning of Adhesive Bonding Processes

Once it has been decided to use adhesive bonding as a joining method, numerous influences have to be considered. This includes communication chains, education of employees as well as storage conditions and supply chains. Beside these factors, the bonding process itself requires particular attention. If the adhesive bonding process is well-planed and understood, many defects can be avoided. In general, the planning of all adhesive bonding process with respect to quality can be divided in six stages. The process chain to implement an adhesive bonding process is given in Figure 1.

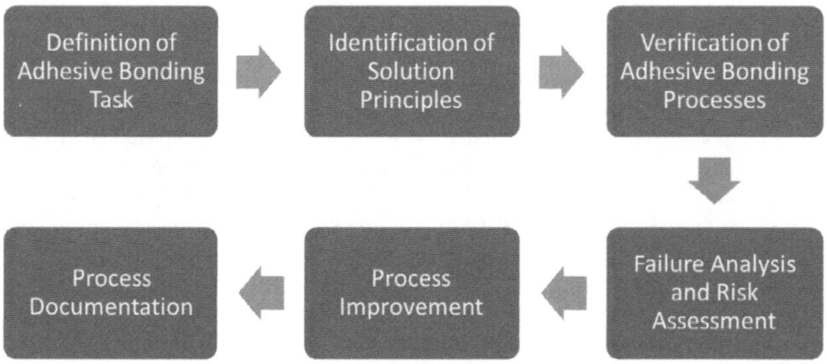

Figure 1: Process for the implementation of an adhesive bonding process

First, a precise definition of the bonding task is required. This include on one hand the bonding task itself (what should be bonded) as well as constructive limitations and all requirements that arise from the production process (cycle times, batch size) and the final application (loading conditions, safety requirements) on the other hand. Based on these boundary conditions, possible bonding process can be defined. This includes the choice of suitable adhesives, surface pre-treatments as well as the application of the adhesive. Then the bonding process has to be verified. Within DIN 2304-1, several ways to verify the bonding process are proposed. If these steps are conducted, the requirements are well-described and the bonding process fulfilling these requirements is defined. However, before the bonding process can be applied, a further critical analysis with respect to the manufacturing conditions has to be conducted. The influences of production parameters have to be investigated and an evaluation of the effect of parameter variations have to be performed. Potential defects have to be identified and a risk assessment has to be performed. This weak point analysis is the basis for the following step, the process improvement, where measures are taken to reduce the risk of unexpected failure. Following this improvements, a process documentation can be provided and the bonding process can be implemented.

3. Quality Management Methods for Adhesive Bonding Processes

Also the approach described above seems quite logical and simple, the practical application is challenging. As adhesive bonding processes are influenced by many parameters, it requires a strategic approach to identify all boundary conditions and parameters so that a well-planned adhesive bonding can be implemented. Some methods of quality management have shown potential to be applied with this process and are described below.

3.1 Step 1: Definition of Adhesive Bonding Task

In order to get a proper definition of the adhesive bonding task, it has shown to be appropriate to use a checklist based on “Five Ws and How” (5W1H). The application of this method is basic in information gathering and enables a detailed description of the bonding task. However, a detailed description has to cover several fields. First of all, the people involved have to be identified and the responsibilities have to be clarified. Then it is also important to give the motivation, why adhesive bonding should be applied and the expectations on this joining method should be specified. Furthermore, it is necessary to describe the basic bonding task as detailed as possible. This includes a description of the materials and the surfaces to be joined as well as a description of planned construction. In addition, the production process should be described. This includes information of cycle times, batch sizes as well as the expected costs as all these factors are essential for the process planning. Additional facts to be considered may be the functions the adhesive bond should have. This may be sealing or insulation functions as well as load bearing functions. However, the main question to be answered may be summarized as follows:

- Who is involved in the process planning?
- Who is responsible for the adhesive bond?
- What materials/parts should be adhesively bonded?
- Why should the materials be bonded adhesively?
- How does the production process look like?
- What production requirements have to be considered?

In a second step, the functions have to be gathered more detailed. The loads that apply in all stages after the actual bonding process (manufacturing process, transport, assembly and operation) have to be considered as well as chemical and physical loads. In addition, requirements concerning stiffness and deformation have to be considered. This information can be gathered by 5WH1 as well and by properly answering the associated questions, it is possible to get a good description of the bonding process, the expectations and the requirements and to derive the requirements on the adhesive, the surface pre-treatment, the application of the adhesive and the curing process as major steps within the adhesive bonding manufacturing chain.

3.2 Step 2: Identification of Solution Principles

Once the process is described, principles solutions can be developed. Within the project, two approaches have shown to be suitable to identify principle solutions for given adhesive bonding tasks.

One suitable tool to identify solution principles is the application of a morphological box as it offers the possibility to generate numerous possible solutions and is easy to implement. Within this method, suitable adhesives and surface pre-treatments can be identified in a first step and, in a second step, associated techniques for the application of the adhesive and the curing can be added. However, as many solution may be possible, a selection of the most significant approaches is difficult.

An alternative approach is 6-3-5 Brainwriting where six participants develop three initial ideas each that fulfil the requirements in a first step. This ideas are continuously detailed by the other participants. This method offers the opportunity to develop and specify solution for the adhesive bonding task. The ideas of all participants are respected and the detailed description of the solution allows a sound selection of suitable solution principle and ease the documentation of the selection. A main disadvantage is, that it is possible, that the proposed solutions overlap and therefore, only a limited number of solutions may appear. Furthermore, it requires basic knowledge of adhesive bonding processes from all participants.

In both cases, the evaluation of the defined processes starts from the selection of possible adhesive fulfilling the requirements. Based on that, suitable surface pre-treatments, application methods as well as curing procedures can be selected.

3.3 Step 4: Verification of Adhesive Bonding Processes

The suitability of the selected bonding processes has to be demonstrated in the next step. According to DIN 2304-1, this may be done by experimental tests, component tests or based on experiences made with adhesive bonding processes that are already approved or a combination of the before-mentioned methods. As the outcome of the bonding process highly depends on the process parameters, in a first step all controllable process parameters have to be identified and the specimens for the experimental evaluation have to be manufactured under realistic conditions. Ishikawa diagrams were identified as a suitable method for parameter identification for each step of the bonding process. The considered parameters can be divided into two categories. On one hand, there are parameters that are well to handle. This includes machines, materials the working environment, the materials as well as the measurements where possible parameters and parameter windows can be defined. On the other hand, the influence of human resources can hardly be captured. Therefore it is essential to determine to derive minimum requirements that have to be fulfilled and to give detailed instructions on how to bond.

If all parameters are identified, design of experiments was found to be a suitable method to reduce the testing effort to verify the adhesive bonding process. It has to be shown, that the chosen process is suitable to fulfil all requirements. If it fails, the failure cause has to be identified and it has to be checked, if parameter variations are suitable to prevent failure. In case parameter variations do not give satisfying solutions, the adhesive bonding process has to be changed to meet the requirements.

3.4 Step 4: Failure Analysis and Risk Assessment

When the adhesive bonding process has been chosen, it is suitable to perform a Failure Mode and Effects Analysis (FMEA). Based on the results of the verification of the process and the determined parameters, an evaluation of the likelihood of appearance a specific fault has to be conducted. This can be done by simulations, experiments or based on experiences from former processes. The probability P is then rated within a group starting from 1 (extremely unlikely) to 5 (frequent). Furthermore, the severity S of the fault has to be rated. This can be done by a rating form 1 (no effect) to 5 (complete failure). The third parameter is the probability to detect the fault. It has to be determined whether a fault will certainly will be detected (1) or whether a fault cannot be detected (5). The numbers can be changed to meet the requirements of the bonding process and as a result, it offers the possibility to prioritize the risk of a certain fault by a risk priority number (RPN) computed as a product of P, S and D. However, it has to be considered, that these RPN numbers may not depict the most critical faults. Therefore, a critical evaluation of the result is essential.

Other techniques for failure analysis like Ishikawa diagrams as well as failure tree analysis have also be considered. These methods can help to identify the fault and their causes, however an evaluation is not really possible and therefore, FMEA was preferred. Based on these results

3.5 Step 5 & Step 6: Process Improvement and Process Documentation

The process parameters that have been identified as critical have to be considered to improve the process. In general, there are three possibilities to improve the process: The probability for a certain fault can be reduced, the severity of the fault can be lowered or the possibility to detect the fault can be improved. When the actions to improve the process are defined, a detailed documentation has to be prepared. Then the process can be implemented.

As a QMS is designed to improve the process continuously, all actions the whole process have to be evaluated regularly.

3. Example - Surface pre-treatment for a test case

As mentioned, the recommendations concerning the suitability of the methods base on the experiences gathered during the project “EcoQuality”. Within this project, two adhesive bonding task with industrial background were considered to develop a basic QMS for adhesive bonding processes. One of the bonding tasks was the adhesive bonding of a pane of toughened safety glass to powder-coated aluminium profiles. Within the project, two different epoxy-based powder-coatings (black and white) were considered.

According to the proposed procedure, in a first step the bonding task was precisely defined based on a questionnaire developed within the project. In a second step, solution principles were identified using a morphological box. Two processes using different adhesive were chosen. On one hand, a pressure sensitive adhesive was applied, on the other hand, a two part silicon-based adhesive was used. In this case, the surface pre-treatment consisted of a cleaning of the surfaces and the application of an activator to improve the adhesion. The adhesive bonding process is shown in Figure 2.

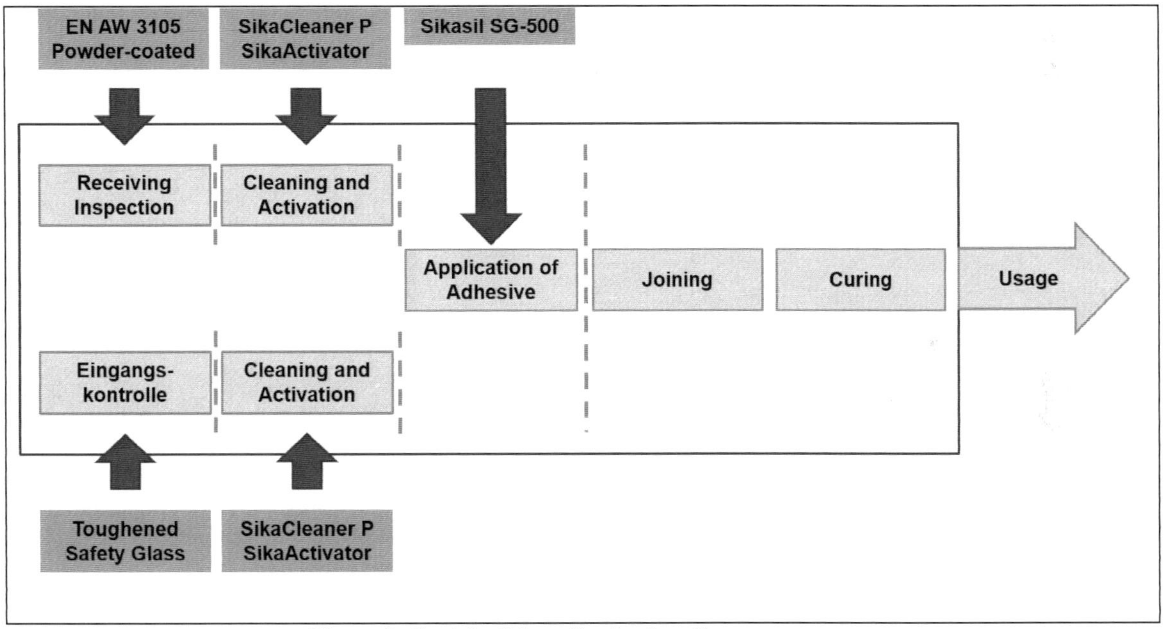

Figure 2: Scetch of the adhesive bonding process

After the process was defined, the parameters influencing each process step were determined using Ishikawa diagrams. As it is a manual manufacturing process, machines have no influence on the outcome of the surface pre-treatment. In contrast, the employee has a large impact as is fully responsible for the process. This is closely connected to the method as the number of cleaning steps or the way how to apply the activator have to be described. Additionally, the materials have a major influence on the surface pre-treatment as well. The level of contamination of the adherends as well as the chemical composition of the surface influence the result of the surface pre-treatment. The cleaner and the activator have an influence, too. On one hand, they can be contaminated or they might be expired. Furthermore, the environmental conditions have to be considered. For example, it has to be investigated, whether contaminations could occur.

Based on the defined influence parameters a worst case scenario was defined. That would mean, the adherends were bonded without any surface pre-treatment. Further investigations considered the lack of activator and the use of another cleaning agent (Isopropanol). Specimen were manufactured with the described fault and compared to the verified process. The test results for all three considered surfaces are shown in Figure 3.

a.) b.)

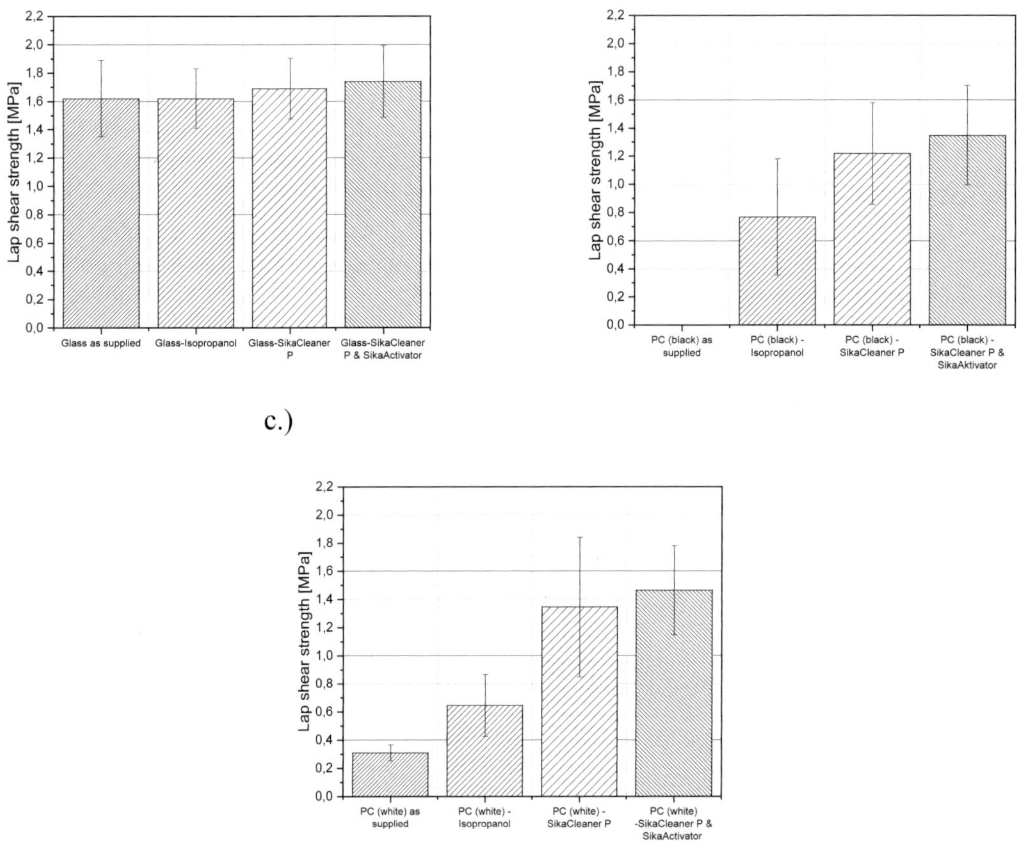

Figure 3: Lap shear test strength of specimen with different surface properties – a.) Adherends made of toughened safety glass; b.) Adherends made of EN AW 3105 with a black epoxy based powder coating and c.) Adherends made of EN AW 3105 with a white epoxy based powder

As the results show, the lap shear strength highly depend on the surface condition of the adherends. The lap shear strength for the adherends made of toughened safety glass are independent from the surface pre-treatment (a.). The adhesive shows cohesive failure in all cases. In contrast, the powder-coated specimens are highly sensitive to the surface conditions. If the specimens are bonded as delivered, they either fail on a low level (c.) in case of the white coating or they cannot be bonded at all in case of black coating (b.). A cleaning step with isopropanol increases the bonding strength in both cases significantly, however the failure mode is still partly adhesive. The recommend SikaCleaner P shows better results and the lap shear strength increase even more. This is caused by a bonding agent added to the cleaner. The failure is mainly cohesive. The application of the activator finally gives cohesive failure.
Based on these results, an FMEA was performed. The three cases considered were "Lack of surface pre-treatment", "Use of wrong cleaner and No activator" and "No activator". The results of the FMEA is given in Table 1. It is possible to detect the absence of the activator as the surface shows streaks on the surface.

Table 1: FMEA for surface pre-treatments

Adherends	Fault	P	S	D	RPN
Glass	No surface pre-treatment	2	1	5	5
Glass	Use of wrong cleaner and no activator	2	1	5	5
Glass	No activator	2	1	1	2
PC (white)	No surface pre-treatment	2	5	5	25
PC (white)	Use of wrong cleaner and no activator	2	4	5	20
PC (white)	No activator	2	2	1	4
PC (black)	No surface pre-treatment	2	5	5	25
PC (black)	Use of wrong cleaner and no activator	2	4	5	20
PC (black)	No activator	2	2	1	4

The results show, that the most critical case is the absence of any surface pre-treatment for the powder-coated aluminium. This would cause a drop in the lap-shear strength and it cannot be detected properly. Wetting measurements were performed, however the results do not give an indication of the missing cleaning step. Although the probability of an occurrence is low, as there are no suitable methods to detect the fault and the effect on the bonding strength is significant, the RPN is high. Therefore, actions has to be taken to improve the process and to prevent the fault. In order to avoid the critical case action can be taken. For example, it is suitable to remove all liquids that might be interchanged with the SikaCleaner P from the work space. Furthermore, it is necessary that the worker confirm the orderly implementation of the surface pre-treatment. This would not affect the severity or the detectability, however the probability of occurrence is reduced and therefore, the process is improved.

4. Conclusions

Within this study, the suitability of established methods for quality assurance for adhesive bonding processes was investigated. A process for the implementation based on the selected methods was proposed and the suitability was demonstrated for a surface pre-treatment. The developed process based on established methods for quality assurance seems to be suitable to identify the critical parameters and allow a clear presentation of the effects and a good basis to establish a QMS. A critical evaluation of the adhesive bonding is essential and the methods proposed can help to identify deficiencies and their causes. However, it is the responsibility of the user to implement the process and to improve the adhesive bonding process continuously.

5. Acknowledgements

The IGF project No: 19.391 N / DVS-No: 08.096 of the research association "Forschungsvereinigung Schweißen und verwandte Verfahren e. V." of the DVS, Aachener Str. 172, 40223 Düsseldorf, Germany, was supported by the AiF within the scope of a program for the fostering of the industrial collective research (IGF) by the Federal Ministry of Economy Affairs and Energy due to a resolution of the German Bundestag. We appreciate this support very much. Furthermore, we would like to thank the committee which accompanied the project for their constructive support during in this project.

Supported by:

Federal Ministry for Economic Affairs and Energy

on the basis of a decision by the German Bundestag

References

[1] SRAJBR, Christian ; TANASIE, Gheorghe ; DILGER, Klaus ; BÖHM, Stefan: *Active Thermography for Quality Assurance of joints in automobile manufacturing*. In: *Welding in the World* 55 (2011), 7-8, S. 90–97

[2] KRYUKOV, Igor ; BÖHM, Stefan: *Prospects and limitations of eddy current shearography for non-destructive testing of adhesively bonded structural joints*. In: *The Journal of Adhesion* 16 (2018), S. 1–13

[3] GOGLIO, Luca ; ROSSETTO, Massimo: *Ultrasonic testing of adhesive bonds of thin metal sheets*. In: *NDT & E International* 32 (1999), Nr. 6, S. 323–331

[4] YONATHAN SUNARSA, Timotius ; ARYAN, Pouria ; JEON, Ikgeun ; PARK, Byeongjin ; LIU, Peipei ; SOHN, Hoon: *A Reference-Free and Non-Contact Method for Detecting and Imaging Damage in Adhesive-Bonded Structures Using Air-Coupled Ultrasonic Transducers*. In: *Materials (Basel, Switzerland)* 10 (2017), Nr. 12

[5] SAVAGE, Gary: *Failure prevention in bonded joints on primary load bearing structures*. In: *Engineering Failure Analysis* 14 (2007), Nr. 2, S. 321–348

AUTHOR INDEX